T0323809

Sufficient Dimension Reduction

Methods and Applications with R

MONOGRAPHS ON STATISTICS AND APPLIED PROBABILITY

Editors: F. Bunea, P. Fryzlewicz, R. Henderson, N. Keiding, T. Louis, R. Smith, and W. Wong

Semialgebraic Statistics and Latent Tree Models
Piotr Zwiernik 146

Inferential Models
Reasoning with Uncertainty
Ryan Martin and Chuanhai Liu 147

Perfect Simulation
Mark L. Huber 148

State-Space Methods for Time Series Analysis
Theory, Applications and Software
Jose Casals, Alfredo Garcia-Hiernaux, Miguel Jerez, Sonia Sotoca, and A. Alexandre Trindade 149

Hidden Markov Models for Time Series
An Introduction Using R, Second Edition
Walter Zucchini, Iain L. MacDonald, and Roland Langrock 150

Joint Modeling of Longitudinal and Time-to-Event Data
Robert M. Elashoff, Gang Li, and Ning Li 151

Multi-State Survival Models for Interval-Censored Data
Ardo van den Hout 152

Generalized Linear Models with Random Effects
Unified Analysis via H-likelihood, Second Edition
Youngjo Lee, John A. Nelder, and Yudi Pawitan 153

Absolute Risk
Methods and Applications in Clinical Management and Public Health
Ruth M. Pfeiffer and Mitchell H. Gail 154

Asymptotic Analysis of Mixed Effects Models
Theory, Applications, and Open Problems
Jiming Jiang 155

Missing and Modified Data in Nonparametric Estimation With R Examples
Sam Efromovich 156

Probabilistic Foundations of Statistical Network Analysis
Harry Crane 157

Multistate Models for the Analysis of Life History Data
Richard J. Cook and Jerald F. Lawless 158

Nonparametric Models for Longitudinal Data with Implementation in R
Colin O. Wu and Xin Tian 159

Multivariate Kernel Smoothing and Its Applications
José E. Chacón and Tarn Duong 160

Sufficient Dimension Reduction Methods and Applications with R
Bing Li 161

For more information about this series please visit:
https://www.crcpress.com/Chapman--HallCRC-Monographs-on-Statistics--
Applied-Probability/book-series/CHMONSTAAPP

Sufficient Dimension Reduction

Methods and Applications with R

Bing Li

CRC Press
Taylor & Francis Group
Boca Raton London New York

CRC Press is an imprint of the
Taylor & Francis Group, an **informa** business

A CHAPMAN & HALL BOOK

CRC Press
Taylor & Francis Group
6000 Broken Sound Parkway NW, Suite 300
Boca Raton, FL 33487-2742

First issued in paperback 2020

ISBN-13: 978-1-4987-0447-2 (hbk)
ISBN-13: 978-0-367-73472-5 (pbk)

Visit the Taylor & Francis Web site at
http://www.taylorandfrancis.com

and the CRC Press Web site at
http://www.crcpress.com

To Yanling

Contents

List of Figures   xiii

List of Tables   xvii

Preface   xix

Author   xxi

1 Preliminaries   **1**
    1.1 Empirical Distribution and Sample Moments   1
    1.2 Principal Component Analysis   2
    1.3 Generalized Eigenvalue Problem   3
    1.4 Multivariate Linear Regression   3
    1.5 Generalized Linear Model   5
        1.5.1 Exponential Family   5
        1.5.2 Generalized Linear Models   6
    1.6 Hilbert Space, Linear Manifold, Linear Subspace   8
    1.7 Linear Operator and Projection   10
    1.8 The Hilbert Space $\mathbb{R}^p(\Sigma)$ 11
    1.9 Coordinate Representation   12
    1.10 Generalized Linear Models under Link Violation   13

2 Dimension Reduction Subspaces   **17**
    2.1 Conditional Independence   17
    2.2 Sufficient Dimension Reduction Subspace   21
    2.3 Transformation Laws of Central Subspace   24
    2.4 Fisher Consistency, Unbiasedness, and Exhaustiveness   25

3 Sliced Inverse Regression   **27**
    3.1 Sliced Inverse Regression: Population-Level Development   27
    3.2 Limitation of SIR   30
    3.3 Estimation, Algorithm, and R-codes   31
    3.4 Application: The Big Mac Index   33

4   Parametric and Kernel Inverse Regression   **37**
   4.1   Parametric Inverse Regression   37
   4.2   Algorithm, R Codes, and Application   39
   4.3   Relation of PIR with SIR   40
   4.4   Relation of PIR with Ordinary Least Squares   42
   4.5   Kernel Inverse Regression   42

5   Sliced Average Variance Estimate   **47**
   5.1   Motivation   47
   5.2   Constant Conditional Variance Assumption   47
   5.3   Sliced Average Variance Estimate   49
   5.4   Algorithm and R-code   52
   5.5   Relation with SIR   55
   5.6   The Issue of Exhaustiveness   56
   5.7   SIR-II   58
   5.8   Case Study: The Pen Digit Data   60

6   Contour Regression and Directional Regression   **63**
   6.1   Contour Directions and Central Subspace   63
   6.2   Contour Regression at the Population Level   65
   6.3   Algorithm and R Codes for CR   67
   6.4   Exhaustiveness of Contour Regression   69
   6.5   Directional Regression   70
   6.6   Representation of Λ_{DR} Using Moments   74
   6.7   Algorithm and R Codes for DR   76
   6.8   Exhaustiveness Relation with SIR and SAVE   77
   6.9   Pen Digit Case Study Continued   79

7   Elliptical Distribution and Predictor Transformation   **83**
   7.1   Linear Conditional Mean and Elliptical Distribution   83
   7.2   Box-Cox Transformation   88
   7.3   Application to the Big Mac Data   92
   7.4   Estimating Equations for Handling Non-Ellipticity   94

8   Sufficient Dimension Reduction for Conditional Mean   **97**
   8.1   Central Mean Subspace   97
   8.2   Ordinary Least Squares   100
   8.3   Principal Hessian Direction   101
   8.4   Iterative Hessian Transformation   104

9   Asymptotic Sequential Test for Order Determination   **107**
   9.1   Stochastic Ordering and Von Mises Expansion   107
   9.2   Von Mises Expansion and Influence Functions   109
   9.3   Influence Functions of Some Statistical Functionals   110
   9.4   Random Matrix with Affine Invariant Eigenvalues   112
   9.5   Asymptotic Distribution of the Sum of Small Eigenvalues   115

9.6 General Form of the Sequential Tests 117
9.7 Sequential Test for SIR 118
9.8 Sequential Test for PHD 124
9.9 Sequential Test for SAVE 126
9.10 Sequential Test for DR 132
9.11 Applications 139

10 Other Methods for Order Determination **141**
10.1 BIC Type Criteria for Order Determination 141
10.2 Bootstrapped Eigenvector Variation 147
10.3 Eigenvalue Magnitude and Eigenvector Variation 150
10.4 Ladle Estimator 152
10.5 Consistency of the Ladle Estimator 156
10.6 Application: Identification of Wine Cultivars 156

11 Forward Regressions for Dimension Reduction **159**
11.1 Outer Product of Gradients 160
11.2 Fisher Consistency of Gradient Estimate 163
11.3 Minimum Average Variance Estimate 167
11.4 Refined MAVE and refined OPG 170
11.5 From Central Mean Subspace to Central Subspace 173
11.6 dOPG and Its Refinement 173
11.7 dMAVE and Its Refinement 178
11.8 Ensemble Estimators 180
11.9 Simulation Studies and Applications 184
11.10 Summary 188

12 Nonlinear Sufficient Dimension Reduction **191**
12.1 Reproducing Kernel Hilbert Space 192
12.2 Covariance Operators in RKHS 193
12.3 Coordinate Mapping 199
12.4 Coordinate of Covariance Operators 200
12.5 Kernel Principal Component Analysis 202
12.6 Sufficient and Central σ-Field for Nonlinear SDR 204
12.7 Complete Sub σ-Field for Nonlinear SDR 206
12.8 Converting σ-Fields to Function Classes for Estimation 208

13 Generalized Sliced Inverse Regression **211**
13.1 Regression Operator 212
13.2 Generalized Sliced Inverse Regression 213
13.3 Exhaustiveness and Completeness 215
13.4 Relative Universality 216
13.5 Implementation of GSIR 217
13.6 Precursors and Variations of GSIR 220
13.7 Generalized Cross Validation for Tuning ε_X and ε_Y 220
13.8 k-Fold Cross Validation for Tuning $\rho_X, \rho_Y, \varepsilon_X, \varepsilon_Y$ 223

13.9 Simulation Studies 225
13.10 Applications 227
    13.10.1 Pen Digit Data 227
    13.10.2 Face Sculpture Data 228

14 Generalized Sliced Average Variance Estimator **233**
14.1 Generalized Sliced Average Variance Estimation 233
14.2 Relation with GSIR 237
14.3 Implementation of GSAVE 239
14.4 Simulation Studies and an Application 248
14.5 Relation between Linear and Nonlinear SDR 251

15 The Broad Scope of Sufficient Dimension Reduction **253**
15.1 Sufficient Dimension Reduction for Functional Data 253
15.2 Sufficient Dimension Folding for Tensorial Data 256
15.3 Sufficient Dimension Reduction for Grouped Data 259
15.4 Variable Selection via Sufficient Dimension Reduction 260
15.5 Efficient Dimension Reduction 262
15.6 Partial Dimension Reduction for Categorical Predictors 264
15.7 Measurement Error Problem 265
15.8 SDR via Support Vector Machine 267
15.9 SDR for Multivariate Responses 268

Bibliography **271**

Index **281**

List of Figures

1.1 Estimating gradient direction under link violation in Generalized Linear Model. 16

2.1 Sufficient Dimension Reduction subspace. 23

3.1 Illustration of unbiasedness of SIR. 29
3.2 Illustration of limitation of Sliced Inverse Regression. 31
3.3 Scatter plot matrix of ten economic variables in the Big Mac data. 34
3.4 Scatter plot of the response versus the first two SIR predictors. 35

4.1 Big Mac index versus the first PIR predictor. 40
4.2 Big Mac index versus the OLS predictor. 43
4.3 Big Mac index versus the first KIR predictor. 45

5.1 Leading eigenvector $\text{var}(X|Y \in J_\ell)$ in the monotone case. 51
5.2 Leading eigenvector of $\text{var}(X|Y_\ell)$ in symmetric case. 52
5.3 Y versus first two SAVE predictor for the Big Mac data. 53
5.4 Y versus first two SAVE predictor for the Big Mac data with first predictor removed. 54
5.5 Y versus X_1 and X_2. 54
5.6 Upper panels: Y versus first two SIR predictors; lower panels: Y versus first two SAVE predictors. 55
5.7 Y versus first two SIR-II predictors for the Big Mac data. 60
5.8 Y versus first two SIR-II predictors for Model (5.8) in Example 5.1. 60
5.9 First three predictors from SIR (upper panel), SAVE (lower-left panel), and SIR-II (lower-right panel). 62

6.1 Illustration of contour directions and central subspace. 64
6.2 Y versus first two CR predictors for the model in Example 5.1. 68
6.3 Reorganization of empirical directions by Directional Regression. 73
6.4 Y versus first two DR predictors for the model in Example 5.1. 77
6.5 Perspective plots for the pen digit case study. 80
6.6 Effect of scale separation by SIR and DR for the pen digit data. 81
6.7 Five handwriting shapes in the pen digit data. 81

7.1   Scatter plot matrix for Box-Cox transformed of the predictor of the
      Big Mac data.   93
7.2   First SIR predictor for the Box-Cox transformed Big Mac data.   93

8.1   OLS predictor for Big Mac data.   101
8.2   PHD predictor for Big Mac data. Left panel: Y versus first predictor;
      right panel: Y versus the second predictor.   103
8.3   Perspective plot for the first three PHD predictors for the pen digit
      data.   103
8.4   Application of IHT to pen digit data.   106

9.1   Boxplots of p-values for sequential test based on SIR as applied to
      Model (9.18).   123
9.2   Boxplots of p-values for sequential test based on PHD as applied to
      Model (9.18).   126
9.3   Boxplots of p-values for sequential test based on SAVE as applied
      to the model in Example 5.1.   132
9.4   Boxplots of p-values for sequential test based on DR for the model
      in Example 5.1.   136

10.1 Bootstrapped eigenvector variation based on SIR for $k = 1, \ldots, 6$, as
      applied to the model in Example 9.1.   149
10.2 Ladle plot for the model in Example 9.1.   153
10.3 Ladle plot for the wine data.   157
10.4 Comparing the 2-d and 3-d plots to see the effect of the third
      sufficient predictor in the wind data.   157

11.1 Comparison of OPG, MAVE, rOPG, rMAVE, rdOPG, rdMAVE,
      reOPG, reMAVE for the model in Example 11.1.   185
11.2 Comparison of OPG, MAVE, rOPG, rMAVE, rdOPG, rdMAVE,
      reOPG, reMAVE for the model in Example 11.2.   186
11.3 Scatter plot matrix for the abalone data.   187
11.4 Age versus the first two predictors from rOPG and reOPG for the
      abalone data.   188

12.1 First three linear principal components and first three kernel principal
      components for the pen digit data.   204

13.1 GSIR for Model I, Scenario A.   226
13.2 First two GSIR predictors for the pen digit data.   228
13.3 Face data representation.   229
13.4 First three GSIR predictors as evaluated at the testing set.   230
13.5 Relation between the true responses and the first three GSIR
      predictors in the face data.   231

14.1 First five sufficient predictors by GSAVE for the pen digit data.   250

15.1 A schematic representation of the functional data collected by EEG
       from a subject.   254

List of Tables

3.1 Spearman's correlation with the response   35

9.1 p-values for sequential tests applied to the Big Mac data   139
9.2 p-values for sequential tests applied to the pen digit data   140

10.1 Percentage (with symbol % omitted) of correct estimation by BIC   146
10.2 Percentage (with symbol % omitted) of correct estimation by ladle
estimator and BIC estimator coupled with the ZMP criterion   155
10.3 p-values for DR-sequential test for the wine data   157

11.1 Types of forward regression estimators   189

13.1 Performance of GSIR (with options $\Lambda_{GSIR}^{(1)}$ and $\Lambda_{GSIR}^{(2)}$) under Models I,
II, III. Scenarios A, B, C, and four different combinations of sample
sizes and dimensions   227

14.1 Comparison of GSAVE and GSIR under Models IV, V, VI and
Scenarios A, B, C   249
14.2 Performance of GSAVE under Models I, II, III and Scenarios A, B,
C   249

Preface

Sufficient Dimension Reduction is a rapidly developing research field that has wide applications in regression diagnostics, data visualization, Machine Learning, Genomics, image processing, pattern recognition, and medicine, which often contain a large number of variables. The purpose of the book is to introduce the basic theories and the main methodologies that have been developed in this field, to explore the key technical machineries that have been proven useful for conducting related research, to provide practical and easy-to-use algorithms and computer codes to implement these methodologies, and to survey the recent advances in the frontier of this field, which has grown too vast to be covered in detail in a single book.

Sufficient Dimension Reduction is a powerful tool to extract the core information hidden in the high-dimensional data, for the purpose of classifying or predicting one or several response variables. The extraction of information is based on the notion of sufficiency, which means a set of functions of the predictors provides all the information needed to understand the response, so that the rest of the predictors can be ignored without loss of information. Sufficiency is derived from conditional independence, a statistical concept that plays the central role in this theory.

Sufficient Dimension Reduction is akin to Principal Component Analysis — they both try to organize the variations in the data in an intelligent and interpretable way. However, Principal Component Analysis organizes the variations in the data itself, according to the magnitudes of variations; whereas Sufficient Dimension Reduction organizes the variations in the predictor according to how much they can explain the response variables. Sufficient Dimension Reduction is also akin to variable selection — they both try to reduce the number of variables that predict the response. However, variable selection tries to reduce the number of coordinates in the predicting vector; whereas Sufficient Dimension Reduction tries to reduce the predictor to a few linear combinations, or a few nonlinear functions, of the coordinates. In other words, variable selection reduces the data to achieve sparsity; Sufficient Dimension Reduction reduces the data to achieve low rank.

Sufficient Dimension Reduction has undergone momentous development in recent years, partly due to the increased demands for techniques to process high-dimensional data, a hallmark of our age of Big Data. The heightened development is also propelled by the increased complexity of the data structure. The classical dimension reduction problem proposed in the early 90's was concerned with a single response variable and a vector of continuous predicting variables; it used linear combinations as the sufficient predictors; its objective was to reduce the predictor in the conditional distribution. Since then, Sufficient Dimension Reduction has expanded in many directions. For example, the predictors and the responses can both be functions or vectors of functions; the predictor can be matrix- or tensor-valued; the predictors can have grouped structures, and can be either continuous or categorical. The sufficient predictors are no longer limited to linear functions; it can be a member of a reproducing kernel Hilbert space. The target of reduction is no longer restricted to the whole conditional distribution; they can be the conditional means, conditional quantiles, conditional variances, or other conditional functionals of the response, according to our primary interests in the study.

The book is organized around four main themes. The first three themes belong to linear Sufficient Dimension Reduction: the inverse regression methods, order determination methods and related asymptotic developments, and the forward regression methods. The last theme is nonlinear Sufficient Dimension Reduction.

Specifically, Chapter 1 introduces the preliminary tools that will be used throughout the book, as well as some backgrounds and motivations. Chapters 2 lays out the basic theoretical framework, such as Sufficient Dimension Reduction subspaces and Fisher consistency. Chapters 3 through 6 develop a variety of inverse regression estimators, such as the Sliced Inverse Regression, the Parametric and the Kernel Inverse Regression, the Sliced Average Variance Estimate, Contour Regression, and Directional Regression. Chapter 6 discusses the key assumption — the elliptical distribution assumption — that underlies these inverse regression methods. Chapter 7 introduces the dimension reduction framework where the conditional mean is of interest. Chapters 8 and 9 cover the order determination methods that determine the number of sufficient predictors to be extracted from the data. Chapter 10, a relatively long chapter, covers the forward regression methods, such as the Outer Product of Gradients, the Minimal Average Variance Estimator, and the Ensemble Estimator. Chapters 12 through 14 cover Nonlinear Sufficient Dimension Reduction, which includes the basic theory, the Generalized Sliced Inverse Regression, and the Generalized Sliced Average Variance Estimator. In the last chapter, Chapter 15, we give an overview of the developments that cannot be explored in detail in the previous chapters, which reveals the current scope and trends of this field.

This book grew out of the lecture notes I wrote when I taught such a course in the Spring of 2014 in the Department of Statistics of the Pennsylvania State University, chaired at the time by Professor D. Hunter. I thank the department for giving me such an opportunity and for the stimulating research environment. My work during this period has been supported by the National Science Foundation grants. My special thanks are due to Professor R. D. Cook, whose many inspiring discussions and collaborations during and before the writing of this book have benefited me greatly. I thank Professor X. Yin for reading a large part of the book. I thank my former students, K.-Y. Lee and W. Luo, for helping to collect and organize some computer codes. My other former students, S. Wang, Y. Dong, and A. Artemiou also contributed to the computing codes. I thank Professor B. Sriperumbudur for his useful discussions with me on the reproducing kernel Hilbert space.

State College   *Bing Li*

Author

Bing Li obtained his Ph.D. from the University of Chicago in 1992. He is a Professor of Statistics at the Pennsylvania State University. His research interests cover Sufficient Dimension Reduction, Statistical Graphical Models, Functional Data Analysis, Machine Learning, Quasilikelihood, Estimating Equations, and Robust Statistics. He is a fellow of the Institute of Mathematical Statistics and a fellow of the American Statistical Association. He is serving as an Associate Editor for *The Annals of Statistics* and the *Journal of the American Statistical Association*.

Chapter 1

Preliminaries

1.1   Empirical Distribution and Sample Moments   1
1.2   Principal Component Analysis   2
1.3   Generalized Eigenvalue Problem   3
1.4   Multivariate Linear Regression   3
1.5   Generalized Linear Model   5
      1.5.1   Exponential Family   5
      1.5.2   Generalized Linear Models   6
1.6   Hilbert Space, Linear Manifold, Linear Subspace   8
1.7   Linear Operator and Projection   10
1.8   The Hilbert Space $\mathbb{R}^p(\Sigma)$ 11
1.9   Coordinate Representation   12
1.10 Generalized Linear Models under Link Violation   13

1.1   Empirical Distribution and Sample Moments

Let X be a random vector defined on a probability space $(\Omega, \mathscr{F}, P)$, taking values in a measurable space $(\Omega_X, \mathscr{F}_X)$. Let $X_1, \ldots, X_n$ be independent copies of X. We assume Ω_X to be a subset of $\mathbb{R}^p$, the p-dimensional Euclidean space, and $\mathscr{F}_X = \{\Omega_X \cap B : B \in \mathscr{R}^p\}$, where $\mathscr{R}^p$ is the Borel σ-field on $\mathbb{R}^p$.

Throughout this book, when there is a sample of n random vectors of p dimension, we always use subscript to indicate subjects, and superscript to indicate components. Thus X_i^k is the kth component of the ith subject. The symbol X_i without a superscript is used to denote the p-dimensional vector $(X_i^1, \ldots, X_i^p)^\top$.

The empirical distribution of X based on $X_1, \ldots, X_n$ is defined to be the measure on $(\Omega_X, \mathscr{F}_X)$ that assigns n^{-1} mass to each X_i. This measure is denoted by F_n. That is,

$$F_n = n^{-1} \sum_{i=1}^{n} \delta_{X_i},$$

1

where δ_{X_i} is a point mass at X_i, defined as the set function

$$\delta_{X_i}(A) = \begin{cases} 1 & \text{if } X_i \in A \\ 0 & \text{if } X_i \notin A \end{cases}.$$

The measure F_n is a random measure, because it depends on the sample $X_1, \ldots, X_n$.

The moments with respect to the measure F_n are called sample moments, and will be indicated by E_n. Thus, for a vector-valued function $f : \Omega_X \to \mathbb{R}^r$,

$$E_n f(X) = \int f(X) dF_n = n^{-1} \sum_{i=1}^{n} f(X_i) = n^{-1} \sum_{i=1}^{n} \begin{pmatrix} f_1(X_i) \\ \vdots \\ f_r(X_i) \end{pmatrix}.$$

The sample covariance matrix and the sample variance matrix can then be defined using E_n, as follows. If $g : \Omega_X \to \mathbb{R}^r$ is another vector-valued function, then $\mathrm{cov}_n(f(X), g(X))$ is defined as

$$E_n[(f(X) - E_n f(X))(g(X) - E_n g(X))^\mathsf{T}],$$

where $(\cdots)^\mathsf{T}$ denote the transpose of a matrix. The sample variance matrix $\mathrm{var}_n[f(X)]$ is then defined to be the sample covariance matrix between $f(X)$ and $f(X)$; that is,

$$\mathrm{var}_n[f(X)] = \mathrm{cov}_n[f(X), f(X)].$$

1.2　Principal Component Analysis

Suppose X is a random vector in $\mathbb{R}^p$. The principal components of X are defined to be the set of linear combinations of X that have the largest variances. Thus, at the population level, the first principal component is defined through the following maximization problem:

$$\text{maximize} \quad \mathrm{var}(\alpha^\mathsf{T} X) \quad \text{subject to } \|\alpha\| = 1.$$

Let α_1 be the solution to the above problem. Then $\alpha_1^\mathsf{T} X$ is called the first principal component at the population level. Let $\Sigma = \mathrm{var}(X)$. Then $\mathrm{var}(\alpha^\mathsf{T} X) = \alpha^\mathsf{T} \Sigma \alpha$, and so α_1 is the first eigenvector of Σ. Similarly, the kth principal component of X is defined by the problem of

$$\begin{aligned} &\text{maximizing} \quad \alpha^\mathsf{T} \Sigma \alpha \\ &\text{subject to} \quad \|\alpha\| = 1, \ \ell = 1, \ldots, k-1, \ \alpha^\mathsf{T} \alpha_\ell = 0. \end{aligned} \tag{1.1}$$

The solution is the kth eigenvector of Σ. The kth principal component at the population level is defined as the random variable $\alpha_k^\mathsf{T} X$.

Intuitively, the random variable $\alpha_1^\mathsf{T} X$ explains the most variation in X; $\alpha_2^\mathsf{T} X$ explains most variation in X left in the orthogonal complement of α_1. In this way, we decompose the variations of X sequentially by orthogonal linear combinations.

At the sample level, suppose that $X_1, \ldots, X_n$ is an independent and identically distributed (i.i.d.) sample of X. Let $\hat{\Sigma} = \mathrm{var}_n(X)$, and let $\hat{\alpha}_1, \ldots, \hat{\alpha}_k$ be the first k eigenvectors of $\hat{\Sigma}$. The first k sample-level principal components of X are

$$\{\hat{\alpha}_\ell^\mathsf{T} X_i : i = 1, \ldots, n\}, \quad \ell = 1, \ldots, k.$$

1.3 Generalized Eigenvalue Problem

Principal Component Analysis is one of many problems that can be formulated as a generalized eigenvalue problem. Let Σ and Λ be symmetric matrix and Λ be positive definite. The generalized eigenvalue problem is defined by the following iterative optimization problem: at the kth step

$$
\begin{aligned}
\text{maximizing} \quad & \alpha^\mathsf{T} \Sigma \alpha \\
\text{subject to} \quad & \alpha^\mathsf{T} \Lambda \alpha = 1, \ \alpha^\mathsf{T} \Lambda \alpha_\ell = 0, \ \ell = 1, \ldots, k-1,
\end{aligned}
\tag{1.2}
$$

where $\alpha_1, \ldots, \alpha_{k-1}$ are the maximizers in the previous $k-1$ steps. This is a generalization of problem (1.1) and can be reduced to it by making the transformation $\beta = \Lambda^{1/2} \alpha$. Then this problem becomes

$$
\begin{aligned}
\text{maximizing} \quad & \beta^\mathsf{T} \Lambda^{-1/2} \Sigma \Lambda^{-1/2} \beta \\
\text{subject to} \quad & \beta^\mathsf{T} \beta = 1, \ \beta^\mathsf{T} \beta_\ell = 0, \ \ell = 1, \ldots, k-1.
\end{aligned}
$$

Thus, the solution to problem (1.2) is $\alpha_k = \Lambda^{-1/2} \beta_k$, where β_k is the kth eigenvector of the symmetric matrix $\Lambda^{-1/2} \Sigma \Lambda^{-1/2}$.

We call α_k the kth eigenvector of the generalized eigenvalue problem (Σ, Λ). We abbreviate the phrase "generalized eigenvalue problem with respect to (Σ, Λ)" as $\mathrm{GEV}(\Sigma, \Lambda)$.

1.4 Multivariate Linear Regression

Let U and V be random vectors in $\mathbb{R}^p$ and $\mathbb{R}^q$. In multivariate linear regression, at the population level, we are interested in minimizing the least squares criterion

$$E\|U - BV\|^2$$

over all matrices in $\mathbb{R}^{p \times q}$. This problem has an explicit solution, which will be useful in discussing many problems in Sufficient Dimension Reduction.

Henceforth, we will say a random vector V is square integrable if $E\|V\|^2 < \infty$. By the Cauchy-Schwarz inequality, this is true if and only if each component of V has finite second moment. In the following, if A is a positive definite matrix, we write $A > 0$.

Theorem 1.1 *Suppose U and V are square integrable with $E(U) = 0$ and $E(V) = 0$ and $\mathrm{var}(V) > 0$. Then $E\|U - BV\|^2$ is uniquely minimized over $\mathbb{R}^{p \times q}$ by*

$$B^* = E(UV^\mathsf{T})[E(VV^\mathsf{T})]^{-1}.$$

PROOF. First, expand $E\|U - BV\|^2$ as

$$\begin{aligned}
E\|U - BV\|^2 &= E\|U - B^*V + B^*V - BV\|^2 \\
&= E\|U - B^*V\|^2 + 2\mathrm{tr}E[(U - B^*V)(B^*V - BV)^\mathsf{T}] + E\|B^*V - BV\|^2,
\end{aligned} \tag{1.3}$$

where $\mathrm{tr}(\cdots)$ stands for the trace of a matrix. The middle term on the right-hand side is 0, because

$$\begin{aligned}
E[(U - B^*V)(B^*V - BV)^\mathsf{T}] &= E[(U - B^*V)V^\mathsf{T}](B^* - B)^\mathsf{T} \\
&= [E(UV^\mathsf{T}) - E(UV^\mathsf{T})](B^* - B)^\mathsf{T} = 0.
\end{aligned}$$

Therefore

$$E\|U - BV\|^2 \geq E\|U - B^*V\|^2$$

for all $B \in \mathbb{R}^{p \times q}$.

To see that the minimizer B^* is unique, we note that if $B \neq B^*$, then the third term on the right-hand side of (1.3) is

$$E\|B^*V - BV\|^2 = \mathrm{tr}[(B^* - B)\mathrm{var}(V)(B^* - B)^\mathsf{T}],$$

which is greater than 0 because $\mathrm{var}(V)$ is positive definite. □

There are several variations of Theorem 1.1 that will also be useful.

Corollary 1.1 *Suppose U and V are square integrable and $\mathrm{var}(V) > 0$. Then the function $E\|U - a - BV\|^2$ is minimized uniquely by*

$$B^* = \mathrm{cov}(U, V)[\mathrm{var}(V)]^{-1}, \quad a^* = EU - B^*EV.$$

PROOF. Let $U_c = U - E(U)$ and $V_c = V - E(V)$. Then

$$E\|U - a - BV\|^2 = E\|U_c - BV_c\|^2 + \|EU - a - BE(V)\|^2$$

By Proposition 1.1 the first term is minimized at

$$B^* = E(U_c V_c^\mathsf{T})(EV_c V_c^\mathsf{T})^{-1} = \mathrm{cov}(U, V)[\mathrm{var}(V)]^{-1}.$$

The second term is 0 if $a^* = E(U) - B^*E(V)$. □

This result is also applicable if we replace the true distribution of (U, V) by its empirical distribution. Let $(U_1, V_1), \ldots, (U_n, V_n)$ be an i.i.d. sample of (U, V).

Corollary 1.2 *If $\mathrm{var}_n(V) > 0$, then the criterion $E_n\|U - a - BV\|^2$ is uniquely minimized by*

$$\hat{B} = \mathrm{cov}_n(U, V)(\mathrm{var}_n V)^{-1}, \quad \hat{a} = E_n U - \hat{B}E_n V.$$

1.5 Generalized Linear Model

Since one of the first ideas of Sufficient Dimension Reduction stems from a study of Generalized Linear Models under link violation (Li and Duan (1989), Li (1991)), it is helpful to review the basic structure and properties of the Generalized Linear Models. For more information on this topic, see McCullagh and Nelder (1989).

1.5.1 Exponential Family

Let Y be a random variable that takes values in $(\Omega_Y, \mathscr{F}_Y)$. We say that the distribution of Y belongs to an exponential family if the probability density function (p.d.f.) of Y has the form $c(\theta)e^{\theta y}$ with respect to some σ-finite measure ν on Ω_Y. This can be rewritten as

$$e^{\theta y - b(\theta)},$$

where $b(\theta) = -\log c(\theta)$. The moment generating function of Y can be easily computed, as follows:

$$M_Y(t) = \int e^t e^{\theta y - b(\theta)} d\nu(y) = e^{b(t+\theta)-b(\theta)} \int e^{(t+\theta)y-b(t+\theta)} d\nu(y) = e^{b(t+\theta)-b(\theta)}.$$

The cumulant generating function, defined as the natural log of the moment generating function, is then

$$C_Y(\theta) = b(t+\theta) - b(\theta).$$

The derivatives of the cumulant generating function evaluated at $t = 0$ generate cumulants, the first two of which are the mean and the variance:

$$\dot{C}_Y(0) = E_\theta(Y), \quad \ddot{C}_Y(0) = \mathrm{var}_\theta(Y). \tag{1.4}$$

See, for example, McCullagh (1987). It follows that

$$\dot{b}(\theta) = E_\theta(Y), \quad \ddot{b}(\theta) = \mathrm{var}_\theta(Y).$$

From the second equality we see that if $\mathrm{var}_\theta(Y) > 0$ for all θ, then $\dot{b}$ is a monotone increasing function, and therefore its inverse $\dot{b}^{-1}$ is a well defined function. If we denote $E_\theta(Y)$ by μ, then

$$\theta = \dot{b}^{-1}(\mu).$$

Moreover, $\mathrm{var}_\theta(Y)$ can be reexpressed in μ as $\ddot{b}(\dot{b}^{-1}(\mu))$. The function $\ddot{b} \circ \dot{b}^{-1}$ characterizes the mean-variance relation in an exponential family, and is called the *variance function*. We denote the variance function by $V(\mu)$.

1.5.2   Generalized Linear Models

Let X be a random vector in $\mathbb{R}^p$ as defined in Section 1.1. In a Generalized Linear Model we assume that Y is related with X by the conditional density

$$f_{Y|X}(y|x) \propto e^{\theta(x)y - b(\theta(x))}, \tag{1.5}$$

where $\theta(x)$ is a function of x. The regression relation between Y and X is modeled through the link function. Note that

$$\theta(x) = \dot{b}^{-1}(E(Y|x)).$$

We model $E(Y|x)$ by

$$E(Y|x) = \mu(\eta), \quad \eta = \alpha + \beta^\mathsf{T} x,$$

where $\mu(\eta)$ is called the mean function and $\eta = \alpha + \beta^\mathsf{T} x$ is called the the linear predictor or the linear index. Usually, we assume $\mu(\cdot)$ to be one-to-one, and its inverse μ^{-1} is called the link function.

Substituting the relation $\theta(x) = \dot{b}^{-1}(\mu(\alpha + \beta^\mathsf{T} x))$ into the conditional density (1.5), we have

$$f_{Y|X}(y|x; \alpha, \beta) \propto \exp\left\{(\dot{b}^{-1} \circ \mu)(\alpha + \beta^\mathsf{T} x)y - b((\dot{b}^{-1} \circ \mu)(\alpha + \beta^\mathsf{T} x))\right\}. \tag{1.6}$$

In Generalized Linear Models, α and β are estimated by maximum likelihood estimation based on the density (1.6). Suppose that $\mathbb{D}_n = \{(X_1, Y_1), \ldots, (X_n, Y_n)\}$ are a sample of i.i.d. observations on (X, Y). Then the joint log likelihood is proportional to

$$\begin{aligned} \ell(\alpha, \beta; \mathbb{D}_n) &= E_n\left\{(\dot{b}^{-1} \circ \mu)(\alpha + \beta^\mathsf{T} X)X - b((\dot{b}^{-1} \circ \mu)(\alpha + \beta^\mathsf{T} X))\right\} \\ &= E_n\left\{(\dot{b}^{-1} \circ \mu)(\gamma^\mathsf{T} \tilde{X})Y - b((\dot{b}^{-1} \circ \mu)(\gamma^\mathsf{T} \tilde{X}))\right\}, \end{aligned} \tag{1.7}$$

where

$$\gamma = \begin{pmatrix} \alpha \\ \beta \end{pmatrix}, \quad \tilde{X} = \begin{pmatrix} 1 \\ X \end{pmatrix}.$$

Differentiate (1.7) with respect to γ to obtain

$$\partial \ell(\gamma; \mathbb{D}_n)/\partial \gamma = E_n\left\{\partial[(\dot{b}^{-1} \circ \mu)(\gamma^\mathsf{T} \tilde{X})Y]/\partial \gamma - \partial[b((\dot{b}^{-1} \circ \mu)(\gamma^\mathsf{T} \tilde{X}))]/\partial \gamma\right\}.$$

This function is called the *score function*, and we denote it by $s(\gamma; \mathbb{D}_n)$. The derivatives in the score function are computed by the chain rule:

$$\frac{\partial(\dot{b}^{-1} \circ \mu)(\gamma^\mathsf{T} \tilde{X})}{\partial \gamma} = \frac{\partial \dot{b}^{-1}(\mu)}{\partial \mu} \frac{\partial \mu}{\partial \eta} \frac{\partial \eta}{\partial \gamma} = \frac{\tilde{X}\dot{\mu}(\gamma^\mathsf{T} \tilde{X})}{\ddot{b}(\dot{b}^{-1}(\mu(\gamma^\mathsf{T} \tilde{X})))} = \frac{\tilde{X}\dot{\mu}(\gamma^\mathsf{T} \tilde{X})}{V(\mu(\gamma^\mathsf{T} \tilde{X}))}.$$

Here, $\dot{\mu}(\eta)$ denote the function $\eta \mapsto \partial \mu/\partial \eta$. Similarly,

$$\frac{\partial b((\dot{b}^{-1} \circ \mu)(\gamma^\mathsf{T} \tilde{X}))}{\partial \gamma} = \frac{\partial b(\theta)}{\partial \theta}\bigg|_{\theta = \dot{b}^{-1}(\mu)} \times \frac{\partial \dot{b}^{-1}(\mu)}{\partial \mu} \frac{\partial \mu}{\partial \eta} \frac{\partial \eta}{\partial \gamma} = \frac{\tilde{X}\dot{\mu}(\gamma^\mathsf{T} \tilde{X})\mu(\gamma^\mathsf{T} \tilde{X})}{V(\mu(\gamma^\mathsf{T} \tilde{X}))}.$$

Hence the score function is written explicitly as

$$s(\gamma; \mathbb{D}_n) = E_n \left\{ \frac{\tilde{X}\dot{\mu}(\gamma^T\tilde{X})[Y - \mu(\gamma^T\tilde{X})]}{V(\mu(\gamma^T\tilde{X}))} \right\}.$$

This is completely specified by the mean function μ, which is our regression model, and the mean-variance relation $V(\mu)$, which is determined by the exponential family.

The parameter γ is usually estimated by the maximum likelihood estimation. Under the exponential family assumption, the log likelihood is concave and differentiable. Thus the maximum likelihood estimate can be found by solving the *likelihood equation*

$$s(\gamma; \mathbb{D}_n) = 0.$$

This is usually solved by the Newton-Raphson algorithm, or the Fisher scoring method. See, for example, Section 2.5.1 of McCullagh and Nelder (1989) for details.

The link function that makes $\theta(x) = \gamma^T\tilde{x}$ is called the natural link, or the canonical link. In other words μ has to make $\dot{b}^{-1} \circ \mu$ the identity mapping, which implies $\mu^{-1} = \dot{b}^{-1}$. Under the natural link the conditional density (1.6) reduces to

$$f_{Y|X}(y|x; \gamma) \propto \exp\left\{(\gamma^T\tilde{x})y - b(\gamma^T\tilde{x})\right\}.$$

The score function reduces to the simple form

$$s(\gamma; \mathbb{D}_n) = E_n\left[\tilde{X}(Y - \mu(\gamma^T\tilde{X}))\right].$$

We now illustrate the Generalized Linear Models by two simple examples.

Example 1.1 Suppose $Y \sim \text{Poisson}(\lambda)$. Then

$$f(y; \theta) \propto \lambda^y e^{-\lambda} = e^{y\log\lambda - \lambda} = e^{\theta y - e^\theta}.$$

Here, λ is the conventional parameter of a Poisson distribution, $\theta = \log\lambda$ is the canonical parameter, and the cumulant generating function of Y is

$$C_Y(t) = e^{\theta + t} - e^\theta.$$

From this we see that

$$\dot{b}^{-1}(\mu) = \log\mu, \quad \ddot{b}(\theta) = e^\theta, \quad V(\mu) = \exp(\log(\mu)) = \mu.$$

The natural link function is $\dot{b}^{-1}(\mu) = \log(\mu)$, and the score function is simply

$$E_n[\tilde{X}(Y - e^{\gamma^T\tilde{x}})] = 0.$$

This model is also known as the log linear regression model. $\qquad\square$

Example 1.2 Suppose, for a fixed p, Y has a binomial distribution $b(n,p)$, where p is a function of x. That is,

$$f(y) = \binom{n}{x} p^y (1-p)^{n-y} \propto e^{y \log \frac{p}{1-p} + n \log(1-p)}.$$

If we let $\theta = \log[p/(1-p)]$, then $n \log(1-p) = -n \log(1+e^\theta)$. The density $f(y)$ can be rewritten as the canonical form

$$f(y) \propto \exp[\theta y - n \log(1+e^\theta)].$$

Hence

$$b(\theta) = n \log(1+e^\theta), \quad \dot{b}(\theta) = n \frac{e^\theta}{1+e^\theta}, \quad \ddot{b}(\theta) = n \frac{e^\theta}{(1+e^\theta)^2}.$$

It follows that

$$\dot{b}^{-1}(\mu) = \log \frac{\mu/n}{1-\mu/n}, \quad (\ddot{b} \circ \dot{b}^{-1})(\mu) = n(\mu/n)(1-\mu/n).$$

Thus the natural link function is $\log \frac{\mu/n}{1-\mu/n}$, which is called the logit function, and the score function is

$$s(\gamma; \mathbb{D}_n) = E_n \left[\tilde{X} \left(Y - n \frac{e^{\gamma^\mathsf{T} \tilde{X}}}{1+e^{\gamma^\mathsf{T} X}} \right) \right].$$

This type of Generalized Linear Model is called the logistic regression.   □

1.6   Hilbert Space, Linear Manifold, Linear Subspace

The theory of Sufficient Dimension Reduction is geometric in nature, where inner product, orthogonality, and projection play a critical role. In this and the next two sections we bring together some geometric concepts and machineries that will be used repeatedly in this book. When developing these concepts we follow this path:

$$\text{group} \rightarrow \text{Abelian group} \rightarrow \text{vector space} \rightarrow \begin{cases} \text{normed space} \rightarrow \text{Banach space} \\ \text{inner product space} \rightarrow \text{Hilbert space} \end{cases}$$

More information about these topics can be found in Kelley (1955) and Conway (1990).

Let $\mathscr{H}$ be a set. Let $+$ be a mapping from $\mathscr{H} \times \mathscr{H}$ to $\mathscr{H}$ such that the following conditions are satisfied:

1. $+(+(g_1, g_2), g_3) = +(g_1, +(g_2, g_3))$;
2. there is a member e of $\mathscr{H}$ such that $+(e, g) = +(g, e) = g$ for all $g \in \mathscr{H}$;
3. for each $g \in \mathscr{H}$, there is a member $f \in \mathscr{H}$ such that $+(g, f) = e$.

The pair $(\mathcal{H},+)$ of the set $\mathcal{H}$ and the operation $+$ is called a group. A group is an Abelian group or commutative group if it satisfies the additional condition

4. $+(g,f) = +(f,g)$ for all $f,g \in \mathcal{H}$.

Usually, we write $+(g,f)$ as $g+f$, write f in statement 3 as $-g$, and write the e in statement 2 as 0. We call it the zero element of $\mathcal{H}$.

Now suppose $(\mathcal{H},+)$ is an Abelian group, and suppose there is a mapping $\cdot$ from $\mathbb{R} \times \mathcal{H}$ to $\mathcal{H}$ such that

5. for any $a,b \in \mathbb{R}$ and $f \in \mathcal{H}$, $\cdot(a,\cdot(b,f)) = \cdot(ab,f)$

6. for any $a \in \mathbb{R}$ and $f,g \in \mathcal{H}$, $\cdot(a,f+g) = \cdot(a,f) + \cdot(a,g)$.

7. for any $a,b \in \mathbb{R}$, $f \in \mathcal{H}$, $\cdot(a+b,f) = \cdot(a,f) + \cdot(b,f)$

8. for any $f \in \mathcal{H}$, $\cdot(1,f) = f$.

Usually, we write $\cdot(a,f)$ as $a \cdot f$ or simply af. An Abelian group $(\mathcal{H},+)$, together with the mapping $\cdot : \mathbb{R} \times \mathcal{H} \to \mathcal{H}$ that satisfies the above conditions, is called a vector space, and is denoted by $(\mathcal{H},+,\cdot)$. For simplicity, we will just say $\mathcal{H}$ is a vector space, without writing $+,\cdot$ explicitly.

For a vector space $\mathcal{H}$, if there is a mapping $u : \mathcal{H} \times \mathcal{H} \to \mathbb{R}$ that satisfies the following conditions:

9. for any $f,g \in \mathcal{H}, u(f,g) = u(g,f)$

10. for any $f,g,h \in \mathcal{H}, u(f+g,h) = u(f,h) + u(g,h)$

11. for any $a \in \mathbb{R}, f,g \in \mathcal{H}, u(af,g) = au(f,g)$,

then u is called a semi-inner product, and $(\mathcal{H},u)$ is called a semi-inner product space. Usually, we write $u(f,g)$ as $\langle f,g \rangle$. If, in addition,

12. for any $f \in \mathcal{H}, u(f,f) = 0$ implies f is the 0 element of $\mathcal{H}$,

then we call u an inner product and $(\mathcal{H},u)$ an inner product space.

Suppose $\mathcal{H}$ is a vector space. If there is a mapping $\rho : \mathcal{H} \to \mathbb{R}$ satisfies the following conditions

9′. for any $a \in \mathbb{R}, f \in \mathcal{H}, \rho(af) = |a|\rho(f)$

10′. for any $f,g \in \mathcal{H}, \rho(f+g) \leq \rho(f) + \rho(g)$

11′. for any $f \in \mathcal{H}, \rho(f) = 0$ implies that f is the 0 element in $\mathcal{H}$.

Then we call ρ a norm in $\mathcal{H}$ and $(\mathcal{H},\rho)$ a normed space. Usually, we write $\rho(f)$ as $\|f\|$.

Suppose $(\mathcal{H},u)$ is an inner product space, then it can be shown that the mapping

$$\mathcal{H} \to \mathbb{R}, \quad f \mapsto \langle f,f \rangle^{1/2}$$

is a norm in $\mathcal{H}$. So an inner product space is a special normed space with its norm defined by $\|f\| = \langle f,f \rangle^{1/2}$.

A sequence $\{f_n\}$ in a normed space $(\mathcal{H}, \|\cdot\|)$ is called a Cauchy sequence if, for any $\varepsilon > 0$, there is an m such that for all $n_1,n_2 > m$, $\|f_{n_1} - f_{n_2}\| < \varepsilon$. A normed space $(\mathcal{H}, \|\cdot\|)$ is said to be complete if every Cauchy sequence in $\mathcal{H}$ converges to a member of $\mathcal{H}$. That is, there is $f \in \mathcal{H}$ such that $\|f_n - f\| \to 0$. A complete normed space is called a Banach space. If an inner product space $(\mathcal{H}, \langle \cdot, \cdot \rangle)$ is complete in terms of the norm $\|f\| = \langle f, f \rangle^{1/2}$, then it is called a Hilbert space.

Suppose $\mathcal{H}$ is a vector space. A subset $\mathcal{S} \subseteq \mathcal{H}$ is called a linear manifold in $\mathcal{H}$ if

1. for any $f, g \in \mathcal{S}$, $f + g \in \mathcal{S}$
2. for $a \in \mathbb{R}$, $f \in \mathcal{S}$, $af \in \mathcal{S}$.

If $\mathcal{S}$ is a linear manifold in $\mathcal{H}$ and $\mathcal{S}$ is closed, then $\mathcal{S}$ is a linear subspace of $\mathcal{H}$.

1.7   Linear Operator and Projection

In later parts of the book, on those topics related to functional data and kernel mapping, we will frequently employ linear operators in Hilbert spaces and their coordinate representations.

Let $\mathcal{H}$ be a Hilbert space with inner product $\langle \cdot, \cdot \rangle$. A linear operator is a mapping $T : \mathcal{H} \to \mathcal{H}$ such that

1. for any $f, g \in \mathcal{H}$, $T(f + g) = T(f) + T(g)$
2. for any $a \in \mathbb{R}$, $T(af) = aT(f)$.

A linear operator T is said to be idempotent if $T^2 = T$. That is, for any $f \in \mathcal{H}$, $T(T(f)) = T(f)$. A linear operator T is self adjoint if, for any $f, g \in \mathcal{H}$,

$$\langle f, Tg \rangle = \langle Tf, g \rangle.$$

If a linear operator $P : \mathcal{H} \to \mathcal{H}$ is both idempotent and self adjoint, then it is called a projection.

If T is a linear operator, then $\ker(T)$, the kernel of T, is the set $\{f \in \mathcal{H} : Tf = 0\}$. It is easy to show that $\ker(T)$ is a linear manifold; it can also be shown that $\ker(T)$ is closed. Therefore $\ker(T)$ is a linear subspace of $\mathcal{H}$. The symbol $\mathrm{ran}(T)$ stands for the range of T, which is the set $\{Tf : f \in \mathcal{H}\}$. It is also easy to verify that $\mathrm{ran}(T)$ is a linear manifold in $\mathcal{H}$, but it may not be closed, and consequently $\mathrm{ran}(T)$ may not be a subspace of $\mathcal{H}$. We use $\overline{\mathrm{ran}}\, T$ to denote the closure of $\mathrm{ran}(T)$, which is always a subspace of $\mathcal{H}$. The range of a projection P is always closed, implying $\mathrm{ran}(P) = \overline{\mathrm{ran}}\,(P)$.

Sometimes we call a projection P the projection on to the subspace $\mathcal{S}$, where $\mathcal{S}$ is $\mathrm{ran}(P)$. Conversely, for any subspace $\mathcal{S}$ of $\mathcal{H}$, there is a unique projection P, whose range is $\mathcal{S}$.

Let $\mathcal{S}$ be a subspace of $\mathcal{H}$ and let f be a member of $\mathcal{H}$. Then there is a unique member of $\mathcal{S}$, say f^*, such that $f - f^* \perp \mathcal{S}$. Furthermore, it can be shown

$$f - f^* \perp \mathcal{S} \Leftrightarrow \|f - f^*\| \leq \|f - g\| \quad \text{for all } g \in \mathcal{S}.$$

The unique existence of f^* defines the mapping

$$\mathcal{H} \to \mathcal{S}, \quad f \mapsto f^*.$$

It can be shown that this mapping is precisely the mapping $P_{\mathcal{S}}$.

1.8 The Hilbert Space $\mathbb{R}^p(\Sigma)$

Now let us consider the special case of $\mathcal{H} = \mathbb{R}^p$. A member v of $\mathbb{R}^p$ is the vector $(v_1, \ldots, v_p)^{\mathsf{T}}$. For $u, v \in \mathbb{R}^p$, define

$$u + v = (u_1 + v_1, \ldots, u_p + v_p)^{\mathsf{T}}.$$

Then $(\mathbb{R}^p, +)$ is an Abelian group. For $v \in \mathbb{R}^p$, $a \in \mathbb{R}$, define

$$\cdot(a, v) = av = (av_1, \ldots, av_p)^{\mathsf{T}}.$$

Then $(\mathbb{R}^p, +, \cdot)$ is a vector space. Let $\Sigma \in \mathbb{R}^{p \times p}$ be a positive definite matrix. For $u, v \in \mathbb{R}^p$, define

$$\langle u, v \rangle = u^{\mathsf{T}} \Sigma v.$$

Then $(\mathbb{R}^p, +, \cdot, \langle \cdot, \cdot \rangle)$ is an inner product space. It is true that any finite-dimensional normed space is complete. Therefore $(\mathbb{R}^p, +, \cdot, \langle \cdot, \cdot \rangle)$ is also a Hilbert space. We will abbreviate this Hilbert space by the simple $\mathbb{R}^p(\Sigma)$.

Let $u_1, \ldots, u_m$ be a collection of vectors in $\mathbb{R}^p$. Let $\mathscr{S}$ be the set

$$\operatorname{span}(u_1, \ldots, u_n) = \{c_1 u_1 + \cdots + c_m u_m : c_1, \ldots, c_m \in \mathbb{R}\}.$$

It is easy to see that this is a linear manifold in $\mathbb{R}^p$. Because this linear manifold has finite dimension, it is also closed. Thus it is a subspace of $\mathbb{R}^p$. We call this subspace the linear span of $u_1, \ldots, u_m$, and write it as $\operatorname{span}(u_1, \ldots, u_m)$. Conversely, any linear subspace of $\mathscr{S}$ is a linear span of a set of vectors in $\mathbb{R}^p$. The projection on to a subspace of the Euclidean space $\mathbb{R}^p$ can be expressed explicitly. Let $\mathscr{S} = \operatorname{span}\{u_1, \ldots, u_m\}$, and let B denote the matrix $(u_1, \ldots u_m)$. We use $P_{\mathscr{S}}$ to denote the projection on to the subspace $\mathscr{S}$. Let

$$P_B(\Sigma) = B(B^{\mathsf{T}} \Sigma B)^{-1} B^{\mathsf{T}} \Sigma.$$

Proposition 1.1 *The linear operator*

$$P_{\mathscr{S}} : \quad \mathbb{R}^p \to \mathbb{R}^p, \quad v \mapsto P_B(\Sigma) v$$

is a projection.

PROOF. We first note that

$$P_B(\Sigma) P_B(\Sigma) = B(B^{\mathsf{T}} \Sigma B)^{-1} (B^{\mathsf{T}} \Sigma B)(B^{\mathsf{T}} \Sigma B)^{-1} B^{\mathsf{T}} \Sigma = P_B(\Sigma).$$

Hence, for any $v \in \mathbb{R}^p(\Sigma)$, $P_{\mathscr{S}}^2(v) = [P_B(\Sigma)]^2 v = P_B(\Sigma) v = P_{\mathscr{S}}(v)$. Thus $P_{\mathscr{S}}$ is idempotent. Moreover, for any $u, v \in \mathbb{R}^p$,

$$\begin{aligned}
\langle P_{\mathscr{S}}(u), v \rangle &= (P_B(\Sigma) u)^{\mathsf{T}} \Sigma v \\
&= u^{\mathsf{T}} \Sigma B(B^{\mathsf{T}} \Sigma B)^{-1} B^{\mathsf{T}} \Sigma v \\
&= u^{\mathsf{T}} \Sigma [B(B^{\mathsf{T}} \Sigma B)^{-1} B^{\mathsf{T}} \Sigma] v \\
&= \langle u, P_{\mathscr{S}}(v) \rangle.
\end{aligned}$$

Thus $P_{\mathscr{S}}$ is self-adjoint. $\qquad \square$

1.9   Coordinate Representation

To implement kernel related methods for in Chapters 12 through 14, we need to use the coordinate representations of a function or a linear operator in finite-dimensional Hilbert spaces. Our notations are adopted from Horn and Johnson (1985). Let $\mathscr{H}$ be a finite-dimensional Hilbert space with spanning system $\mathscr{B} = \{b_1, \ldots, b_m\}$. Then any member f of $\mathscr{H}$ can be written as $c_1 b_1 + \cdots + c_m b_m$, where $c_1, \ldots, c_m \in \mathbb{R}$. The vector $(c_1, \ldots, c_m)^{\mathsf{T}}$ is called the coordinate of f relative to the spanning system $\mathscr{B}$. In the cases where $b_1, \ldots, b_m$ are linearly dependent, a member of $\mathscr{H}$ can have many coordinate representations, but this does not concern us because the function it represents is unique. The coordinate of f relative to a spanning system $\mathscr{B}$ is written as $[f]_{\mathscr{B}}$. Thus we can write $f = [f]_{\mathscr{B}}^{\mathsf{T}} b_{1:m}$, where $b_{1:m} = (b_1, \ldots, b_m)^{\mathsf{T}}$.

The matrix of inner products $\{\langle b_i, b_j \rangle_{\mathscr{H}} : i, j = 1, \ldots, m\}$ is called the Gram matrix of $\mathscr{B}$, and is written as $G_{\mathscr{B}}$. We can represent the inner product between two members of $\mathscr{H}$ using their coordinates and the Gram matrix as follows

$$\langle f, g \rangle_{\mathscr{H}} = \sum_{i=1}^{m} \sum_{j=1}^{m} ([f]_{\mathscr{B}})_i ([g]_{\mathscr{B}})_j \langle b_i, b_j \rangle_{\mathscr{H}} = [f]_{\mathscr{B}}^{\mathsf{T}} G_{\mathscr{B}} [g]_{\mathscr{B}},$$

where, for example, $([f]_{\mathscr{B}})_i$ stands for the ith component of the vector $[f]_{\mathscr{B}}$.

Let $\mathscr{H}_1$ and $\mathscr{H}_2$ be two finite dimensional Hilbert spaces spanned by the subsets $\mathscr{B}_1 = \{b_1^{(1)}, \ldots, b_{m_1}^{(1)}\} \subseteq \mathscr{H}_1$ and $\mathscr{B}_2 = \{b_1^{(2)}, \ldots, b_{m_2}^{(2)}\} \subseteq \mathscr{H}_2$, respectively. Let $A : \mathscr{H}_1 \to \mathscr{H}_2$ be a linear operator and f be a member of $\mathscr{H}_1$. Then Af is a member of $\mathscr{H}_2$ and its coordinate representation relative to $\mathscr{B}_2$ is obtained by the following calculation:

$$\begin{aligned} Af &= A \left(\sum_{i=1}^{m_1} ([f]_{\mathscr{B}_1})_i b_i^{(1)} \right) \\ &= \sum_{i=1}^{m_1} ([f]_{\mathscr{B}_1})_i A b_i^{(1)} \\ &= \sum_{i=1}^{m_1} ([f]_{\mathscr{B}_1})_i \sum_{j=1}^{m_2} ([A b_i^{(1)}]_{\mathscr{B}_2})_j b_j^{(2)}. \end{aligned}$$

We see that the coordinate of Af relative to $\mathscr{B}_2$ is simply the vector

$$\left\{ \sum_{i=1}^{m_1} ([f]_{\mathscr{B}_1})_i ([A b_i^{(1)}]_{\mathscr{B}_2})_j : j = 1, \ldots, m_2 \right\}.$$

Motivated by this relation, we write the matrix

$$\begin{pmatrix} ([A b_1^{(1)}]_{\mathscr{B}_2})_1 & \cdots & ([A b_{m_1}^{(1)}]_{\mathscr{B}_2})_1 \\ \vdots & & \vdots \\ ([A b_1^{(1)}]_{\mathscr{B}_2})_{m_2} & \cdots & ([A b_{m_1}^{(1)}]_{\mathscr{B}_2})_{m_2} \end{pmatrix}$$

as ${}_{\mathscr{B}_2}[A]_{\mathscr{B}_1}$, and call it the coordinate representation of the linear operator A. Using this notation we can conveniently write

$$[Af]_{\mathscr{B}_2} = ({}_{\mathscr{B}_2}[A]_{\mathscr{B}_1})[f]_{\mathscr{B}_1}.$$

Carrying the logic in this notation further, let $\mathscr{H}_3$ be a third finite-dimensional Hilbert space with spanning system $\mathscr{B}_3 = \{b_1^{(3)}, \ldots, b_{m_3}^{(3)}\}$. Let $A_1 : \mathscr{H}_1 \to \mathscr{H}_2$ and $A_2 : \mathscr{H}_2 \to \mathscr{H}_3$ be linear operators, and let f be a member of $\mathscr{H}_1$. Then

$$\begin{aligned}
_{\mathscr{B}_3}[A_2 A_1 f]_{\mathscr{B}_1} &= {}_{\mathscr{B}_3}[A_2(A_1 f)]_{\mathscr{B}_1} \\
&= ({}_{\mathscr{B}_3}[A_2]_{\mathscr{B}_2})[A_1 f]_{\mathscr{B}_1} \\
&= ({}_{\mathscr{B}_3}[A_2]_{\mathscr{B}_2})({}_{\mathscr{B}_2}[A_1]_{\mathscr{B}_1})([f]_{\mathscr{B}_1}).
\end{aligned}$$

In the meantime, we have

$$_{\mathscr{B}_3}[(A_2 A_1) f]_{\mathscr{B}_1} = ({}_{\mathscr{B}_3}[A_2 A_1]_{\mathscr{B}_1})[f]_{\mathscr{B}_1}.$$

Comparing these two equations we have

$$_{\mathscr{B}_3}[A_2 A_1]_{\mathscr{B}_1} = ({}_{\mathscr{B}_3}[A_2]_{\mathscr{B}_2})({}_{\mathscr{B}_2}[A_1]_{\mathscr{B}_1}).$$

1.10 Generalized Linear Models under Link Violation

As a prelude to Sufficient Dimension Reduction, we present a case study of a property of the Generalized Linear Model when the link function μ^{-1} (or equivalently, μ) is misspecified. This property was discovered by Li and Duan (1989), and is a starting point (and arguably the starting point) of Sufficient Dimension Reduction. This property states that the maximum likelihood estimate of α, β is Fisher consistent even if the link function is misspecified, provided there is some symmetry in X.

Recall that, under the canonical link, the log likelihood for the Generalized Linear Model is $(\alpha + \beta^{\mathsf{T}} X) Y - b(\alpha + \beta^{\mathsf{T}} X)$, where b is a monotone increasing function, which means that b is a convex function. The maximum likelihood estimation pertains to maximizing $E_n[(\alpha + \beta^{\mathsf{T}} X) Y - b(\alpha + \beta^{\mathsf{T}} X)]$. At the population level, we maximize the function

$$E[(\alpha + \beta^{\mathsf{T}} X) Y - b(\alpha + \beta^{\mathsf{T}} X)].$$

Let (α_0, β_0) represent the true values of (α, β). Then, by the well known theory of maximum likelihood estimation (see, for example, Lehmann and Casella (1998a), Theorem 3.2), the above function is uniquely maximized at (α_0, β_0). What is interesting is that, as shown in Li and Duan (1989), even when the function b is misspecified, the maximizer of the above function (with incorrectly specified b) still gives the correct direction of β_0, as long as X has linear conditional expectation.

Specifically, let us replace b by c, an arbitrary convex function. Equivalently, one can regard this replacement as replacing the true mean function μ by an arbitrary monotone function. That is, let

$$R(\alpha, \beta) = E[(\alpha + \beta^{\mathsf{T}} X) Y - c(\alpha + \beta^{\mathsf{T}} X)]. \tag{1.8}$$

The main result of Li and Duan (1989) is that, if (α_1, β_1) minimizes $R(\alpha, \beta)$, then, under the condition that $E(X|\beta_0^\mathsf{T}X)$ is a linear function of $\beta_0^\mathsf{T}X$, β_1 is proportional to β_0. Since $R_0(\alpha,\beta)$ is the expectation of the true log likelihood, (α_0,β_0) is the true parameter. Hence β_1 is proportional to the true parameter β_0 regardless of whether the link function is correctly specified, provided that X has a linear conditional mean given $\beta_0^\mathsf{T}X$. Since this assumption will appear frequently in this book, we give it a formal definition.

Assumption 1.1 *We say that X satisfies the linear conditional mean assumption with respect to β if $E(X|\beta^\mathsf{T}X)$ is a linear function of $\beta^\mathsf{T}X$.*

It can be shown that if this condition is satisfied for all $\beta \in \mathbb{R}^p$, then X has an elliptically contoured distribution, and vice versa. See Eaton (1986). The consequence of the linear conditional mean assumption is that $E(X|\beta^\mathsf{T}X)$ can be expressed as the projection in $\mathbb{R}^p(\Sigma)$.

Lemma 1.1 *Suppose $\beta^\mathsf{T}\Sigma\beta > 0$ and X satisfies Assumption 1.1, then*

$$E(X - EX|\beta^\mathsf{T}X) = P_\beta^\mathsf{T}(\Sigma)(X - EX).$$

PROOF. Denote EX by μ. Because $E(X|\beta^\mathsf{T}X)$ is linear in $\beta^\mathsf{T}X$, we have

$$E(X|\beta^\mathsf{T}X) = c + D\beta^\mathsf{T}X$$

where $c \in \mathbb{R}^p$ and $D \in \mathbb{R}^{p \times d}$. Take unconditional expectation on both sides to obtain

$$\mu = c + D\beta^\mathsf{T}\mu.$$

Substituting this into the previous line, we have $E(Z|\beta^\mathsf{T}Z) = D\beta^\mathsf{T}Z$. Multiply both sides of this equation from the right-hand side by $Z^\mathsf{T}\beta$, to obtain

$$E(Z|\beta^\mathsf{T}Z)Z^\mathsf{T}\beta = D\beta^\mathsf{T}ZZ^\mathsf{T}\beta.$$

Taking unconditional expectation on both sides, we have $\Sigma\beta = D\beta^\mathsf{T}\Sigma\beta$, which implies the desired relation. □

We now prove the result of Li and Duan (1989).

Proposition 1.2 *Suppose the following conditions hold.*

1. The probability density function of $Y|X$ is proportional to

$$\exp\{(\alpha_0 + \beta_0^\mathsf{T}x)y - b(\alpha_0 + \beta_0^\mathsf{T}x)\};$$

2. Assumption 1.1 is satisfied for β_0;

3. $\beta^\mathsf{T}\Sigma\beta > 0$;

4. c is a strictly convex function; $(\alpha + \beta^\mathsf{T}X)Y - c(\alpha + \beta^\mathsf{T}X)$ is integrable for all $\alpha \in \mathbb{R}$ and $\beta \in \mathbb{R}^p$.

If (α_1,β_1) is the maximizer of $R_1(\alpha,\beta)$, then $\beta_1 \propto \beta_0$.

PROOF. Condition 1 implies

$$E(Y|X) = \mu(\alpha_0 + \beta_0^\mathsf{T} X) = \dot{b}(\alpha_0 + \beta_0^\mathsf{T} X).$$

This also implies $E(Y|X) = E(Y|\beta_0^\mathsf{T} X)$. By definition,

$$R(\alpha, \beta) = E[(\alpha + \beta^\mathsf{T} X)Y] - E[c(\alpha + \beta^\mathsf{T} X)].$$

The first term on the right-hand side is

$$
\begin{aligned}
E[(\alpha + \beta^\mathsf{T} X)Y] &= E[(\alpha + \beta^\mathsf{T} X)E(Y|X)] \\
&= E[(\alpha + \beta^\mathsf{T} X)E(Y|\beta_0^\mathsf{T} X)] \\
&= E[E(\alpha + \beta^\mathsf{T} X|\beta_0^\mathsf{T} X)Y] \\
&= E[a + b^\mathsf{T} E(X|\beta_0^\mathsf{T} X)Y].
\end{aligned}
$$

By Lemma 1.1,

$$E(X|\beta_0^\mathsf{T} X) = P_{\beta_0}^\mathsf{T}(\Sigma)(X - \mu) + \mu.$$

Hence

$$
\begin{aligned}
E[(\alpha + \beta^\mathsf{T} X)Y] &= E[\alpha + \beta^\mathsf{T} P_{\beta_0}^\mathsf{T}(\Sigma)(X - \mu) + \beta^\mathsf{T} \mu)Y] \\
&= E[\alpha + \beta^\mathsf{T} P_{\beta_0}^\mathsf{T}(\Sigma)X - \beta^\mathsf{T} P_{\beta_0}^\mathsf{T}(\Sigma)\mu + \beta^\mathsf{T} \mu)Y] \\
&= E[\alpha + \beta^\mathsf{T} \mu - \beta^\mathsf{T} P_{\beta_0}^\mathsf{T}(\Sigma)\mu + \beta^\mathsf{T} P_{\beta_0}^\mathsf{T}(\Sigma)X)Y] \\
&= E[\gamma + \delta\beta_0^\mathsf{T} X)Y],
\end{aligned}
$$

where $\gamma = \alpha + \beta^\mathsf{T} \mu - \beta^\mathsf{T} P_{\beta_0}^\mathsf{T}(\Sigma)\mu$, $\delta = \beta^\mathsf{T} \Sigma \beta_0 / (\beta_0^\mathsf{T} \Sigma \beta_0)$. Meanwhile, by Jensen's inequality,

$$
\begin{aligned}
E[c(\alpha + \beta^\mathsf{T} X)] &= E[E(c(\alpha + \beta^\mathsf{T} X)|\beta_0^\mathsf{T} X)] \\
&\geq E[c(E(\alpha + \beta^\mathsf{T} X)|\beta_0^\mathsf{T} X)] \\
&= E[c(\gamma + \delta\beta_0^\mathsf{T} X)].
\end{aligned}
$$

So $R(\alpha, \beta) \leq R(\gamma, \delta\beta_0)$ for all $\alpha \in \mathbb{R}$, $\beta \in \mathbb{R}^p$. Because c is strictly convex, R has a unique maximizer, leading to $\beta_1 = \delta\beta_0$. $\qquad\square$

   To gain more insight into this somewhat surprising result, in Figure 1.1 we plot a surface S, representing $E(Y|X) = E(Y|\beta^\mathsf{T} X)$, and a plain P, representing the least-squares estimate of the surface S. The surface varies only in one direction, which corresponds to the direction of β. We see that, as long as the distribution of X is symmetric about the direction in which the surface varies (as indicated by the dotted rectangle at the bottom), the gradient of the plain shares the same direction with the gradient of the surface. Thus, if we are interested in the direction, but not the magnitude, of the gradient, then we need not have an accurate estimate of the surface, because a simple model such as the linear model provides a reasonably good estimate of the gradient direction, even though it is a poor estimate of the surface.

This result means that we can borrow the symmetry in the distribution of X to estimate the direction of the gradient using a rough estimate of the surface. This is significant because, especially in a high-dimensional setting, it is not easy to estimate a surface accurately as it involves high-dimensional smoothing, which causes what is known as "the curse of dimensionality". However, fitting a least-squares plane does not involve any smoothing, and thus it provides a reasonably good estimate of the gradient direction in the high-dimensional setting. The same can be said of any other simple parametric regression model, such as the quadratic regression model. This is one of the most important principles of Sufficient Dimension Reduction, and leads to many useful estimators described in the subsequent chapters.

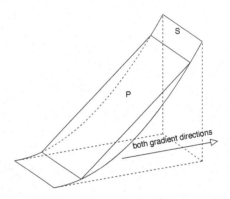

Figure 1.1 *Estimating gradient direction under link violation in Generalized Linear Model.*

Chapter 2

Dimension Reduction Subspaces

2.1 Conditional Independence   17
2.2 Sufficient Dimension Reduction Subspace   21
2.3 Transformation Laws of Central Subspace   24
2.4 Fisher Consistency, Unbiasedness, and Exhaustiveness   25

2.1 Conditional Independence

In this section we review the definitions and basic properties of conditional independence, which are crucial for Sufficient Dimension Reduction. Our development is in terms of conditional independence of σ-fields — rather than that of random variables — which is more general than can be found in most text books. Besides the generality, it is in fact easier to discuss conditional independence in terms of σ-fields than in terms of random variables. After developing the properties of conditional independence in the general setting, we then discuss the special cases where the relevant σ-fields are generated by random variables. For more information on conditional independence of σ-fields, see Hoffmann-Jorgensen (1994), page 460.

As in Chapter 1, let $(\Omega, \mathscr{F}, P)$ be a probability space. Let $\mathscr{G}_1$ and $\mathscr{G}_2$ be two sub σ-fields of $\mathscr{F}$. We say that $\mathscr{G}_1$ and $\mathscr{G}_2$ are independent, and write $\mathscr{G}_1 \perp\!\!\!\perp \mathscr{G}_2$, if, for any $A \in \mathscr{G}_1$ and $B \in \mathscr{G}_2$, A and B are independent; that is, $P(A \cap B) = P(A)P(B)$. Now let $\mathscr{G}_3$ be a third σ-field and, for any $A \in \mathscr{F}$, let $P(A|\mathscr{G}_3)$ be the conditional probability of A given $\mathscr{G}_3$; that is, $P(A|\mathscr{G}_3)$ is a mapping from Ω that (i) is measurable with respect to $\mathscr{G}_3$; (ii) satisfies $\int_G P(A|\mathscr{G}_3)dP = P(G \cap A)$ for all $G \in \mathscr{G}_3$. We say that $\mathscr{G}_1$ and $\mathscr{G}_2$ are conditional independent given $\mathscr{G}_3$, and write $\mathscr{G}_1 \perp\!\!\!\perp \mathscr{G}_2|\mathscr{G}_3$ if, for every $A \in \mathscr{G}_1$ and $B \in \mathscr{G}_2$, we have

$$P(A \cap B|\mathscr{G}_3) = P(A|\mathscr{G}_3)P(B|\mathscr{G}_3), \quad a.s.\ P.$$

Let $\mathscr{P} = \{A \cap B : A \in \mathscr{G}_1, B \in \mathscr{G}_2\}$. It is easy to show that $\mathscr{P}$ is a π-system. Let $(\mathscr{G}_1, \mathscr{G}_2)$ — or simply $\mathscr{G}_1, \mathscr{G}_2$ — denote the σ-field generated by $\mathscr{P}$. The next proposition gives an equivalent definition of conditional independence.

Proposition 2.1 *Let $\mathscr{G}_1$, $\mathscr{G}_2$, and $\mathscr{G}_3$ be sub-σ-fields of $\mathscr{F}$. Then the following statements are equivalent:*

1. $\mathscr{G}_1 \perp\!\!\!\perp \mathscr{G}_2 | \mathscr{G}_3$;

2. $P(A|\mathscr{G}_2, \mathscr{G}_3) = P(A|\mathscr{G}_3)$ a.s. P for each $A \in \mathscr{G}_1$.

Before proving this proposition, we first prove a proposition that will be used many times in later development.

Proposition 2.2 *Let X, Y, and Z be random variables defined on $(\Omega, \mathscr{F}, P)$, where the product XY is integrable with respect to P. Then*

$$E(XE(Y|Z)) = E(E(X|Z)Y) = E(E(X|Z)E(Y|Z)). \tag{2.1}$$

PROOF. This is because

$$
\begin{aligned}
E(XE(Y|Z)) &= E(E(XE(Y|Z)|Z)) \\
&= E(E(X|Z)E(Y|Z)) \\
&= E(E(E(X|Z)Y|Z)) \\
&= E(E(X|Z)Y).
\end{aligned}
\qquad \square
$$

We are now ready to prove Proposition 2.1.

PROOF OF PROPOSITION 2.1. $1 \Rightarrow 2$. It suffices to show that, for any $A_1 \in \mathscr{G}_1$, $P(A_1|\mathscr{G}_3)$ is the conditional probability of A_1 given $(\mathscr{G}_2, \mathscr{G}_3)$. Since $P(A_1|\mathscr{G}_3)$ is measurable with respect to $(\mathscr{G}_2, \mathscr{G}_3)$, it suffices to show that, for any $B \in (\mathscr{G}_2, \mathscr{G}_3)$ we have

$$\int_B P(A_1|\mathscr{G}_3)dP = P(A_1 \cap B).$$

Because $\mathscr{P} = \{A_2 A_3 : A_2 \in \mathscr{G}_2, A_3 \in \mathscr{G}_3\}$ is a π system generating $(\mathscr{G}_2, \mathscr{G}_3)$, by the π-λ Theorem (see, for example, Billingsley (1995), page 42), it suffices to show the above inequality for all $B \in \mathscr{P}$. Let $B = A_2 \cap A_3$ where $A_2 \in \mathscr{G}_2$, $A_3 \in \mathscr{G}_3$. Then

$$\int_B P(A_1|\mathscr{G}_3)dP = E[I_{A_2}I_{A_3}E(I_{A_1}|\mathscr{G}_3)]$$

By Proposition 2.1, the right-hand side can be written as

$$E[E(I_{A_2}I_{A_3}|\mathscr{G}_3)E(I_{A_1}|\mathscr{G}_3)] = E[I_{A_3}E(I_{A_2}|\mathscr{G}_3)E(I_{A_1}|\mathscr{G}_3)].$$

By 1, we have $E(I_{A_2}|\mathscr{G}_3)E(I_{A_1}|\mathscr{G}_3) = E(I_{A_2}I_{A_1}|\mathscr{G}_3)$. Hence the right-hand side above reduces to $E[I_{A_3}E(I_{A_2}I_{A_1}|\mathscr{G}_3)]$, which, by Proposition 2.1, is the same as

$$E[E(I_{A_3}|\mathscr{G}_3)I_{A_2}I_{A_1}] = E[I_{A_3}I_{A_2}I_{A_1}] = E(I_{A_1}I_B) = P(A_1 \cap B),$$

as desired.

$2 \Rightarrow 1$. The relation in statement 1 is equivalent to

$$E(I_{A_1}I_{A_2}|\mathscr{G}_3) = E(I_{A_1}|\mathscr{G}_3)E(I_{A_2}|\mathscr{G}_3)$$

for all $A_1 \in \mathscr{G}_1$ and $A_2 \in \mathscr{G}_2$. Because $\mathscr{G}_3 \subseteq (\mathscr{G}_2, \mathscr{G}_3)$, the left-hand side is

$$E[E(I_{A_1} I_{A_2} | \mathscr{G}_2, \mathscr{G}_3) | \mathscr{G}_3] = E[E(I_{A_1} I_{A_2} | \mathscr{G}_2, \mathscr{G}_3) | \mathscr{G}_3]$$
$$= E[I_{A_2} E(I_{A_1} | \mathscr{G}_2, \mathscr{G}_3) | \mathscr{G}_3].$$

By statement 2, $E(I_{A_1} | \mathscr{G}_2, \mathscr{G}_3) = E(I_{A_1} | \mathscr{G}_3)$. Hence the right-hand side reduces to $E(I_{A_2} | \mathscr{G}_3) E(I_{A_1} | \mathscr{G}_3)$. □

Conditional independence satisfies a set of axioms — called semigraphoid axioms — which will be extremely useful for our discussions. These axioms were first proved in Dawid (1979) as properties for conditional independence. They were proposed as axioms for constructing graphical models and for making causal inference by Pearl and Verma (1987) and Pearl et al. (1989). While in its original form the semigraphoid axioms were stated in terms of random variables, here we state them in terms of sub σ-fields of $\mathscr{F}$, which makes the statements more general and the proof more transparent.

Theorem 2.1 *Let* $\mathscr{G}_1, \ldots, \mathscr{G}_4$ *be sub-σ-fields of* $\mathscr{F}$. *Then the following statements hold true:*

1. $\mathscr{G}_1 \perp\!\!\!\perp \mathscr{G}_2 | \mathscr{G}_3 \Rightarrow \mathscr{G}_2 \perp\!\!\!\perp \mathscr{G}_1 | \mathscr{G}_3$;

2. $\mathscr{G}_1 \perp\!\!\!\perp (\mathscr{G}_2, \mathscr{G}_3) | \mathscr{G}_4 \Rightarrow \mathscr{G}_1 \perp\!\!\!\perp \mathscr{G}_2 | \mathscr{G}_4$;

3. $\mathscr{G}_1 \perp\!\!\!\perp (\mathscr{G}_2, \mathscr{G}_3) | \mathscr{G}_4 \Rightarrow \mathscr{G}_1 \perp\!\!\!\perp \mathscr{G}_2 | (\mathscr{G}_3, \mathscr{G}_4)$;

4. $\mathscr{G}_1 \perp\!\!\!\perp \mathscr{G}_2 | (\mathscr{G}_3, \mathscr{G}_4), \ \mathscr{G}_1 \perp\!\!\!\perp \mathscr{G}_3 | \mathscr{G}_4 \Rightarrow \mathscr{G}_1 \perp\!\!\!\perp (\mathscr{G}_2, \mathscr{G}_3) | \mathscr{G}_4$.

PROOF. *1.* Let $A_1 \in \mathscr{G}_1, A_2 \in \mathscr{G}_2$. Then

$$P(A_1 \cap A_2 | \mathscr{G}_3) = P(A_1 | \mathscr{G}_3) P(A_2 | \mathscr{G}_3)$$

if and only if

$$P(A_2 \cap A_1 | \mathscr{G}_3) = P(A_2 | \mathscr{G}_3) P(A_1 | \mathscr{G}_3).$$

2. Since $\mathscr{G}_1 \perp\!\!\!\perp (\mathscr{G}_2, \mathscr{G}_3) | \mathscr{G}_4$, we have, for any $A_1 \in \mathscr{G}_1$ and $B = A_2 \cap A_3$ where $A_2 \in \mathscr{G}_2$ and $A_3 \in \mathscr{G}_3$, we have $P(A_1 \cap B | \mathscr{G}_4) = P(A_1 | \mathscr{G}_4) P(B | \mathscr{G}_4)$. Take $A_3 = \Omega$. Then $B = A_2$ and we have $P(A_1 \cap A_2 | \mathscr{G}_4) = P(A_1 | \mathscr{G}_4) P(A_2 | \mathscr{G}_4)$. Since this is true for all $A_1 \in \mathscr{G}_1$ and $A_2 \in \mathscr{G}_2$, we have $\mathscr{G}_1 \perp\!\!\!\perp \mathscr{G}_2 | \mathscr{G}_4$.

3. We need to show that, for any $A_1 \in \mathscr{G}_1, A_2 \in \mathscr{G}_2$, we have

$$E(I_{A_1} I_{A_2} | \mathscr{G}_3, \mathscr{G}_4) = E(I_{A_1} | \mathscr{G}_3, \mathscr{G}_4) E(I_{A_2} | \mathscr{G}_3, \mathscr{G}_4).$$

We will show that the right-hand side is the conditional expectation on the left-hand side. Since the right-hand side is measurable with respect to $(\mathscr{G}_3, \mathscr{G}_4)$, we only need to show that, for any $B \in (\mathscr{G}_3, \mathscr{G}_4)$,

$$E[I_B E(I_{A_1} | \mathscr{G}_3, \mathscr{G}_4) E(I_{A_2} | \mathscr{G}_3, \mathscr{G}_4)] = P(A_1 \cap A_2 \cap B).$$

Since $\mathscr{P}$ is a π-system generating $(\mathscr{G}_3, \mathscr{G}_4)$, by the π-λ Theorem, it suffices to show that the above equality holds for all $B = A_3 \cap A_4$ where $A_3 \in \mathscr{G}_3$ and $A_4 \in \mathscr{G}_4$. For such B, the left-hand side becomes

$$E[I_{A_3}I_{A_4}E(I_{A_1}|\mathscr{G}_3,\mathscr{G}_4)E(I_{A_2}|\mathscr{G}_3,\mathscr{G}_4)].$$

Because $\mathscr{G}_1 \perp\!\!\!\perp (\mathscr{G}_2,\mathscr{G}_3)|\mathscr{G}_4$, by statement 2 we have $\mathscr{G}_1 \perp\!\!\!\perp \mathscr{G}_3|\mathscr{G}_4$. Hence, by Proposition 2.1, $E(I_{A_1}|\mathscr{G}_3,\mathscr{G}_4) = E(I_{A_1}|\mathscr{G}_4)$ and the above becomes

$$\begin{aligned}
E[I_{A_3}I_{A_4}E(I_{A_1}|\mathscr{G}_4)E(I_{A_2}|\mathscr{G}_3,\mathscr{G}_4)] &= E[I_{A_4}E(I_{A_1}|\mathscr{G}_4)E(I_{A_2}I_{A_3}|\mathscr{G}_3,\mathscr{G}_4)] \\
&= E[E(I_{A_4}E(I_{A_1}|\mathscr{G}_4)I_{A_2}I_{A_3}|\mathscr{G}_3,\mathscr{G}_4)] \\
&= E[E(I_{A_1}|\mathscr{G}_4)I_{A_4}I_{A_2}I_{A_3}].
\end{aligned}$$

By Proposition 2.1, the right-hand side can be rewritten as

$$E[I_{A_1}E(I_{A_4}I_{A_2}I_{A_3}|\mathscr{G}_4)] = E[I_{A_1}I_{A_4}E(I_{A_2}I_{A_3}|\mathscr{G}_4)] = E[E(I_{A_1}I_{A_4}|\mathscr{G}_4)E(I_{A_2}I_{A_3}|\mathscr{G}_4)].$$

By $\mathscr{G}_1 \perp\!\!\!\perp (\mathscr{G}_2,\mathscr{G}_3)|\mathscr{G}_4$ again, the right-hand side can be rewritten as

$$E[E(I_{A_1}I_{A_4}I_{A_2}I_{A_3}|\mathscr{G}_4)] = E(I_{A_1}I_{A_4}I_{A_2}I_{A_3}) = P(A_1 \cap A_2 \cap B).$$

This proves statement 3.

4. We need to show that, for any $A_1 \in \mathscr{G}_1$, $B \in (\mathscr{G}_2,\mathscr{G}_3)$, we have

$$E(I_{A_1}I_B|\mathscr{G}_4) = E(I_{A_1}|\mathscr{G}_4)E(I_B|\mathscr{G}_4).$$

We will show that the right-hand side is the conditional expectation on the left-hand side, for which it suffices to show that, for any $A_4 \in \mathscr{G}_4$,

$$E(I_{A_4}E(I_{A_1}|\mathscr{G}_4)E(I_B|\mathscr{G}_4)) = E(I_{A_4}I_{A_1}I_B). \tag{2.2}$$

By the π-λ Theorem, it suffices to verify this equality for all $B = A_2 A_3$ where $A_2 \in \mathscr{G}_2$ and $A_3 \in \mathscr{G}_3$. For such a set B, the left-hand side is

$$E(I_{A_4}E(I_{A_1}|\mathscr{G}_4)E(I_{A_2}I_{A_3}|\mathscr{G}_4)) = E(E(I_{A_1}|\mathscr{G}_4)E(I_{A_4}I_{A_2}I_{A_3}|\mathscr{G}_4)). \tag{2.3}$$

By Proposition 2.1, the condition $\mathscr{G}_1 \perp\!\!\!\perp \mathscr{G}_3|\mathscr{G}_4$, and Proposition 2.1, we can rewrite the right-hand side of (2.3) as

$$\begin{aligned}
E(E(I_{A_1}|\mathscr{G}_4)I_{A_4}I_{A_2}I_{A_3}) &= E(E(I_{A_1}|\mathscr{G}_3,\mathscr{G}_4)I_{A_4}I_{A_2}I_{A_3}) \\
&= E(E(I_{A_1}|\mathscr{G}_3,\mathscr{G}_4)E(I_{A_4}I_{A_2}I_{A_3}|\mathscr{G}_3,\mathscr{G}_4)) \\
&= E(E(I_{A_1}|\mathscr{G}_3,\mathscr{G}_4)E(I_{A_2}|\mathscr{G}_3,\mathscr{G}_4)I_{A_4}I_{A_3}).
\end{aligned}$$

By the condition $\mathscr{G}_1 \perp\!\!\!\perp \mathscr{G}_2|\mathscr{G}_3,\mathscr{G}_4$, the right-hand side can be rewritten as

$$\begin{aligned}
E(E(I_{A_1}I_{A_2}|\mathscr{G}_3,\mathscr{G}_4)I_{A_4}I_{A_3}) &= E(E(I_{A_1}I_{A_2}I_{A_4}I_{A_3}|\mathscr{G}_3,\mathscr{G}_4)) \\
&= E(I_{A_1}I_{A_2}I_{A_4}I_{A_3}) \\
&= E(I_{A_1}I_{A_4}I_B),
\end{aligned}$$

which is the right-hand side of (2.2). □

Now let us consider the conditional independence of random elements. Suppose X_1, ..., X_4 are random elements defined on $(\Omega, \mathscr{F}, P)$ taking values in $(\Omega_{X_1}, \mathscr{F}_{X_1}), \ldots, (\Omega_{X_4}, \mathscr{F}_{X_4})$, respectively. Let $\sigma(X_i)$ be the sub-σ-field of $\mathscr{F}$ generated by X_i; that is, $\sigma(X_i) = X_i^{-1}(\mathscr{F}_{X_i})$. We say that X_1 and X_2 are independent if $\sigma(X_1) \perp\!\!\!\perp \sigma(X_2)$. In this case we write $X_1 \perp\!\!\!\perp X_2$. We say that X_1 and X_2 are independent given X_3 if $\sigma(X_1) \perp\!\!\!\perp \sigma(X_2) | \sigma(X_3)$. In this case we write $X_1 \perp\!\!\!\perp X_2 | X_3$. We say that X_1 and X_2 are conditionally independent given (X_3, X_4) if

$$\sigma(X_1) \perp\!\!\!\perp \sigma(X_2) | (\sigma(X_3), \sigma(X_4)).$$

In this case we write $X_1 \perp\!\!\!\perp X_2 | (X_3, X_4)$.

The following corollaries state the special cases of Proposition 2.1 and Theorem 2.1 for random elements.

Corollary 2.1 *Let X_1, X_2, and X_3 be random elements defined on $(\Omega, \mathscr{F}, P)$. Then the following statements are equivalent:*

1. $X_1 \perp\!\!\!\perp X_2 | X_3$;
2. $P(A | X_2, X_3) = P(A | X_3)$ a.s. P for each $A \in \sigma(X_1)$.

Corollary 2.2 *Let $X_1, \ldots, X_4$ be random variables defined on $(\Omega, \mathscr{F}, P)$. Then the following statements hold true:*

1. $X_1 \perp\!\!\!\perp X_2 | X_3 \Rightarrow X_2 \perp\!\!\!\perp X_1 | X_3$;
2. $X_1 \perp\!\!\!\perp (X_2, X_3) | X_4 \Rightarrow X_1 \perp\!\!\!\perp X_2 | X_4$;
3. $X_1 \perp\!\!\!\perp (X_2, X_3) | X_4 \Rightarrow X_1 \perp\!\!\!\perp X_2 | (X_3, X_4)$;
4. $X_1 \perp\!\!\!\perp X_2 | (X_3, X_4), X_1 \perp\!\!\!\perp X_3 | X_4 \Rightarrow X_1 \perp\!\!\!\perp (X_2, X_3) | X_4$.

2.2 Sufficient Dimension Reduction Subspace

Now let $X : \Omega \to \mathbb{R}^p$ be a random vector measurable with respect to $\mathscr{R}^p$, the Borel σ-field in $\mathbb{R}^p$, and let $Y : \Omega \to \mathbb{R}$ be a random variable measurable with respect to $\mathscr{R}$, the Borel σ-field in $\mathbb{R}$. Sufficient Dimension Reduction is concerned with the situations where the distribution of Y given X depends on X only through a set of linear combinations of X. That is, there is a matrix $\beta \in \mathbb{R}^{p \times r}$, where $r \leq p$, such that

$$Y \perp\!\!\!\perp X | \beta^\mathsf{T} X. \tag{2.4}$$

This relation is unchanged if we replace β by any βA, where $A \in \mathbb{R}^{r \times r}$ is any nonsingular matrix; that is,

$$Y \perp\!\!\!\perp X | \beta^\mathsf{T} X \Leftrightarrow Y \perp\!\!\!\perp X | (\beta A)^\mathsf{T} X.$$

This is because there is a one-to-one relation between $\beta^\mathsf{T} X$ and $A^\mathsf{T} \beta^\mathsf{T} X = (\beta A)^\mathsf{T} X$. Thus the identifiable parameter in (2.4) is the space spanned by the columns of β, rather than β itself. We denote the column space of β by span(β).

The conditional independence relation (2.4) is extracted from useful regression models such as the Generalized Linear Model, where the conditional density of $f_{Y|X}$ depends on X through $\beta^{\mathsf{T}}X$, where $\beta \in \mathbb{R}^p$. That is,

$$f_{Y|X}(y|x) = h(y, \beta^{\mathsf{T}}x). \tag{2.5}$$

for some function h. See Li (1991). It is easy to see that (2.5) implies (2.4). However, (2.4) is much more general than (2.5) as specified by the Generalized Linear Model in Chapter 1, because (i) β does not have to be vector, and (ii) no assumption is imposed on the form of h. Some other popular regression models can also be stated, at least in form, as special cases of the Sufficient Dimension Reduction problem (2.4). For example, the Single Index Model is of the form

$$Y = f(\beta^{\mathsf{T}}X) + \varepsilon,$$

where $\beta \in \mathbb{R}^p$ ad $X \perp\!\!\!\perp \varepsilon$. See Ichimura (1993). Similarly, the Multiple Index Model

$$Y = f(\beta_1^{\mathsf{T}}X, \ldots, \beta_d^{\mathsf{T}}X) + \varepsilon,$$

where $\beta_1, \ldots, \beta_d \in \mathbb{R}^p$ and $\varepsilon \perp\!\!\!\perp X$, is also a special case of (2.4). See, for example, Yin et al. (2008).

The original form of Sufficient Dimension Reduction proposed by Li (1991) is

$$Y = h(\beta^{\mathsf{T}}X, \varepsilon), \tag{2.6}$$

where $\varepsilon \perp\!\!\!\perp X$. It is again easy to see that (2.6) implies (2.4). The form (2.4) was proposed in Cook (1994). See also Cook (1998).

If condition (2.4) is satisfied for β, then we call span(β) a Sufficient Dimension Reduction subspace, or SDR subspace. Conversely, a subspace $\mathscr{S}$ of $\mathbb{R}^p$ is an SDR subspace if (2.4) is satisfied for any β such that $\mathscr{S} = \text{span}(\beta)$. Obviously, SDR subspace always exists, because $Y \perp\!\!\!\perp X|X$ always holds, which means $\mathbb{R}^p$ is an SDR subspace. Furthermore, the SDR subspace is not unique, because of the following fact.

Proposition 2.3 *If $\mathscr{S}_1$ is an SDR subspace and $\mathscr{S}_2$ is any subspace of $\mathbb{R}^p$ that contains $\mathscr{S}_1$, then $\mathscr{S}_2$ is also an SDR subspace.*

PROOF. Let β_1 and β_2 be matrices such that span(β_1) $= \mathscr{S}_1$ and span(β_2) $= \mathscr{S}_2$. Because the σ-fields $(X, \beta_2^{\mathsf{T}}X)$ and $\sigma(X)$ are identical, we have $Y \perp\!\!\!\perp (X, \beta_2^{\mathsf{T}}X)|\beta_1^{\mathsf{T}}X$. By Theorem 2.1, statement (iii), we have

$$Y \perp\!\!\!\perp X|(\beta_1^{\mathsf{T}}X, \beta_2^{\mathsf{T}}X).$$

However, since the σ-fields $(\beta_1^{\mathsf{T}}X, \beta_2^{\mathsf{T}}X)$ and $\beta_2^{\mathsf{T}}X$ are the same, we have $Y \perp\!\!\!\perp X|\beta_2^{\mathsf{T}}X$, which means $\mathscr{S}_2$ is an SDR subspace. $\qquad\square$

Since the SDR subspace is not unique, we naturally prefer the smallest SDR subspace, which achieves maximum dimension reduction. Under some mild assumptions, if $\mathscr{S}_1$ and $\mathscr{S}_2$ are SDR subspaces, then so is $\mathscr{S}_1 \cap \mathscr{S}_2$. The assumptions for this to hold is very weak: it was shown in Yin et al. (2008) that, if X is supported by an M-set (or matching set), then this condition is satisfied. Since the result is rather technical we omit its proof and take it for granted in this book. That is, we make the following assumption.

Assumption 2.1 *Let $\mathfrak{A}$ be the class of all SDR subspaces. We assume that*

$$\cap\{\mathscr{S} : \mathscr{S} \in \mathfrak{A}\} \in \mathfrak{A}.$$

That is, the intersection of all SDR subspaces is itself an SDR subspace.

Under this assumption we define the intersection of all SDR subspaces as the target of estimation in Sufficient Dimension Reduction.

Definition 2.1 *Suppose Assumption 2.1. The subspace*

$$\cap\{\mathscr{S} : \mathscr{S} \in \mathfrak{A}\}$$

is called central SDR subspace or the central subspace, and is written as $\mathscr{S}_{Y|X}$. The dimension d of $\mathscr{S}_{Y|X}$ is called the structural dimension.

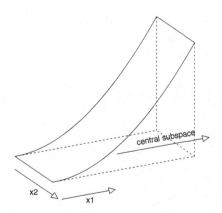

Figure 2.1 *Sufficient Dimension Reduction subspace.*

Figure 2.1 illustrates the idea of the central subspace. The surface in the figure represents the dependence of $E(Y|X)$ on X, where $X = (X_1, X_2)$ is a random vector in $\mathbb{R}^2$. In this case, $E(Y|X_1, X_2)$ is a function of X_1 alone; that is, $E(Y|X_1, X_2) = E(Y|X_1)$.

Under the additional assumption that the distribution of Y given X is determined by the conditional mean $E(Y|X)$, such as the regression model

$$Y = f(X) + \varepsilon,$$

where $X \perp\!\!\!\perp \varepsilon$, the central subspace is spanned by $(1,0)$, as indicated by the long arrow in the plot. The random variable $\beta^\mathsf{T} X$, where $\beta = (1,0)^\mathsf{T}$, carries all the information about the conditional distribution of $Y|X$.

2.3   Transformation Laws of Central Subspace

A useful property of the central subspace is that it transforms equivariantly under affine transformation of X (see, for example, Cook (1998)). Using this property we can work with standardized X in estimation of $\mathscr{S}_{Y|X}$, as we usually do in Sufficient Dimension Reduction.

Theorem 2.2 *If $A \in \mathbb{R}^{p \times p}$ be a nonsingular matrix and $b \in \mathbb{R}^p$, then*

$$\mathscr{S}_{Y|X} = A^\mathsf{T} \mathscr{S}_{Y|AX+b}.$$

PROOF. Let

$$\mathfrak{A}_{Y|X} = \{\mathscr{S} : \mathscr{S} \text{ is an SDR space of } Y \text{ versus } X\},$$
$$A^{-\mathsf{T}} \mathfrak{A}_{Y|X} = \{A^{-\mathsf{T}} \mathscr{S} : \mathscr{S} \in \mathfrak{A}_{Y|X}\},$$
$$\mathfrak{A}_{Y|AX+b} = \{\mathscr{S} : \text{ is an SDR subspace of } Y \text{ versus } AX + b\},$$

where, for a matrix A, $A^{-\mathsf{T}}$ denote $(A^{-1})^\mathsf{T}$ or $(A^\mathsf{T})^{-1}$. We first show that

$$A^{-\mathsf{T}} \mathfrak{A}_{Y|X} = \mathfrak{A}_{Y|AX+b}. \tag{2.7}$$

Let $\mathscr{S}' \in A^{-\mathsf{T}} \mathfrak{A}_{Y|X}$. Then $\mathscr{S}' = A^{-\mathsf{T}} \mathscr{S}$, where $\mathscr{S} \in \mathfrak{A}_{Y|X}$. Let β be a matrix such that $\mathrm{span}(\beta) = \mathscr{S}$. Then $Y \perp\!\!\!\perp X | \beta^\mathsf{T} X$, which implies

$$Y \perp\!\!\!\perp | \beta^\mathsf{T} A^{-1}[(AX + b) - b] \Rightarrow Y \perp\!\!\!\perp X | \beta^\mathsf{T} A^{-1}(AX + b)$$
$$\Rightarrow Y \perp\!\!\!\perp X | (A^{-\mathsf{T}} \beta)^\mathsf{T} (AX + b)$$
$$\Rightarrow Y \perp\!\!\!\perp (AX + b) | (A^{-\mathsf{T}} \beta)^\mathsf{T} (AX + b).$$

The last line means $\mathrm{span}(A^{-\mathsf{T}} \beta) = A^{-\mathsf{T}} \mathscr{S} = \mathscr{S}' \in \mathfrak{A}_{AX+b}$. Hence $A^{-\mathsf{T}} \mathfrak{A}_{Y|X} \subseteq \mathfrak{A}_{Y|AX+b}$. Next, let $\mathscr{S}' \in \mathfrak{A}_{Y|AX+b}$ and let β' be a matrix such that $\mathrm{span}(\beta') = \mathscr{S}'$. Then

$$Y \perp\!\!\!\perp AX + b | (\beta')^\mathsf{T} (AX + b) \Rightarrow Y \perp\!\!\!\perp X | (A^\mathsf{T} \beta')^\mathsf{T} X.$$

So $\mathrm{span}(A^\mathsf{T} \beta') \in \mathfrak{A}_{Y|X}$, which implies $\mathfrak{A}_{Y|AX+b} \subseteq A^{-\mathsf{T}} \mathscr{S}_{Y|X}$. Thus we have proved (8.3).

The relation (8.3) can be equivalently written as

$$\{\mathscr{S} : \mathscr{S} \in \mathfrak{A}_{Y|X}\} = A^\mathsf{T} \{\mathscr{S}' : \mathscr{S}' \in \mathfrak{A}_{Y|AX+b}\} = \{A^\mathsf{T} \mathscr{S}' : \mathscr{S}' \in \mathfrak{A}_{Y|AX+b}\}.$$

Take intersection of the classes of sets on both sides to obtain

$$\mathscr{S}_{Y|X} = \cap\{\mathscr{S} : \mathscr{S} \in \mathfrak{A}_{Y|X}\}$$
$$= \cap\{A^{\mathsf{T}}\mathscr{S}' : \mathscr{S}' \in \mathfrak{A}_{Y|AX+b}\}$$
$$= A^{\mathsf{T}} \cap \{\mathscr{S}' : \mathscr{S}' \in \mathfrak{A}_{Y|AX+b}\} = A^{\mathsf{T}}\mathscr{S}_{Y|AX+b},$$

which is the desired equality.   □

Typically, to estimate the central subspace $\mathscr{S}_{Y|X}$, we start with an estimator of a positive semidefinite matrix M such that span(M) is a subspace of $\mathscr{S}_{Y|Z}$, where $Z = \Sigma^{-1/2}(X - \mu)$, $\Sigma = \text{var}(X)$, and $\mu = E(X)$. Let $s \leq d$ be the rank of M, and let $v_1, \ldots, v_s$ be the vectors of M corresponding the nonzero eigenvalues. Then we use the sample-level counterpart of $\mathscr{S} = \text{span}(v_1, \ldots, v_s)$ as estimator of $\mathscr{S}_{Y|Z}$. The space $\Sigma^{-1/2}\mathscr{S}$ is a subspace of $\Sigma^{-1/2}\mathscr{S}_{Y|Z}$ which, by Theorem 2.2, is the central subspace $\mathscr{S}_{Y|X}$.

Another type of transformation is that on the response Y. Because

$$Y \perp\!\!\!\perp X | \beta^{\mathsf{T}}X \Rightarrow g(Y) \perp\!\!\!\perp X | \beta^{\mathsf{T}}X,$$

any SDR subspace of Y versus X is also a subspace of $g(Y)$ versus X. Let $\mathfrak{A}$ be the collection of all SDR subspaces for Y versus X, and $\mathfrak{B}$ be the collection of all SDR subspaces of $g(Y)$ versus X. Then $\mathfrak{A} \subseteq \mathfrak{B}$. It follows that $\cap\{\mathscr{S} : \mathscr{S} \in \mathfrak{A}\} \supseteq \{\mathscr{S} : \mathscr{S} \in \mathfrak{B}\}$. Thus we have proved the following theorem.

Theorem 2.3 *For any measurable function $g(Y)$, we have $\mathscr{S}_{g(Y)|X} \subseteq \mathscr{S}_{Y|X}$.*

2.4   Fisher Consistency, Unbiasedness, and Exhaustiveness

In estimation theory, a statistic can often be represented as a mapping defined on the family of all distributions of a random element W to the parameter space. Specifically, let $\mathfrak{F}$ be the family of all distributions of W, let Θ be the parameter space, and let F_n be the empirical distribution based on an i.i.d. sample $W_1, \ldots, W_n$ of observations on W. For example, in our context, W is (X, Y) and W_i is (X_i, Y_i). A statistical functional T is a mapping from $\mathfrak{F}$ to some space $\mathbb{S}$. Many statistics can be represented as $T(F_n)$. Let $\theta_0 \in \Theta$ be the parameter to be estimated.

Definition 2.2 *We say that a statistic $T(F_n)$ is Fisher consistent for estimating θ_0 if $T(F_0) = \theta_0$.*

For example, suppose $T(F) = E_F(X)$, and we are interested in estimating $\mu_0 = E_{F_0}(X)$. Then $T(F_n) = \bar{X}$ and $T(F_0) = \mu_0$. So $\bar{X}$ is Fisher consistent estimator of μ_0. Similarly, if $\Sigma(F) = \text{var}_F(X)$, then $\Sigma(F_n)$, the sample covariance matrix of $X_1, \ldots, X_n$, is a Fisher consistent estimate of true covariance matrix $\text{var}_{F_0}(X)$.

In the context of Sufficient Dimension Reduction, let $\mathfrak{F}$ represent the collection of all distributions of (X, Y), let F_n represent the empirical distribution based on an i.i.d. sample $(X_1, Y_1), \ldots, (X_n, Y_n)$ of (X, Y), and let F_0 be the true distribution of (X, Y). The parameter space is the collection of all p dimensional subspace of $\mathbb{R}^p$. This is called the Grassman manifold $\mathbb{G}_{p,d}$. The central subspace $\mathscr{S}_{Y|X}$ is a member of $\mathbb{G}_{p,d}$. An estimator of the central subspace typically takes the form $M(F_n)$, where M is a mapping from $\mathfrak{F}$ to $\mathbb{R}^{p \times r}$. Following Ye and Weiss (2003), we call the estimator $M(F_n)$ a *candidate matrix*.

Definition 2.3 *We say that an estimator $M(F_n)$ is an unbiased estimator of $\mathscr{S}_{Y|X}$ if* $\mathrm{span}(M(F_0)) \subseteq \mathscr{S}_{Y|X}$. *We say that $M(F_n)$ is an exhaustive estimator of $\mathscr{F}_{Y|X}$ if* $\mathrm{span}(M(F_0)) \supseteq \mathscr{S}_{Y|X}$. *We say that $M(F_n)$ is a Fisher consistent estimate of $\mathscr{S}_{Y|X}$ if* $\mathrm{span}(M(F_0)) = \mathscr{S}_{Y|X}$.

Chapter 3

Sliced Inverse Regression

3.1 Sliced Inverse Regression: Population-Level Development   27
3.2 Limitation of SIR   30
3.3 Estimation, Algorithm, and R-codes   31
3.4 Application: The Big Mac Index   33

3.1 Sliced Inverse Regression: Population-Level Development

Sliced Inverse Regression (SIR), introduced by Li (1991), is the first and most commonly known Sufficient Dimension Reduction estimator. The term "inverse regression" refers to the conditional expectation $E(X|Y)$. The word "inverse" is used because, in usual regression analysis, what is of interest is the conditional mean $E(Y|X)$. The word "slice" refers to the fact that we estimate the conditional mean $E(X|Y)$ by taking an interval of Y.

Similar to the development in Section 1.10, here we also require linearity of conditional mean of the form $E(X|\beta^\mathsf{T}X)$, more general in form than Assumption 1.1.

Assumption 3.1 *Let $\beta \in \mathbb{R}^{p \times d}$ be a matrix such that* $\mathrm{span}(\beta) = \mathscr{S}_{Y|X}$. *We assume that $E(X|\beta^\mathsf{T}X)$ is a linear function of the d-dimensional random vector $\beta^\mathsf{T}X$.*

The next lemma is a generalization of Lemma 1 in Section 1.10; its proof is omitted.

Lemma 3.1 *Suppose $\beta^\mathsf{T}\Sigma\beta$ is positive definite. Then*

$$E[X - E(X)|\beta^\mathsf{T}X] = P_\beta^\mathsf{T}(\Sigma)[X - E(X)].$$

A random vector is said to have an elliptical distribution if there is a positive definite matrix A such that the density of X depends on x only through $x^\mathsf{T}Ax$; that is,

$$f_X(x) = h(x^\mathsf{T}Ax)$$

for some function $h : \mathbb{R} \to \mathbb{R}$. The following result was proved in Eaton (1986).

Lemma 3.2 *If X is integrable and has an elliptical distribution then $E(X|B^\mathsf{T}X)$ is linear in B for any matrix B. If $E(X|v^\mathsf{T}X)$ is linear in $v^\mathsf{T}X$ for each $v \in \mathbb{R}^p$, then X has an elliptical distribution.*

We now prove that SIR is unbiased.

Theorem 3.1 *Suppose X is square-integrable and $\Sigma = \mathrm{var}(X)$ is nonsingular. Then, under Assumption 3.1,*

$$\Sigma^{-1}[E(X|Y) - E(X)] \in \mathscr{S}_{Y|X}.$$

PROOF. First, assume $E(X) = 0$. Then

$$
\begin{aligned}
E(X|Y) &= E(E(X|\beta^\mathsf{T}X, Y)|Y)\\
&= E(E(X|\beta^\mathsf{T}X)|Y)\\
&= E(P_\beta^\mathsf{T}(\Sigma)X|Y)\\
&= P_\beta^\mathsf{T}(\Sigma)E(X|Y),
\end{aligned}
\tag{3.1}
$$

where the second equality holds because $Y \perp\!\!\!\perp X|\beta^\mathsf{T}X$; the third equality follows from Assumption 3.1 and Lemma 3.1. Because $P_\beta^\mathsf{T}(\Sigma) = \Sigma P_\beta(\Sigma)\Sigma^{-1}$, the right-hand side of (3.1) can be rewritten as $\Sigma P_\beta(\Sigma)\Sigma^{-1}E(X|Y)$, and consequently,

$$E(X|Y) = \Sigma P_\beta(\Sigma)\Sigma^{-1}E(X|Y).$$

Hence $\Sigma^{-1}E(X|Y) \in \mathrm{span}(P_\beta(\Sigma)) = \mathscr{S}_{Y|X}$. In the case where $E(X) \neq 0$, we simply apply the above result to $X - E(X)$. □

The above result shows that, under the linear conditional mean assumption, we can recover the gradient direction of the regression function $E(Y|X)$, or more generally the conditional distribution $F_{Y|X}$, by estimating the inverse conditional expectation $E(X|Y)$. This is significant because, to estimate the forward conditional moment $E(Y|X)$ nonparametrically, we need to smooth over a p-dimensional space; but to estimate $E(X|Y)$ nonparametrically, we only need to smooth over 1-dimensional space – the space of Y, which is much more accurate than high-dimensional smoothing. This is similar to the phenomenon observed in Section 1.10, where the symmetry in the distribution of X allows us to use linear regression to recover the gradient direction of $E(Y|X)$, thus avoiding fitting a high-dimensional surface nonparametrically.

Since we do not know the true β, Assumption 3.1 cannot be verified. Hence, in practice, we replace Assumption 3.1 by the stronger assumption that X has an elliptical distribution. We can intuitively verify the elliptical distribution condition by looking at the scatter plot matrix of the components of X, as we do in Section 3.4, Figure 3.3. Even though elliptical shape of the distribution of (X_i, X_j) for all $i \neq j$ does not imply the ellipticity of the joint distribution of $X = (X_1, \ldots, X_p)$, the scatter plot matrix is often a simple and effective way of detecting non-ellipticity.

Figure 3.1 illustrates how the vector $E(X|Y) - E(X)$ recovers the central subspace in the special case where $\Sigma = I_p$. When $E(Y|X) = E(Y|\beta^\mathsf{T}X)$, the vector $E(X|Y) - E(X)$ is aligned with the direction of β as long as the distribution of X is symmetric about the direction of β, which is guaranteed by the linear conditional mean condition.

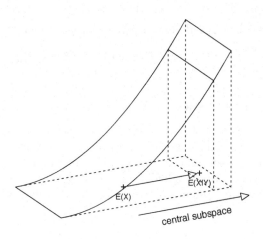

Figure 3.1 *Illustration of unbiasedness of SIR.*

Corollary 3.1 *Under the assumptions of Theorem 3.1,*

$$\text{span}(\Sigma^{-1}\text{cov}[E(X|Y)]\Sigma^{-1}) \subseteq \mathscr{S}_{Y|X}.$$

PROOF. Let $U = E(X|Y) - E(X)$. We need to span$[\Sigma^{-1}E(UU^{\mathsf{T}})\Sigma^{-1}] = \mathscr{S}_{Y|X}$. Since Σ is nonsingular, it is equivalent to

$$\text{span}[\Sigma^{-1}E(UU^{\mathsf{T}})] \subseteq \mathscr{S}_{Y|X} \Rightarrow \text{span}[E(UU^{\mathsf{T}})] \subseteq \Sigma\mathscr{S}_{Y|X}$$
$$\Rightarrow (\Sigma\mathscr{S}_{Y|X})^{\perp} \subseteq \text{span}[E(UU^{\mathsf{T}})]^{\perp}.$$

Now let $v \in (\Sigma\mathscr{S}_{Y|X})^{\perp}$. Because, by Theorem 3.1, $U \in \Sigma\mathscr{S}_{Y|X}$, we have $(\Sigma\mathscr{S}_{Y|X})^{\perp} \subseteq$ span$(U)^{\perp}$. Hence $v \perp U$, which implies $E(v^{\mathsf{T}}U)^2 = 0$, or $v^{\mathsf{T}}E(UU^{\mathsf{T}})v = 0$. Hence $v \in \text{span}[E(UU^{\mathsf{T}})]^{\perp}$. □

Let Λ_{SIR} denote the matrix $\text{cov}[E(X|Y)]$. This corollary implies that we can use the column space of

$$\mathscr{S}_{\text{SIR}} = \text{span}(\Sigma^{-1}\Lambda_{\text{SIR}}\Sigma^{-1})$$

to recover at least a part of the central subspace. This can be formulated as solving a generalized eigenvalue problem, as described in Section 1.3. That is, $\mathscr{S}_{\text{SIR}}$ is spanned by the set of eigenvectors $\{v : \Lambda_{\text{SIR}}v = \lambda\Sigma v, \lambda > 0\}$ in GEV$(\Lambda_{\text{SIR}}, \Sigma)$. This means we first solve the standard eigenvalue problem $\mathrm{GEV}(\Sigma^{-1/2}\Lambda_{\mathrm{SIR}}\Sigma^{-1/2}, I_p)$:

$$A = \{u : \Sigma^{-1/2}\Lambda_{\mathrm{SIR}}\Sigma^{-1/2} = \lambda v, \lambda > 0\},$$

and then recover $\mathscr{S}_{\mathrm{SIR}}$ by the set of transformed eigenvectors $\{\Sigma^{-1/2}u : u \in A\}$.

We can easily extend this result as follows using the relation $\mathscr{S}_{g(Y)|X} \subseteq \mathscr{S}_{Y|X}$.

Corollary 3.2 *If* $E(X|\beta^{\mathsf{T}}X)$ *is linear in* $\beta^{\mathsf{T}}X$ *then for any measurable function* $g(Y)$,

$$\Sigma^{-1}[E(X|g(Y)) - E(X)] \in \mathscr{S}_{Y|X}.$$

3.2   Limitation of SIR

While the theory of the above section guarantees that $\mathscr{S}_{\mathrm{SIR}}$ is always a subspace of $\mathscr{S}_{Y|X}$, it says nothing about whether it can recover the entire central subspace or merely a proper subspace thereof. In this section we use an example to demonstrate the cases where SIR can fail to recover $\mathscr{S}_{Y|X}$ fully.

Suppose X is a p-dimensional random vector with $E(X) = 0$ and

$$Y = f(X_1) + \varepsilon,$$

where $\varepsilon \perp\!\!\!\perp X$, $\Sigma = I_p$, X has a spherical distribution, and f is symmetric about 0. In this case the central subspace is spanned by $(1, 0, \ldots, 0)$. We will show that $E(X|Y) = 0$. First, because f is symmetric, $f(-X_1) = f(X_1)$; so (Y, X_1) and $(Y, -X_1)$ have the same distribution. Hence

$$E(X_1|Y) = E(-X_1|Y), \quad \text{which implies} \quad E(X_1|Y) = 0.$$

For $i \neq 1$, because X has a spherical distribution and $X \perp\!\!\!\perp \varepsilon$, we have

$$(X_1, \ldots, X_p, \varepsilon) \overset{\mathscr{D}}{=} (X_1, \ldots, -X_i, \ldots, X_p, \varepsilon)$$

where $\overset{\mathscr{D}}{=}$ means the two sides of the equality have the same distribution. Hence

$$E(X_i|f(X_1) + \varepsilon) = E(-X_i|f(X_1) + \varepsilon), \quad \text{which implies} \quad E(X_i|Y) = 0.$$

Thus we conclude that $E(X|Y) = 0$.

We see that, in this special case, $\mathscr{S}_{\mathrm{SIR}} = \{0\}$. Even though the assertion $\mathscr{S}_{\mathrm{SIR}} \subseteq \mathscr{S}_{Y|X}$ is not violated, $\mathscr{S}_{\mathrm{SIR}}$ is useless as it does not provide any information about the central subspace.

The situation is illustrated by Figure 3.2, where the U-shaped surface represents $f(X_1)$. The conditional expectation $E(X|Y)$ is at the center of the set $A \cup B$, which is at the center of distribution of X, as represented by "+" in the figure. At the same time unconditional mean $E(X)$ is also at the center of the distribution of X, located at the same point represented by "+". Hence $E(X) - E(X|Y) = 0$. This limitation is one of the motivations of developing other SDR methods, such as the Sliced Average Variance Estimator (SAVE, Cook and Weisberg (1991)), Contour Regression (Li et al. (2005)), and Directional Regression (Li and Wang (2007)).

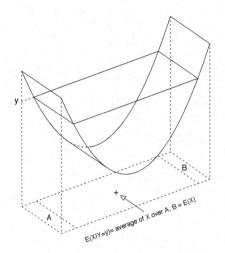

Figure 3.2 *Illustration of limitation of Sliced Inverse Regression.*

3.3   Estimation, Algorithm, and R-codes

Let $(X_1, Y_1), \ldots, (X_n, Y_n)$ be a sample of independent observations on (X, Y). We first describe the estimation procedure for SIR at the population level. Let $J_1, \ldots, J_h$ be intervals in Ω_Y. Let

$$g(Y) = \sum_{\ell=1}^{h} \ell I(Y \in J_\ell).$$

That is, $g(Y)$ takes the value ℓ if Y falls in the ℓ interval. By Corollary 3.2, we have

$$E[X - E(X)|g(Y)] \in \Sigma \mathscr{S}_{Y|X}.$$

Let $Z = \Sigma^{-1/2}(X - EX)$. Let

$$\Lambda = \text{var}[E(Z|g(Y))].$$

Suppose Λ has rank r, which is at most d, the dimension of $\mathscr{S}_{Y|X}$. Let $v_1, \ldots, v_r$ be the eigenvectors of Λ corresponding to its nonzero eigenvalues. Then, by Theorem 2.2 of Chapter 2, $u_k = \Sigma^{-1/2}v_i$, $k = 1, \ldots, r$, belong the the central subspace $\mathscr{S}_{Y|X}$. The random variables

$$u_1^{\mathsf{T}}(X - E(X)), \ldots, u_r^{\mathsf{T}}(X - E(X))$$

are called sufficient predictors, and are the result of the SDR.

For estimation, we mimic the above process at the sample level. We summarize the estimation procedure as the following algorithm.

Algorithm 3.1 Sliced Inverse Regression

1. Compute the sample mean and sample variance:

$$\hat{\mu} = E_n(X), \quad \hat{\Sigma} = \text{var}_n(X).$$

and compute the standardized random vectors

$$Z_i = \hat{\Sigma}^{-1/2}(X_i - \hat{\mu}), \quad i = 1, \dots, n.$$

2. Approximate $E[Z|g(Y) \in J_\ell]$ or $E(Z|Y \in J_\ell)$ by

$$E_n(Z|Y \in J_\ell) = \frac{E_n[ZI(Y \in J_\ell)]}{E_n[I(Y \in J_\ell)]}, \quad \ell = 1, \dots, h.$$

3. Approximate $\text{var}[E(Z|g(Y))]$ by

$$\hat{\Lambda} = \sum_{i=\ell}^{h} E[I(Y \in J_\ell)] E_n(Z|Y \in J_\ell) E_n(Z^T|Y \in J_\ell). \tag{3.2}$$

4. Let $\hat{v}_1, \dots, \hat{v}_r$ be the first r eigenvectors of $\hat{\Lambda}$, and let $\hat{\beta}_k = \hat{\Sigma}^{-1/2} \hat{v}_k$, $k = 1, \dots, r$. The sufficient predictors are

$$\hat{\beta}_k^T(X_1 - \hat{\mu}), \dots, \hat{\beta}_k^T(X_n - \hat{\mu}), \quad k = 1, \dots, r.$$

Let S_{ik} represent the random variable $\hat{u}_k^T(X_i - \hat{\mu})$, and let $S_i = (S_{i1}, \dots, S_{ir})^T$. We now have a sample of the lower dimensional predictor $S_1, \dots, S_n$, which serves as a compressed version of the high-dimensional predictor $X_1, \dots, X_n$. Under the premise of $r = d$, we can perform statistical analysis of Y versus S without losing information about the relation between Y and X. For example, we can perform regression analysis or classification based on the sample $(S_1, Y_1), \dots (S_n, Y_n)$.

There are several issues that we will resolve in later chapters. For example, in the above algorithm r is assumed known, but in practice it must be estimated. Also, in practice, we often choose J_ℓ so that each interval has roughly an equal number of observations. The above algorithm is implemented by the following R-codes.

1. Function to compute power of a matrix   This function computes the `alpha` power of a matrix a, which must be a symmetric matrix.

```
matpower = function(a,alpha){
a = round((a + t(a))/2,7); tmp = eigen(a)
return(tmp$vectors%*%diag((tmp$values)^alpha)%*%t(tmp$vectors))}
```

2. Function to discretize Y   This function computes $g(Y)$ from Y. The input y is the original sample of response; h is the number of slices. The code divides the sample of Y roughly evenly.

```
discretize=function(y,h){
```

```
n=length(y);m=floor(n/h)
y=y+.00001*mean(y)*rnorm(n)
yord = y[order(y)]
divpt=numeric();for(i in 1:(h-1)) divpt = c(divpt,yord[i*m+1])
y1=rep(0,n);y1[y<divpt[1]]=1;y1[y>=divpt[h-1]]=h
for(i in 2:(h-1)) y1[(y>=divpt[i-1])&(y<divpt[i])]=i
return(y1)}
```

3. Function to compute $\hat{\beta}$ This function computes the vectors $\hat{\beta}_1,\ldots,\hat{\beta}_r$. The input x is a matrix of dimension $n \times p$, whose rows are X_i. r is the dimension of $\mathscr{S}_{\text{SIR}}$. The choice of r will be discussed in a later chapter.

```
sir=function(x,y,h,r,ytype){
p=ncol(x);n=nrow(x)
signrt=matpower(var(x),-1/2)
xc=t(t(x)-apply(x,2,mean))
xst=xc%*%signrt
if(ytype=="continuous") ydis=discretize(y,h)
if(ytype=="categorical") ydis=y
yless=ydis;ylabel=numeric()
for(i in 1:n) {if(var(yless)!=0) {ylabel=
            c(ylabel,yless[1]);yless=yless[yless!=yless[1]]}}
ylabel=c(ylabel,yless[1])
prob=numeric();exy=numeric()
for(i in 1:h) prob=c(prob,length(ydis[ydis==ylabel[i]])/n)
for(i in 1:h) exy=rbind(exy,apply(xst[ydis==ylabel[i],],2,mean))
sirmat=t(exy)%*%diag(prob)%*%exy
return(signrt%*%eigen(sirmat)$vectors[,1:r])}
```

3.4   Application: The Big Mac Index

In this section we illustrate SIR using a data set involving 10 economic variables from 45 countries. The data set is taken from the *Arc Software* of the University of Minnesota, which can be found at the website

        `http://www.stat.umn.edu/arc/software.html`

The detailed description of the data set can be found at the above website. The 10 variables are

```
X1.  Min labor to buy 1 kg bread
X2.  Lowest cost of 10k public transit
X3.  Electrical engineers' annual salary
X4.  Tax rate paid by engineer
X5.  Annual cost of 19 services
X6.  Primary teacher salary
X7.  Tax rate paid by primary teacher
X8.  Average days vacation per year
X9.  Average hours worked per year
Y.   Min labor to buy a Big Mac and fries
```

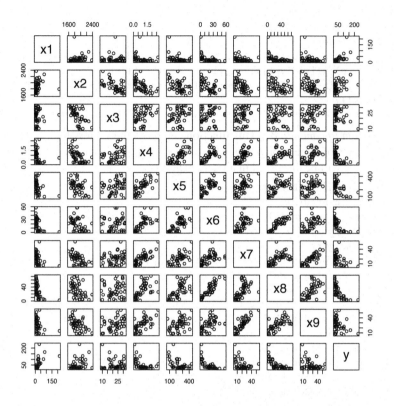

Figure 3.3 *Scatter plot matrix of ten economic variables in the Big Mac data.*

The first 9 variables are taken as predictors and the last one is taken as the response. The Big Mac index is sometimes used by economists as a informal measure of a country's purchasing-power parity (PPP). The goal of our study is to find the linear combinations of the above variables that best predict the Big Mac index. To explore the basic shape of the multivariate data, we present in Figure 3.3 the scatter plot matrix. We see that the random vector $(X_2, \ldots, X_9)$ roughly follows an elliptical distribution, but the joint distributions of the first variable with the other variables are skewed. In practice, we often make a transformation before dimension reduction to make the predictor distribution roughly elliptical. This will be discussed in a later chapter. For now, we carry out the dimension reduction without transformation. Also, we can see that Y clearly depends on X, especially X_6 and X_8.

We apply SIR to this data set, using 8 slices of roughly equal sizes. The scatter plot of Y versus the first and the second SIR predictor are presented in Figure 3.4, which demonstrates a strong nonlinear relation between Y and $\hat{\beta}_1^T X$, and hardly any relation between Y and X_2. This indicates the dimension of the central subspace is 1.

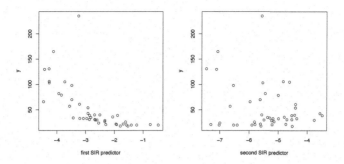

Figure 3.4 *Scatter plot of the response versus the first two SIR predictors.*

Also observe that there appears to be a stronger relation between Y and $\hat{\beta}_1^T X$ than between Y and other individual predictors. To confirm this, in Table 3.1 we present the Spearman's correlation between Y versus $X_1, \ldots, X_9$ and $\hat{\beta}^T X$. We use Spearman's correlation rather than Pearson's correlation because the dependence of Y on these variables is clearly nonlinear.

Table 3.1 *Spearman's correlation with the response*

X_1	0.607	X_7	−0.426
X_2	0.238	X_8	−0.797
X_3	0.159	X_9	−0.348
X_4	−0.607	$\hat{\beta}_1^T X$	−0.894
X_5	−0.457	$\hat{\beta}_2^T X$	0.036
X_6	−0.827		

For further development of Sliced Inverse Regression, see, for example, Fang and Zhu (1996) and Chen and Li (1998). In particular, Fang and Zhu (1996) extends SIR by replacing slice averages with kernel regression estimate (of X versus Y), and studied its asymptotic behavior when the kernel bandwidth decreases to 0 as $n \to \infty$. Chen and Li (1998) recast SIR as a minimization problem, which is influential on the development of related methods, such as the parametric inverse regression Bura and Cook (2001) and the canonical correlation method Fung et al. (2002), which will be developed in the next chapter.

Chapter 4

Parametric and Kernel Inverse Regression

4.1 Parametric Inverse Regression   37
4.2 Algorithm, R Codes, and Application   39
4.3 Relation of PIR with SIR   40
4.4 Relation of PIR with Ordinary Least Squares   42
4.5 Kernel Inverse Regression   42

4.1 Parametric Inverse Regression

As we have seen in the last section, what makes SIR unbiased is the fact that the set of sliced means $\{E(X|Y \in J_\ell) : \ell = 1, \ldots h\}$ are vectors in $\Sigma \mathscr{S}_{Y|X}$. It is then natural to speculate that perhaps performing parametric regression of X versus a set of functions would also produce vectors in $\Sigma \mathscr{S}_{Y|X}$ because, after all, $E(X|Y \in J_\ell)$ is nothing but the projection of X on to the indicator functions $I(Y \in J_h)$. This is the intuition behind the Parametric Inverse Regression (PIR) introduced by Bura and Cook (2001) and the canonical correlation (CANCOR) estimator proposed by Fung et al. (2002). PIR and CANCOR are closely related. However, the former is proposed as an estimator as the dimension d of the central subspace; whereas the latter is an estimator of the central subspace itself. Moreover, PIR uses a different re-scaling matrix than CANCOR. In this section we describe the procedure of regression X on Y parametrically to estimate the central subspace, and refer to it as PIR because the procedure is in the same spirit as Sliced Inverse Regression.

Let $f_1, \ldots, f_m$ be a set of functions of y. For example, these can be

$$\{1, y, y^2, \ldots, y^m\}.$$

We perform regression of X on $\{f_1(y), \ldots, f_m(y)\}$ which, at the population level, means we minimize the objective function

$$E\|X - \alpha_0 - \alpha_1 f_1(Y) - \cdots - \alpha_m f_m(Y)\|^2 \tag{4.1}$$

among all $\alpha_0, \ldots, \alpha_m \in \mathbb{R}^p$. Let

$$F(Y) = (f_1(Y), \ldots, f_m(Y))^\mathsf{T}, \quad B = (\alpha_1, \ldots, \alpha_m).$$

Then the objective function (4.1) can be rewritten as $E\|X - \alpha_0 - BF(Y)\|^2$. By Proposition 1.1 of Chapter 1 we know the solution to this optimization problem is

$$B = \mathrm{cov}[X, F(Y)][\mathrm{var}(F(Y))]^{-1}, \quad \alpha_0 = EX - BE[F(Y)]. \tag{4.2}$$

Before proceeding further, we state without proof a simple generalization of Proposition 2.1.

Lemma 4.1 *Suppose U, V, and W are random variables, random vectors, or random matrices, and the dimensions of U and V are such that the product UV is defined, and suppose the conditional and unconditional expectations involved are defined. Then*

$$E[UE(V|W)] = E[E(U|W)V] = E[E(U|W)E(V|W)].$$

Just as expected, the next theorem shows that the parametric regression of X on $f_0, \ldots, f_m$ does produce vectors in $\Sigma \mathscr{S}_{Y|X}$. Henceforth, because several covariance matrices are involved, we will denote Σ by Σ_{XX} to distinguish it from the covariance matrix between, say, X and $F(Y)$.n

Theorem 4.1 *Suppose X and $F(Y)$ are square integrable and Assumption 3.1 of Chapter 3 holds. Then*

$$\mathrm{span}\{\Sigma_{XX}^{-1}\mathrm{cov}(X, F(Y))\} \subseteq \mathscr{S}_{Y|X}$$

PROOF. Without loss of generality, assume $E(X) = 0$ and $E[F(Y)] = 0$ (otherwise we can reset X to be $X - E(X)$ and $F(Y)$ to be $F(Y) - E[F(Y)]$ in the following proof). Then $\mathrm{cov}[X, F(Y)] = E[XF^\mathsf{T}(Y)]$. By Lemma 4.1,

$$E[XF^\mathsf{T}(Y)] = E\{XE[F^\mathsf{T}(Y)|Y]\} = E[E(X|Y)F^\mathsf{T}(Y)].$$

Because $Y \perp\!\!\!\perp X | \beta^\mathsf{T}X$, we have

$$E[E(X|Y)F^\mathsf{T}(Y)] = E\{E[E(X|\beta^\mathsf{T}X, Y)|Y]F^\mathsf{T}(Y)\} = E\{E[E(X|\beta^\mathsf{T}X)|Y]F^\mathsf{T}(Y)\}.$$

By Assumption 3.1 and Lemma 1.1,

$$E\{E[E(X|\beta^\mathsf{T}X)|Y]F^\mathsf{T}(Y)\} = P_\beta^\mathsf{T}(\Sigma_{XX})E[E(X|Y)F^\mathsf{T}(Y)].$$

By Lemma 4.1 again

$$P_\beta^\mathsf{T}(\Sigma_{XX})E[E(X|Y)F^\mathsf{T}(Y)] = P_\beta^\mathsf{T}(\Sigma_{XX})E\{XE[F^\mathsf{T}(Y)|Y]\} = P_\beta^\mathsf{T}(\Sigma_{XX})E[XF^\mathsf{T}(Y)],$$

which implies $\mathrm{span}\{\mathrm{cov}(X, F(Y))\} \subseteq \Sigma \mathscr{S}_{Y|X}$. $\square$

Let Σ_{XF} and Σ_{FF} denote $\mathrm{cov}[X, F(Y)]$ and $\mathrm{var}[F(Y)]$, respectively. The above theorem implies that, for any matrix with full row-rank, $\mathrm{span}(\Sigma_{XX}^{-1}\Sigma_{XF}A) \subseteq \mathscr{S}_{Y|X}$. In particular, if we take $A = \Sigma_{FF}^{-1}$, then span($\Sigma_{XX}^{-1}B) \subseteq \mathscr{S}_{Y|X}$, where B is the regression coefficient matrix B in (4.2). Similarly, if we take $A = \Sigma_{FF}^{-1}\Sigma_{FX}\Sigma_{XX}^{-1}$, then we have

$$\text{span}(\Sigma_{XX}^{-1}\Sigma_{XF}\Sigma_{FF}^{-1}\Sigma_{FX}\Sigma_{XX}^{-1}) \subseteq \mathscr{S}_{Y|X}.$$

In other words, we need to solve the generalized eigenvalue problem

$$\Sigma_{XF}\Sigma_{FF}^{-1}\Sigma_{FX}v = \lambda\Sigma_{XX}v.$$

Or, using the notation of Section 1.3, we solve the problem GEV($\Sigma_{XF}\Sigma_{FF}^{-1}\Sigma_{FX}, \Sigma_{XX}$).

4.2 Algorithm, R Codes, and Application

Let $(X_1, Y_1), \ldots, (X_n, Y_n)$ be an i.i.d. sample of (X, Y). As in the case of SIR, to develop the estimation procedure all we need to do is to replace the various covariance matrices by their sample estimate. We summarize the algorithm as follows.

Algorithm 4.1 Parametric Inverse Regression

---

1. Compute $\hat{\Sigma} = \text{var}_n(X)$, $\hat{\mu} = E_n(X)$. Standardize $X_1, \ldots, X_n$ as

$$Z_i = \hat{\Sigma}^{-1/2}(X_i - \hat{\mu}), \quad i = 1, \ldots, n.$$

2. Select functions $f_1, \ldots, f_m$. For example

$$\tilde{f}_i(Y) = \frac{f_i(Y) - E_n f_i(Y)}{\sqrt{\text{var}_n[f_i(Y)]}}, \quad i = 1, \ldots, m.$$

Form the random vector

$$F(Y) = (\tilde{f}_1(Y), \ldots, \tilde{f}_m(Y))^{\mathsf{T}}.$$

3. Compute $\hat{\Sigma}_{ZF} = \text{cov}_n[Z, F(Y)]$ and $\hat{\Sigma}_{FF} = \text{var}_n[F(Y)]$. Compute $\hat{v}_1, \ldots, \hat{v}_r$, the first r eigenvectors of $\hat{\Sigma}_{ZF}\hat{\Sigma}_{FF}^{-1}\hat{\Sigma}_{FZ}$.
4. Use

$$\hat{\beta}_1 = \hat{\Sigma}_{XX}^{-1/2}\hat{v}_1, \ldots, \hat{\beta}_r = \hat{\Sigma}_{XX}^{-1/2}\hat{v}_r.$$

as the estimates of a set of vectors in the central subspace.

Below is an R-code for calculating $\hat{\beta}_1, \ldots, \hat{\beta}_r$ for a given r, using the polynomial basis functions $f_i(Y) = Y^i$, $i = 1, , \ldots, 3$.

```
pir=function(x,y,m,r){
xc=t(t(x)-apply(x,2,mean))
signrt=matpower(var(x),-1/2)
xstand=xc%*%signrt
f=numeric();ystand=(y-mean(y))/sd(y)
```

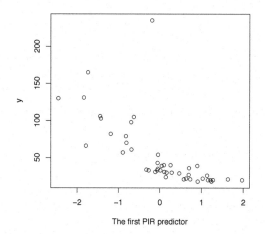

Figure 4.1 *Big Mac index versus the first PIR predictor.*

```
for(i in 1:m) f=cbind(f, ystand^i)
sigxf=cov(xstand,f);sigff=var(f)
cand=sigxf%*%solve(sigff)%*%t(sigxf)
return(signrt%*%eigen(symmetry(cand))$vectors[,1:r])}
```

We now apply PIR to the Big Mac data set, using the polynomial basis $f_i(Y) = Y^i$, $i = 1,2,3$. The Spearman's correlation between Y and the first PIR predictor is -0.886. Figure 4.1 is the scatter plot of Y versus the first PIR direction.

4.3   Relation of PIR with SIR

In this section we show that SIR is in fact a special case of PIR. To see this, consider the following basis functions

$$f_i(Y) = I(Y \in J_i), \quad i = 1,\ldots,h. \tag{4.3}$$

Before proving this result, we first prove a lemma concerning the Moore-Penrose inverse of a matrix, which is denoted by $A^\dagger$.

Lemma 4.2 *Suppose $\pi = (\pi_1,\ldots,\pi_h)^\mathsf{T}$, where $\pi_i \geq 0$ for $i = 1,\ldots,h$ and $\sum_{i=1}^h = 1$. Then*

$$[\mathrm{diag}(\pi) - \pi\pi^\mathsf{T}]^\dagger = \mathrm{diag}(\pi)^{-1} - 1_p 1_p^\mathsf{T}.$$

PROOF. Because

$$\mathrm{diag}(\pi) - \pi\pi^\mathsf{T} = \mathrm{diag}(\pi)^{1/2}[I_p - \mathrm{diag}(\pi)^{-1/2}\pi\pi^\mathsf{T}\mathrm{diag}(\pi)^{-1/2}]\mathrm{diag}(\pi)^{1/2},$$

we have

$$(\text{diag}(\pi) - \pi\pi^\mathsf{T})^\dagger = \text{diag}(\pi)^{-1/2}(I_p - vv^\mathsf{T})^\dagger \text{diag}(\pi)^{-1/2},$$

where $v = \text{diag}(\pi)^{-1/2}\pi$. Since

$$v^\mathsf{T}v = \pi^\mathsf{T}\text{diag}(\pi)^{-1}\pi = \pi^\mathsf{T}1_p = 1,$$

the matrix $(I_p - vv^\mathsf{T})^{-1}$ is the projection on to $\text{span}(v)^\perp$. It follows that $(I_p - vv^\mathsf{T})^\dagger = I_p - vv^\mathsf{T}$. Hence

$$\begin{aligned}
(\text{diag}(\pi) - \pi\pi^\mathsf{T})^\dagger &= \text{diag}(\pi)^{-1/2}(I_p - vv^\mathsf{T})\text{diag}(\pi)^{-1/2} \\
&= \text{diag}(\pi)^{-1} - \text{diag}(\pi)^{-1}\pi\pi^\mathsf{T}\text{diag}(\pi)^{-1} \\
&= \text{diag}(\pi)^{-1} - 1_p 1_p^\mathsf{T},
\end{aligned}$$

as desired.   □

We now prove the main theorem of this section. In the following, for a matrix A, $A._j$ stands for its jth column.

Theorem 4.2 *Suppose f_i, $i = 1, \ldots, m$ are chosen as the basis (4.3), then PIR reduces to solving the generalized eigenvalue problem*

$$\text{GEV}(\text{var}[E(X|g(Y))], \text{var}(X)),$$

where $g(Y) = \sum_{i=1}^h iI(Y \in J_i)$ is the discretized version of Y defined in Section 3.3.
PROOF. Let $\tilde{f}_i(Y) = I(Y \in J_i) - P(Y \in J_i)$. Then

$$\begin{aligned}
(\Sigma_{XF})._j &= E[X\tilde{f}_i(Y)] \\
&= E\{X[I(Y \in J_j) - P(Y \in J_j)]\} \\
&= P(Y \in J_i)[E(X|Y \in J_j) - E(X)].
\end{aligned}$$

Let π denote the vector $\{P(Y \in J_i)\}_{i=1}^p$, and let G denote the matrix

$$(E(X|Y \in J_1), \ldots, E(X|Y \in J_h)).$$

Then Σ_{XF} can be written as $G\text{diag}(\pi)$. In the meantime,

$$(\Sigma_{FF})_{ij} = \text{cov}(I(Y \in J_i), I(Y \in J_j)) = P(Y \in J_i \cap J_j) - P(Y \in J_i)P(Y \in J_j).$$

Since $J_i \cap J_j = \varnothing$, the first term on the right-hand side is $\delta_{ij}P(Y \in J_i)$ where δ_{ij} is the Kronecker δ function. Consequently,

$$\Sigma_{FF} = \text{diag}(\pi) - \pi\pi^\mathsf{T}.$$

By the discussion at the end of Section 4.1, PIR is determined by the eigenvalue problem

$$\text{GEV}(\Sigma_{XF}\Sigma_{FF}^\dagger\Sigma_{FX}, \Sigma_{XX}).$$

However, by Lemma 4.2,

$$\text{diag}(\pi)\Sigma_{FF}^{\dagger}\text{diag}(\pi) = \text{diag}(\pi)[\text{diag}(\pi)^{-1} - 1_p 1_p^{\mathsf{T}}]\text{diag}(\pi)$$
$$= \text{diag}(\pi) - \pi\pi^{\mathsf{T}}.$$

So

$$G\,\text{diag}(\pi)\Sigma_{FF}^{\dagger}\text{diag}(\pi)\,G^{\mathsf{T}} = G\text{diag}(\pi)G^{\mathsf{T}} - G\pi\pi^{\mathsf{T}}G^{\mathsf{T}}. \tag{4.4}$$

Note that

$$G\pi = \sum_{i=1}^{h} E(X|Y \in J_i)P(Y \in J_i) = \sum_{i=1}^{h} E(XI(Y \in J_h)) = E(X)$$
$$G\text{diag}(\pi)G^{\mathsf{T}} = \sum_{i=1}^{h} P(Y \in J_i)E(X|Y \in J_i)E(X^{\mathsf{T}}|Y \in J_i)$$
$$= E[E(X|g(X))E(X^{\mathsf{T}}|g(X))],$$

where $g(Y) = \sum_{i=1} iI(Y \in J_i)$ is the discretized version of Y as defined in Section 3.3. Hence (4.4) is in fact the matrix $\text{var}[E(X|g(Y)]$. $\qquad\square$

4.4   Relation of PIR with Ordinary Least Squares

Ordinary Least Squares (OLS) is another important method for Sufficient Dimension Reduction. Its unbiasedness as a dimension reduction estimator was first established by Li and Duan (1989). We have, in effect, demonstrated this result in Section 1.10 to illustrate how we can borrow the symmetry in the distribution of X to recover the central subspace without fitting an accurate regression model. We now show that this can also be derived from the PIR.

Let the set of basis functions for PIR be the singleton $f_1(y) = y$. Then

$$\Sigma_{XF} = \text{cov}(X,Y), \quad \Sigma_{FF} = \text{var}(Y).$$

Thus $\Sigma_{XF}\Sigma_{FF}^{-1}\Sigma_{FX}$ is rank-1 matrix, and the only eigenvector of the problem $\text{GEV}(\Sigma_{XF}\Sigma_{FF}^{-1}\Sigma_{FX}, \Sigma_{XX})$ corresponding to its nonzero eigenvalue is proportional to the vector $\Sigma_{XX}^{-1}\Sigma_{XY}$, which is precisely the population-level expression of the Ordinary Least Squares estimate. At the sample level, we use the OLS estimate

$$\hat{\beta} = \hat{\Sigma}_{XX}^{-1}\hat{\Sigma}_{XY}$$

to estimate the central subspace $\mathscr{S}_{Y|X}$.

Figure 4.2 shows the result of OLS applied the Big Mac data set. The Spearman's correlation between Y and the first PIR predictor is -0.817.

4.5   Kernel Inverse Regression

Another extension of SIR is to replace slice averages of X by kernel smoothing (Fang and Zhu (1996)). Let $\kappa : \Omega_Y \to \mathbb{R}$ be a kernel function. For example, consider the Gaussian kernel function

$$\kappa_b(u) = b^{-1}\exp(-u^2/b).$$

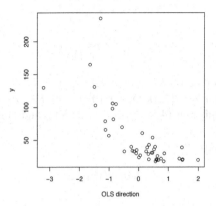

Figure 4.2 *Big Mac index versus the OLS predictor.*

where $b > 0$ is the bandwidth. Instead of approximating $E(X|Y \in J_i)$, we estimate $E(X|Y = y)$ by

$$\hat{E}(X|Y = y) = \frac{E_n[X\kappa_b(Y - y)]}{E_n[\kappa_b(Y - y)]}.$$

As a form of regularization, Fang and Zhu (1996) proposed to replace the denominator by $\max\{E_n[\kappa_b(Y - y)], \varepsilon\}$ where $\varepsilon > 0$ is a tuning constant. That is, we use

$$\tilde{E}(X|Y = y) = \frac{E_n[X\kappa_b(Y - y)]}{\max\{E_n[\kappa_b(Y - y)], \varepsilon\}}$$

to estimate $E(X|Y = y)$. In order to give ε an appropriate scale, we take it to be $\delta E_n\{E_n[\kappa_b(Y - \tilde{Y})]\}$, where

$$E_n\{E_n[\kappa_b(Y - \tilde{Y})]\} = n^{-2}\sum_{i=1}^{n}\sum_{j=1}^{n}\kappa_b(Y_i - Y_j). \tag{4.5}$$

We then approximate $\text{var}[E(X|Y)]$ by

$$\text{var}_n[\tilde{E}(X|Y)] = n^{-1}\sum_{i=1}^{n}[\tilde{E}(X|Y_i) - E_n(X)][\tilde{E}(X|Y_i) - E_n(X)]^{\mathsf{T}}.$$

The central subspace $\mathscr{S}_{Y|X}$ is then estimated by solving the generalized eigenvalue problem

$$\text{GEV}(\text{var}_n[\tilde{E}(X|Y)], \text{var}_n(X)).$$

This method is called the kernel inverse regression (KIR). The kernel estimate can be calculated efficiently by first computing the Gram matrix, as described in the following algorithm.

Algorithm 4.2 Kernel Inverse Regression

1. Standardize X_i and Y_i to Z_i as before and standardize Y_i to be

$$\tilde{Y}_i = \frac{Y_i - E_n(Y)}{\sqrt{\operatorname{var}_n(Y)}}.$$

2. Compute the Gram matrix $K = \{\kappa_b(Y_i - Y_j)\}_{i,j=1}^n$.
3. Compute the vector $V = \{E_n[\kappa_b(Y_i - Y_j)|Y_i]\}_{i=1}^n$ by $K 1_n/n$.
4. Compute the number in (4.5) by $C = n^{-2} 1_n^{\mathsf{T}} K 1_n$, and compute the vector $W = \{\max(V_i, \delta C)\}_{i=1}^n$.
5. Compute $\operatorname{var}_n[\tilde{E}(X|\tilde{Y})]$ as

$$K \operatorname{diag}(W)^{-2} K.$$

6. Compute the first r eigenvectors $v_1, \ldots, v_r$ of the matrix in Step 5.
7. Compute $\hat{\beta}_i = \hat{\Sigma}^{-1/2} v_i$, $i = 1, \ldots, r$.

An R code to implement the above algorithm is given below.

```
kir=function(x,y,b,eps,r){
gker=function(b,y){
n=length(y);k1=y%*%t(y);k2=matrix(diag(k1),n,n)
return((1/b)*exp(-(k2+t(k2)-2*k1)/(2*b^2)))}
xc=t(t(x)-apply(x,2,mean))
signrt=matpower(var(x),-1/2)
xst=xc%*%signrt
f=numeric();yst=(y-mean(y))/sd(y)
kern=gker(b,yst)
mea=mean(c(kern%*%rep(1,n)))
den=apply(cbind(kern%*%rep(1,n),rep(eps*mea,n)),1,max)
scale=eigen(kern)$values[1]
exy=(kern%*%xst)*(1/den);mat=t(exy)%*%exy
return(signrt%*%eigen(mat)$vectors[,1:r])}
```

This algorithm requires the tuning constants b and δ. This can be done by Cross-Validation or Generalized Cross-Validation. But for inverse regression we propose a simpler method. Since inverse regression does not fit a forward model of Y versus X, it does not suffer from the issue of over fitting. Thus it is entirely reasonable to use the strength of the dependence between Y and $\hat{\beta}^{\mathsf{T}} X$ to evaluate the tuning parameters. Since this relation is generally nonlinear, we can use the Spearman's correlation to measure their dependence.

More specifically, recall that the Spearman's correlation between two samples $U_1, \ldots, U_n$ and $V_1, \ldots, V_n$ is defined as the sample correlation of the rank of the two samples. That is, if $R(U_i)$ and $R(V_i)$ are be the ranks of U_i and V_i, respectively, then

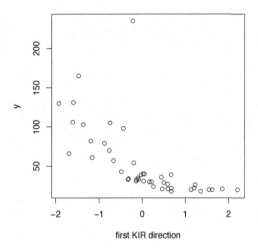

Figure 4.3 *Big Mac index versus the first KIR predictor.*

the Spearman's correlation between the samples is

$$\text{scor}_n(U,V) = \frac{\text{cov}_n[R(U),R(V)]}{\sqrt{\text{var}_n[R(U)]\text{var}_n[R(V)]}}.$$

Let $\hat{\beta}_1(b,\delta)$ be the first KIR direction for a fixed b and δ, and let $\hat{Y}(b,\delta) = [\hat{\beta}(b,\delta)]^{\mathsf{T}}X$ be the prediction of Y, we maximize

$$\text{scor}_n[\hat{Y}(b,\delta),Y]$$

over a grid of (b,δ), say

$$b = 0.1, 0.2, \ldots, 1, \quad \delta = 0.1, 0.2, \ldots, 2.$$

The maximum Spearman's correlation is achieved at $b = 0.2$, $\delta = 1.1$, and the Spearman's correlation is 0.897. Figure 4.3 shows the scatter plot of Y versus the first KIR predictor.

Chapter 5

Sliced Average Variance Estimate

5.1   Motivation   47
5.2   Constant Conditional Variance Assumption   47
5.3   Sliced Average Variance Estimate   49
5.4   Algorithm and R-code   52
5.5   Relation with SIR   55
5.6   The Issue of Exhaustiveness   56
5.7   SIR-II   58
5.8   Case Study: The Pen Digit Data   60

5.1   Motivation

As we explained in Section 3.2, SIR cannot recover any vector in the central subspace $\mathscr{S}_{Y|X}$ if the regression function is symmetric about 0. The same applies to all the first-order methods described in Chapter 4. More generally, if the regression surface is symmetric about 0 along some directions, then those directions cannot be recovered by the first-order methods. In other words, these methods can only recover a proper subset of the central subspace under such circumstances. In practice, this also means that, even when the regression surface is not exactly symmetric but is roughly symmetric about 0 along certain directions, the first-order methods cannot estimate these parts of the central subspace accurately. To remedy the situation, methods based on the second-order conditional moments, such as $\mathrm{var}(X|Y)$ and $E(XX^{\mathsf{T}}|Y)$, have been developed. These methods are called the "second-order methods". The first such method, called the Sliced Average Variance Estimate (SAVE), was proposed by Cook and Weisberg (1991), which is the subject of this chapter.

5.2   Constant Conditional Variance Assumption

The following assumption, which is similar in nature to the linear conditional mean assumption (Assumption 3.1 of Chapter 3), is important for all the second-order methods.

Assumption 5.1 $\mathrm{var}(X|\beta^\mathsf{T}X)$ *is a nonrandom matrix.*

Henceforth we refer to this assumption as the constant conditional variance assumption with respect to β, and abbreviate it by CCV(β). Echoing this abbreviation, we write the linear conditional mean condition in Assumption 3.1 as LCM(β). As the next Proposition shows, CCV is satisfied by multivariate Gaussian distribution.

Proposition 5.1 *Suppose* $X \in \mathbb{R}^p$ *has a multivariate Gaussian distribution with a nonsingular covariance matrix* Σ, *and* $B \in \mathbb{R}^{p \times r}$, $r < p$, *is a matrix with full column rank. Then the matrix* $\mathrm{var}(X|B^\mathsf{T}X)$ *is nonrandom.*

PROOF. Let $V = B^\mathsf{T}X$ and let A be any matrix such that (A, B) is a nonsingular $p \times p$ matrix. Let $U = A^\mathsf{T}X$. Then

$$\begin{pmatrix} U \\ V \end{pmatrix} = \begin{pmatrix} A^\mathsf{T}X \\ B^\mathsf{T}X \end{pmatrix} = \begin{pmatrix} A^\mathsf{T} \\ B^\mathsf{T} \end{pmatrix} X \equiv CX.$$

Let $\Lambda = C\Sigma C^\mathsf{T}$. Let $\Lambda_{UV}, \Lambda_{UU}, \Lambda_{VV}, \Lambda_{VU}$ be blocks of Λ as partitioned by U and V. Let $\mu_U = E(U)$, $\mu_V = E(V)$. Then, the conditional distribution of U given V is

$$U|V \sim N(\mu_U - \Lambda_{UV}\Lambda_{VV}^{-1}\mu_V, \ \Lambda_{UU} - \Lambda_{UV}\Lambda_{VV}^{-1}\Lambda_{VU}).$$

Hence

$$\mathrm{var}(X|B^\mathsf{T}X) = \mathrm{var}[C^{-1}(U^\mathsf{T}, V^\mathsf{T})^\mathsf{T}|V] = C^{-1}\begin{pmatrix} \Lambda_{UU} - \Lambda_{UV}\Lambda_{VV}^{-1}\Lambda_{VU} & 0 \\ 0 & 0 \end{pmatrix}C^{-\mathsf{T}},$$

which is a nonrandom matrix.   □

The following consequence of the CCV assumption will be used repeatedly in later development.

Corollary 5.1 *If* $X \in \mathbb{R}^p$ *satisfies LCM(β) and CCV(β), then*

$$\mathrm{var}(X|\beta^\mathsf{T}X) = \Sigma Q_\beta(\Sigma),$$

where $Q_\beta(\Sigma) = I - P_\beta(\Sigma)$.

PROOF. By Lemma 3.1 of Section 3.1 and CCV(β),

$$\begin{aligned}
\Sigma = \mathrm{var}(X) &= \mathrm{var}[E(X|\beta^\mathsf{T}X)] + E[\mathrm{var}(X|\beta^\mathsf{T}X)] \\
&= \mathrm{var}[P_\beta^\mathsf{T}(\Sigma)X] + \mathrm{var}(X|\beta^\mathsf{T}X) \\
&= P_\beta^\mathsf{T}(\Sigma)\Sigma P_\beta(\Sigma) + \mathrm{var}(X|\beta^\mathsf{T}X) \\
&= \Sigma P_\beta(\Sigma) + \mathrm{var}(X|\beta^\mathsf{T}X),
\end{aligned}$$

where the last equality holds because $\Sigma P_\beta(\Sigma) = P_\beta^\mathsf{T}(\Sigma)\Sigma$ and $P_\beta(\Sigma)^2 = P_\beta(\Sigma)$. Now subtract both sides of the equations by the matrix Σ to complete the proof.   □

5.3 Sliced Average Variance Estimate

At the population level, SAVE is based on the following generalized eigenvalue problem

$$\text{GEV}(\Sigma - \text{var}(X|Y), \Sigma).$$

Henceforth, we say that a matrix A is a basis matrix of a subspace $\mathscr{S}$ if the columns of A form a basis of $\mathscr{S}$; that is, A is of full column rank and $\text{span}(A) = \mathscr{S}$. The next theorem shows that SAVE is unbiased.

Theorem 5.1 *Suppose β is basis matrix for $\mathscr{S}_{Y|X}$, and X satisfies LCM(β) and CCV(β). Then*

$$\text{span}(\Sigma - \text{var}(X|Y)) \subseteq \Sigma \mathscr{S}_{Y|X}.$$

PROOF. First note that

$$\begin{aligned}
\text{var}(X|Y) &= E[\text{var}(X|\beta^\mathsf{T}X, Y)|Y] + \text{var}[E(X|\beta^\mathsf{T}X, Y)|Y] \\
&= E[\text{var}(X|\beta^\mathsf{T}X)|Y] + \text{var}[E(X|\beta^\mathsf{T}X)|Y],
\end{aligned} \tag{5.1}$$

where the second equality follows from $Y \perp\!\!\!\perp X | \beta^\mathsf{T}X$. By Lemma 3.1 of Section 3.1 and Corollary 5.1,

$$E(X|\beta^\mathsf{T}X) = P_\beta^\mathsf{T}(\Sigma)X, \quad \text{var}(X|\beta^\mathsf{T}X) = \Sigma Q_\beta(\Sigma).$$

Substitute these into the right-hand side of (5.1) to obtain

$$\begin{aligned}
\text{var}(X|Y) &= \Sigma Q_\beta(\Sigma) + P_\beta^\mathsf{T}(\Sigma)\text{var}(X|Y)P_\beta(\Sigma) \\
&= \Sigma - \Sigma P_\beta(\Sigma) + P_\beta^\mathsf{T}(\Sigma)\text{var}(X|Y)P_\beta(\Sigma).
\end{aligned}$$

Subtract both sides of the above equation by Σ, and use the relation $\Sigma P_\beta(\Sigma) = P_\beta^\mathsf{T}(\Sigma)\Sigma P_\beta(\Sigma)$ to obtain

$$\begin{aligned}
\text{var}(X|Y) - \Sigma &= P_\beta^\mathsf{T}(\Sigma)[\text{var}(X|Y) - \Sigma]P_\beta(\Sigma) \\
&= \Sigma P_\beta(\Sigma)\Sigma^{-1}[\text{var}(X|Y) - \Sigma]P_\beta(\Sigma),
\end{aligned}$$

where the right-hand side is a member of $\Sigma \mathscr{S}_{Y|X}$, as desired.   □

Let $Z = \Sigma^{-1/2}(X - \mu)$. Applying the above theorem to Z yields

$$\text{span}(I_p - \text{var}(Z|Y)) \subseteq \mathscr{S}_{Y|Z}. \tag{5.2}$$

Let $J_1, \ldots, J_h$ be a set of intervals partitioning Ω_Y. Let $\tilde{Y}$ be the discretized Y according to this partition; that is,

$$\tilde{Y} = \sum_{\ell=1}^h \ell I(Y \in J_\ell). \tag{5.3}$$

Applying (5.2) to $\tilde{Y}$, and evoking Theorem 2.3 of Chapter 2, we have

$$I_p - \text{var}(X|\tilde{Y}) \subseteq \mathscr{S}_{\tilde{Y}|X} \subseteq \mathscr{S}_{Y|X}.$$

Let $\Lambda_{\text{SAVE}} = E[I_p - \text{var}(Z|\tilde{Y})]^2$. The next proposition shows that $I_p - \text{var}(Z|\tilde{Y}) \subseteq \mathscr{S}_{Y|Z}$ if and only if

$$\text{span}(\Lambda_{\text{SAVE}}) \subseteq \mathscr{S}_{Y|Z}. \tag{5.4}$$

Proposition 5.2 *If U is a random matrix whose entries are square integrable, and $\mathscr{S}$ is a subspace of $\mathbb{R}^p$, then the following statements are equivalent.*

1. $\text{span}(U) \subseteq \mathscr{S}$ *a.s. P;*
2. $\text{span}(E(UU^{\top})) \subseteq \mathscr{S}$.

PROOF. We first show that

$$b \perp U \ a.s. \ P \Leftrightarrow b \perp \text{span}[E(UU^{\top})]. \tag{5.5}$$

This is because

$$\begin{aligned} b \perp U \ a.s. \ P \ &\Leftrightarrow b^{\top}U = 0 \ a.s. \ P \\ &\Leftrightarrow b^{\top}UU^{\top}b = 0 \ a.s. \ P \\ &\Leftrightarrow E(b^{\top}UU^{\top}b) = 0 \\ &\Leftrightarrow b \perp \text{span}[E(UU^{\top})]. \end{aligned}$$

Now statement *1* holds if and only if $\mathscr{S}^{\perp} \subseteq \text{span}(U)^{\perp}$ *a.s. P*, which, by (5.5), is equivalent to $\mathscr{S}^{\perp} \subseteq \text{span}[E(UU^{\top})]^{\perp}$. This is equivalent to statement *2*. □

By (5.4) and the relation between $\mathscr{S}_{Y|X}$ and $\mathscr{S}_{Y|Z}$ stated in Theorem 2.2 of Chapter 2, we have

$$\Sigma^{-1/2}\text{span}(\Lambda_{\text{SAVE}}) \subseteq \mathscr{S}_{Y|Z}.$$

It follows that, if r is the rank of Λ_{SAVE} (which is necessarily no more than d), and if $v_1, \ldots, v_r$ are the first r eigenvectors of Λ_{SAVE}, then

$$\text{span}(\Sigma^{-1/2}v_1, \ldots, \Sigma^{-1/2}v_r) \subseteq \mathscr{S}_{Y|X}. \tag{5.6}$$

The algorithm for SAVE is derived by mimicking this relation.

To illustrate the intuition behind the relation (5.6), we plot in Figure 5.1 a typical regression surface with a slice of Y projected on the X domain, as indicated by A in the plot. For easy illustration we assume $\Sigma = I_p$. Unlike in SIR, which takes the slice average, in SAVE we take the leading eigenvectors of the matrix

$$\sum_{\ell=1}^{h} P(Y \in J_{\ell})[I_p - \text{var}(X|Y \in J_{\ell})]^2. \tag{5.7}$$

as the estimate of the central subspace. A generic term is this sum, ignoring the coefficient, is of the form $[I_p - \text{var}(X|Y \in J_{\ell})]^2$. The eigenspace of (5.7) is contributed by the eigenvectors of matrices of this form. Since $[I_p - \text{var}(X|Y \in J_{\ell})]^2$ and $[I_p - \text{var}(X|Y \in J_{\ell})]$ have the same eigenvectors, it suffices to consider the leading eigenvectors of $I_p - \text{var}(X|Y \in J_{\ell})$, which are orthogonal to the leading vectors of $\text{var}(X|Y \in J_{\ell})$. Thus, SAVE would estimate the directions in $\mathscr{S}_{Y|X}$ if the leading eigenvectors of $\text{var}(X|Y \in J_{\ell})$ are all orthogonal to the central subspace. This is shown by Figure 5.1 to be the case. The set $\{Y \in J_{\ell}\}$ projected on the X-space is the region A in the figure. If the distribution of X is symmetric, then any such slice is oriented

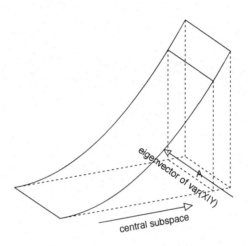

Figure 5.1 *Leading eigenvector* $\mathrm{var}(X|Y \in J_\ell)$ *in the monotone case.*

in the orthogonal direction in which the regression surface changes; thus its leading eigenvectors must be orthogonal to the central subspace.

This also explains why SAVE can overcome the difficulty encountered by SIR, namely, when the regression surface is symmetric about $E(X)$, SIR cannot fully recover the central subspace. To further illustrate this point, we plot in Figure 5.2 a regression surface that is symmetric about 0 along the central subspace. In this case, the set $\{Y \in J_\ell\}$ in the conditional variance $\mathrm{var}(X|Y \in J_\ell)$ maps to the union of two symmetric pieces about 0, as indicated by A and B in the diagram. While the average of of X over $A \cup B$ is 0, the principal component of the variance X over $A \cup B$ is aligned with the two rectangles, as indicated in Figure 5.2.

Then what is the role placed by the CCV assumption? Consider the situation where the two symmetric pieces A and B in Figure 5.2 are so far apart that the leading vector of the covariance of X conditioning on $X \in A \cup B$ becomes aligned with the line connecting the centers of A and B. In this case the leading eigenvector of $I_p - \mathrm{var}(X|Y \in J_\ell)$ would be orthogonal to the central subspace. The CCV assumption prevents this from happening. From this picture we also see that regardless of CCV, the leading eigenvector of $\mathrm{var}(X|Y)$ is either orthogonal or aligned with the central subspace.

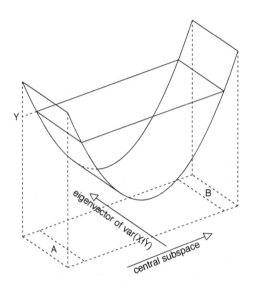

Figure 5.2 *Leading eigenvector of* $\mathrm{var}(X|Y_\ell)$ *in symmetric case.*

5.4   Algorithm and R-code

The population-level development in the previous section motivates the following algorithm. Let $(X_1, Y_1), \ldots, (X_n, Y_n)$ be an independent sample of (X, Y).

Algorithm 5.1 Sliced Inverse Variance Estimation

---

1. Standardize $X_1, \ldots, X_n$ to obtain $\hat{Z}_i$, as in Algorithm 4.2.
2. Discretize Y as $\tilde{Y} = \sum_{\ell=1}^{h} \ell I(Y \in J_\ell)$ where $\{J_1, \ldots, J_\ell\}$ is a collection of intervals that partition $(Y_{(1)}, Y_{(n)})$, $Y_{(i)}$ denoting the order statistic.
3. For each slice J_ℓ, calculate the sample conditional variance of Z given $Y \in J_\ell$:

$$\mathrm{var}_n(\hat{Z}|\tilde{Y} = \ell) = E_n[\hat{Z}\hat{Z}^\mathsf{T} I(\tilde{Y} = \ell)]/E_n[I(\tilde{Y} = \ell)].$$

4. Calculate the sample version of Λ_{SAVE}:

$$\hat{\Lambda}_{\mathrm{SAVE}} = h^{-1}\sum_{\ell=1}^{h} E_n I(\tilde{Y} = \ell)[I_p - \mathrm{var}_n(\hat{Z}|\tilde{Y} = \ell)]^2.$$

5. Compute the first r eigenvalues, say $v_1, \ldots, v_r$, of $\hat{\Lambda}_{\mathrm{SAVE}}$, and compute $u_\ell = \hat{\Sigma}^{-1/2}v_\ell$, $\ell = 1, \ldots, s$. The sufficient predictors are $u_\ell^\mathsf{T}(X_i - E_n X)$, where $i = 1, \ldots, n$, $\ell = 1, \ldots, s$.

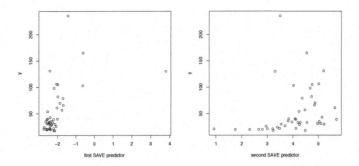

Figure 5.3 *Y versus first two SAVE predictor for the Big Mac data.*

We now present the R-code for implementing SAVE, which calls the function for discretizing Y and the function computing the power of a matrix given in Chapter 3.

```
save=function(x,y,h,r,ytype){
p=ncol(x);n=nrow(x)
signrt=matpower(var(x),-1/2)
xc=t(t(x)-apply(x,2,mean))
xst=xc%*%signrt
if(ytype=="continuous") ydis=discretize(y,h)
if(ytype=="categorical") ydis=y
yless=y;ylabel=numeric()
for(i in 1:n) {if(var(yless)!=0) {ylabel=c(ylabel,yless[1]);
                                  yless=yless[yless!=yless[1]]}}
ylabel=c(ylabel,yless[1])
prob=numeric()
for(i in 1:h) prob=c(prob,length(ydis[ydis==ylabel[i]])/n)
vxy = array(0,c(p,p,h))
for(i in 1:h) vxy[,,i] = var(xst[ydis==ylabel[i],])
savemat=0
for(i in 1:h){
savemat=savemat+prob[i]*(vxy[,,i]-diag(p))%*%(vxy[,,i]-diag(p))}
return(signrt%*%eigen(savemat)$vectors[,1:r])}
```

We applied SAVE to the Big Mac data set. Figure 5.3 shows the plots for the response versus the first two SAVE predictors.

These plots are quite different from the sufficient predictor plots obtained by SIR, as shown in Figure 3.4 of Chapter 3. However, note that the plot for Y versus the second SAVE predictor resembles that for Y versus the first SIR predictor. Thus, both methods captures the main pattern in the data within the first two eigenvectors of the candidate matrix, but SAVE captures the main pattern by its second eigenvector. One reason for this is that SAVE, which depends on the second moment, is rather sensitive to outliers. If we remove the first predictor from the Big Mac data (minimum labor to buy 1 kilogram of bread), which contains an outlier, then the sufficient plots are more similar to those produced by SIR, as shown in Figure 5.4.

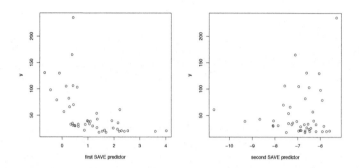

Figure 5.4 *Y versus first two SAVE predictor for the Big Mac data with first predictor removed.*

Because the second sample conditional moment usually requires larger sample sizes to converge, we use a smaller number of slices. In the above example, the number of slices is 2, as compared with 8 for SIR. As a rule, the number of slices used for SAVE are much smaller than that for SIR.

As stated in Section 5.1, an advantage of SAVE is it can recover vectors in the central subspace when the regression surface is symmetric about $E(X)$ along certain directions, which methods like SIR cannot. In the next example, we give a numerical demonstration of this property.

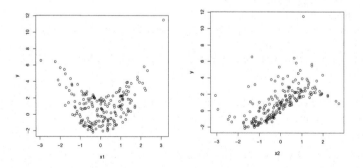

Figure 5.5 *Y versus X_1 and X_2.*

Example 5.1 Let X be a 10-dimensional random vector, distributed as $N(0, I_{10})$, and ε be a random variable independent of X, with distribution $N(0, \sigma^2)$ where $\sigma = 0.2$.

Let

$$Y = X_1^2 + 2\sin(X_2) + \varepsilon. \tag{5.8}$$

Let $(X_1, Y_1), \ldots, (X_{200}, Y_{200})$ be i.i.d. copies of (X, Y). In this case the central subspace is $\mathrm{span}(e_1, e_2)$, where e_i is a 10-dimensional vector whose components are 0 except the ith component, which is 1. Figure 5.5 shows the scatter plot for Y versus X_1 and Y versus X_2.

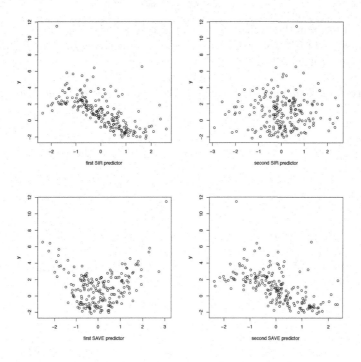

Figure 5.6 *Upper panels: Y versus first two SIR predictors; lower panels: Y versus first two SAVE predictors.*

Since the regression surface is symmetric about $E(X) = 0$ along the direction e_1, we do not expect SIR to estimate this direction well. However, SAVE, which involves second conditional sample moment, should be able to estimate this direction. Figure 5.6 shows the scatter plots for Y versus the first two SIR predictors (upper panels), and Y versus the first two SAVE predictors. These plots show that SIR does not estimate e_1 well, but estimates e_2 well, whereas SAVE gives satisfactory estimates of both directions in the central subspace.

5.5 Relation with SIR

We have seen that SAVE can recover vectors in the central subspace missed SIR due to symmetry. At the population level this is reflected by the fact that the subspace of SAVE is larger than the subspace of SIR, as we will show in this section. Let

$$\mathscr{S}_{\mathrm{SIR}} = \Sigma^{-1/2}\mathrm{span}\{\mathrm{cov}[E(Z|\tilde{Y})]\}, \quad \mathscr{S}_{\mathrm{SAVE}} = \Sigma^{-1/2}\{E[I_p - \mathrm{var}(Z|\tilde{Y})]^2\},$$

where $\tilde{Y}$ is the discretized Y as defined before. We have shown that under $\mathrm{LCM}(\beta)$ and $\mathrm{CCV}(\beta)$

$$\mathscr{S}_{\mathrm{SIR}} \subseteq \mathscr{S}_{Y|X}, \quad \mathscr{S}_{\mathrm{SAVE}} \subseteq \mathscr{S}_{Y|X}.$$

The next theorem shows that $\mathscr{S}_{\mathrm{SAVE}}$ is in general a larger space than $\mathscr{S}_{\mathrm{SIR}}$, and this is true without $\mathrm{LCM}(\beta)$ and $\mathrm{CCV}(\beta)$ — of course, without these conditions none of them is guaranteed to be in $\mathscr{S}_{Y|X}$.

Theorem 5.2 *If all moments involved in the following proof are finite, then*

$$\mathscr{S}_{\mathrm{SIR}} \subseteq \mathscr{S}_{\mathrm{SAVE}}.$$

PROOF. Because $\mathrm{var}(Z) = E[\mathrm{var}(Z|\tilde{Y})] + \mathrm{var}[E(Z|\tilde{Y})]$, we have

$$\mathrm{var}[E(Z|\tilde{Y})] = I_p - E\mathrm{var}(Z|\tilde{Y}) = E[I_p - \mathrm{var}(Z|\tilde{Y})].$$

Let $U = I_p - \mathrm{var}(Z|\tilde{Y})$. We now show that

$$\mathrm{span}(EU) \subseteq \mathrm{span}(EUU^{\mathsf{T}}), \text{ or equivalently, } \mathrm{span}(EUU^{\mathsf{T}})^{\perp} \subseteq \mathrm{span}(EU)^{\perp}. \quad (5.9)$$

Let $b \in E(UU^{\mathsf{T}})$. By (5.5), we know that $b \perp U$ a.s. $P \Leftrightarrow b \perp E(UU^{\mathsf{T}})$. Hence $b^{\mathsf{T}}U = 0$ a.s. P, which implies $b^{\mathsf{T}}E(U)$, or $b \perp E(U)$, proving (5.9). Consequently,

$$\mathrm{span}\{\mathrm{var}[E(Z|\tilde{Y})]\} \subseteq \mathrm{span}\{E[I_p - \mathrm{var}(Z|\tilde{Y})]\} \subseteq \mathrm{span}\{E[I_p - \mathrm{var}(Z|\tilde{Y})]^2\},$$

which implies

$$\Sigma^{-1/2}\mathrm{span}\{\mathrm{var}[E(Z|\tilde{Y})]\} \subseteq \Sigma^{-1/2}\mathrm{span}\{E[I_p - \mathrm{var}(Z|\tilde{Y})]^2\},$$

as desired. □

5.6 The Issue of Exhaustiveness

We now ask a deeper question: under what conditions is SAVE *exhaustive* — that is, $\mathscr{S}_{\mathrm{SAVE}} = \mathscr{S}_{Y|X}$? This question can be answered using a method similar to Li et al. (2005), and the condition turns out to be rather mild. Interestingly, this condition has not been published in the literature. We first prove two lemmas.

Lemma 5.1 *Let $\mathscr{S}$ be subspace of $\mathbb{R}^p$, and let G be positive semidefinite such that $\mathrm{span}(G) \subseteq \mathscr{S}$. The following statements are equivalent:*

1. there is a basis $v_1, \ldots, v_d$ of $\mathscr{S}$ such that $V^{\mathsf{T}}GV > 0$, where V is the $p \times d$ matrix $(v_1, \ldots, v_d)$;

2. *for all $v \in \mathscr{S}$, $v \neq 0$, we have $v^{\mathsf{T}} G v > 0$;*

3. $\operatorname{span}(G) = \mathscr{S}$.

PROOF. $2 \Rightarrow 1$. If statement *1* is not true, then there is a $w \in \mathbb{R}^d$, $w \neq 0$, such that $w^{\mathsf{T}} V^{\mathsf{T}} G V w = 0$. So $v^{\mathsf{T}} G v = 0$ for $v = V w$. Because V is a basis matrix of $\mathscr{S}$, $w \in \mathscr{S}$, and $w \neq 0$, we have $V w \neq 0$. Hence statement 2 is not true.

$1 \Rightarrow 2$. Let $v \in \mathscr{S}$, $v \neq 0$. Then $v = \sum_{i=1}^{d} \alpha_i v_i$, where $\|\alpha\| \neq 0$, α being the vector $(\alpha_1, \ldots, \alpha_d)^{\mathsf{T}}$. Hence

$$v^{\mathsf{T}} G v = \alpha^{\mathsf{T}} (V^{\mathsf{T}} G V) \alpha > 0.$$

Thus statement *2* holds.

$2 \Rightarrow 3$. If statement *2* is not true then there exists $v \in \mathscr{S}$, $v \neq 0$, such that $v^{\mathsf{T}} G v = 0$. So statement *1* is not true.

$3 \Rightarrow 2$. If statement *2* is not true, then there exists $v \in \mathscr{S}$, $v \neq 0$ such that $v^{\mathsf{T}} G v = 0$. Then $v \perp \operatorname{span}(G)$ and statement *2* is not true. $\qquad\square$

Lemma 5.2 *Suppose U and Y are random variables such that $E(U) = 0$, $\operatorname{var}(U) = 1$. Then, unless $E(U|Y)$ and $\operatorname{var}(U|Y)$ are both degenerate, we have*

$$E[1 - \operatorname{var}(U|Y)]^2 > 0.$$

PROOF. If $E(U|Y)$ is degenerate, then by assumption $\operatorname{var}(U|Y)$ is not degenerate. Hence $1 - \operatorname{var}(U|Y)$ is nondegenerate. By Jensen's inequality

$$E[1 - \operatorname{var}(U|Y)]^2 > \{E[1 - \operatorname{var}(U|Y)]\}^2 \geq 0.$$

If $E(U|Y)$ nondegenerate, then $\operatorname{var}[E(U|Y)] > 0$. But because

$$1 = \operatorname{var}(U) = E[\operatorname{var}(U|Y)] + \operatorname{var}[E(U|Y)],$$

we have $E[\operatorname{var}(U|Y)] < 1$. Hence

$$E[1 - \operatorname{var}(U|Y)]^2 \geq (E[1 - \operatorname{var}(U|Y)]) > 0.$$

This completes the proof. $\qquad\square$

The next theorem gives a sufficient condition for SAVE to be exhaustive.

Theorem 5.3 *Suppose $\mathrm{LCM}(\beta)$ and $\mathrm{CCV}(\beta)$, and $E(Z) = 0$, $\operatorname{var}(Z) = I_p$. If, for each $v \in \mathscr{S}_{Y|Z}$, at least one of random variables*

$$E(v^{\mathsf{T}} Z|Y), \quad \operatorname{var}(v^{\mathsf{T}} Z|Y)$$

is nondegenerate, then

$$\operatorname{span}\{E[I_p - \operatorname{var}(Z|Y)]^2\} = \mathscr{S}_{Y|Z}, \tag{5.10}$$

and consequently, $\mathscr{S}_{\mathrm{SAVE}} = \mathscr{S}_{Y|X}$.

Proof. Since, under $\text{LCM}(\beta)$ and $\text{CCV}(\beta)$,

$$\text{span}\{E[I_p - \text{var}(Z|Y)]^2\} \subseteq \mathscr{S}_{Y|Z},$$

by Lemma 5.1, it suffices to show that for all $v \in \mathscr{S}_{Y|Z}$

$$v^\mathsf{T} E[I_p - \text{var}(Z|Y)]^2 v > 0.$$

Without loss of generality, we assume that $\|v\| = 1$. The left-hand side of the above inequality can be rewritten as

$$E\{v^\mathsf{T}[I_p - \text{var}(Z|Y)]^2 v\} = E\{v^\mathsf{T}[I_p - \text{var}(Z|Y)](vv^\mathsf{T} + I - vv^\mathsf{T})[I_p - \text{var}(Z|Y)]v\}.$$

Because $\|v\| = 1$, the matrix vv^T is a projection, and so $I_p - vv^\mathsf{T}$ is positive semi-definite. Consequently, in terms of the Louwner's ordering

$$
\begin{aligned}
E\{v^\mathsf{T}[I_p - \text{var}(Z|Y)]^2 v\} &\geq E\{v^\mathsf{T}[I_p - \text{var}(Z|Y)]vv^\mathsf{T}[I_p - \text{var}(Z|Y)]v\} \\
&\geq E\{v^\mathsf{T}[I_p - \text{var}(Z|Y)]v\}^2 \\
&\geq E[1 - \text{var}(v^\mathsf{T}Z|Y)]^2.
\end{aligned}
$$

Since at least one of $E(v^\mathsf{T}Z|Y)$ and $\text{var}(v^\mathsf{T}Z|Y)$ is nondegenerate, we have, by Lemma 5.2,

$$E[1 - \text{var}(v^\mathsf{T}Z|Y)]^2 > 0,$$

which proves (5.10). Now multiply both sides of (5.10) by $\Sigma^{-1/2}$ to show that $\mathscr{S}_{\text{SAVE}} = \mathscr{S}_{Y|X}$.   □

The sufficient condition "at least one of $E(v^\mathsf{T}Z|Y)$ and $\text{var}(v^\mathsf{T}Z|Y)$ is degenerate" can be understood as follows. If $v \in \mathscr{S}_{Y|Z}$, then $v^\mathsf{T}Z$ explains part of the variation in Y; that is, Y has extra variation due to $v^\mathsf{T}Z$ even after accounting for its variation due to the rest of the vectors in $\mathscr{S}_{Y|Z}$. Then $v^\mathsf{T}Z$ should also vary with Y. Intuitively, if $v^\mathsf{T}Z$ varies with Y, then chances are either the conditional mean or the conditional variance of $v^\mathsf{T}Z$ given Y depends on Y.

5.7 SIR-II

A method related to SAVE is the SIR-II proposed in Li (1991). Let $\tilde{Y}$ be the discretized Y defined by (5.3). Then SIR-II is based on the following population-level matrix

$$\Lambda_{\text{SIRII}} = E[\text{var}(Z|\tilde{Y}) - E(\text{var}(Z|\tilde{Y}))]^2. \tag{5.11}$$

The following theorem proves the unbiasedness of SIR-II.

Theorem 5.4 *Suppose that Z is the standardized X with $E(Z) = 0$ and $\text{var}(Z) = I_p$, and that $\text{LCM}(\beta)$ and $\text{CCV}(\beta)$ are satisfied, where β is any basis matrix of $\mathscr{S}_{Y|Z}$. Then*

$$\text{span}(\Lambda_{\text{SIRII}}) \subseteq \mathscr{S}_{Y|Z}. \tag{5.12}$$

Consequently, $\Sigma^{-1/2}\text{span}(\Lambda_{\text{SIRII}}) \subseteq \mathscr{S}_{Y|X}$.

PROOF. Because $\text{var}(Z) = E[\text{var}(Z|\tilde{Y})] + \text{var}[E(Z|\tilde{Y})]$, (5.11) can be rewritten as

$$\begin{aligned}
\Lambda_{\text{SIRII}} &= E\{\text{var}(Z|\tilde{Y}) - I_p + \text{var}[E(Z|\tilde{Y})]\}^2 \\
&= E\{\text{var}(Z|\tilde{Y}) - I_p\}^2 + \{E[\text{var}(Z|\tilde{Y})] - I_p\}\text{var}[E(Z|\tilde{Y})]\} \\
&\quad + \text{var}[E(Z|\tilde{Y})]\{E[\text{var}(Z|\tilde{Y})] - I_p\} + \{\text{var}[E(Z|\tilde{Y})]\}^2.
\end{aligned}$$

Substituting the relation $E[\text{var}(Z|\tilde{Y})] - I_p = -\text{var}[E(Z|\tilde{Y})]$ into the right-hand side of the above equality, we have

$$\Lambda_{\text{SIRII}} = E\{\text{var}(Z|\tilde{Y}) - I_p\}^2 - \{\text{var}[E(Z|\tilde{Y})]\}^2 = \Lambda_{\text{SAVE}} - (\Lambda_{\text{SIR}})^2.$$

By Theorem 3.1 and Theorem 5.1, $\text{span}(\Lambda_{\text{SIR}}) \subseteq \mathscr{S}_{Y|Z}$ and $\text{span}(\Lambda_{\text{SAVE}}) \subseteq \mathscr{S}_{Y|Z}$. Hence (5.11) holds, which implies $\Sigma^{-1/2}\text{span}(\Lambda_{\text{SIRII}}) \subseteq \mathscr{S}_{Y|X}$. $\square$

At the sample level, SIR-II can be implemented using the matrices $\hat{\Lambda}_{\text{SIR}}$ and $\hat{\Lambda}_{\text{SAVE}}$ for SIR and SAVE, whose algorithms have been described in Chapter 3 and this chapter. Below is an R-code to implement SIR-II.

```
sirii=function(x,y,h,r,ytype="continuous"){
p=ncol(x);n=nrow(x)
signrt=matpower(var(x),-1/2)
xst=t(t(x)-apply(x,2,mean))%*%signrt
if(ytype=="continuous") ydis=discretize(y,h)
if(ytype=="categorical") ydis=y
yless=y;ylabel=numeric()
for(i in 1:n) {if(var(yless)!=0) {ylabel=c(ylabel,yless[1]);
                                 yless=yless[yless!=yless[1]]}}
ylabel=c(ylabel,yless[1])
prob=numeric();exy=numeric()
for(i in 1:h) prob=c(prob,length(ydis[ydis==ylabel[i]])/n)
for(i in 1:h) exy=rbind(exy,apply(xst[ydis==ylabel[i],],2,mean))
sirmat=t(exy)%*%diag(prob)%*%exy
vxy = array(0,c(p,p,h))
for(i in 1:h) vxy[,,i] = var(xst[ydis==ylabel[i],])
savemat=0
for(i in 1:h){
savemat=savemat+prob[i]*(vxy[,,i]-diag(p))%*%(vxy[,,i]-diag(p))}
siriimat=savemat-sirmat%*%t(sirmat)
return(signrt%*%eigen(siriimat)$vectors[,1:r])}
```

We now apply SIR-II to the Big Mac data. Figure 5.7 shows the scatter plots of Y versus the first two SIR-II predictors. As in SAVE, we use two slices for Y.

These plots are similar to the scatter plots for the SAVE predictors in Figure 5.3, though the monotone trend in the second predictor is less pronounced. Like SAVE, SIR-II also involves the second conditional moment. Thus we expect it can also recover the symmetric pattern. To see this, we applied SIR-II to the same model in

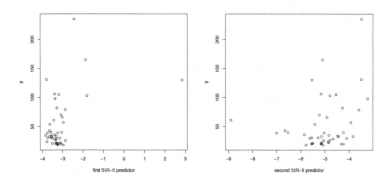

Figure 5.7 *Y versus first two SIR-II predictors for the Big Mac data.*

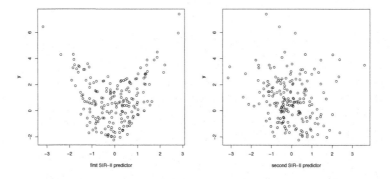

Figure 5.8 *Y versus first two SIR-II predictors for Model (5.8) in Example 5.1.*

Example 5.1. Figure 5.8 shows the response Y versus the first two SIR-II predictors. We see that SIR-II recovers the first vector very well, but has missed monotone trend completely in the second plot. This is not surprising given that SIR-II depends on the random vector, $\text{var}(Z|Y) - E[\text{var}(Z|Y)]$, which excludes the first conditional moment $E(Z|Y)$.

5.8   Case Study: The Pen Digit Data

In this section, we apply SIR, SAVE, and SIR-II on a data set concerning the identification of hand-written digits $\{0, 1, \ldots, 9\}$. The data set contains handwritten digits by 44 subjects, each of whom were asked to write 250 random digits. Eight pairs of 2-dimensional locations were recorded on each digit, yielding a 16-dimensional feature vector. The 44 subjects are divided into two groups of size 30 and 14, in which the first formed the training set (of sample size 7,494) and the second formed the test set (of sample size 3,498). The data set is available in the UCI machine-learning repository at

ftp://ftp.ics.uci.edu/pub/machine-learning-databases/pendigits/

(see also Zhu and Hastie 2003).

For clear demonstration, we focus on the three digits 0, 6, 9. This reduces the sample size to 2,219 for the training set and 1,035 for the test set. To avoid the issue of overfitting, we develop the basis matrix $\hat{\beta}$ for the central subspace based on the training set, and then apply it to the test set to obtain the sufficient predictors. That is, let $(X_1^{(1)}, Y_1^{(1)}), \ldots, (X_m^{(1)}, Y_m^{(1)})$ represent the training data, and let $(X_1^{(2)}, Y_1^{(2)}), \ldots, (X_n^{(2)}, Y_n^{(2)})$ represent the test set. We apply the three SDR methods to $(X_1^{(1)}, Y_1^{(1)}), \ldots, (X_m^{(1)}, Y_m^{(1)})$ to obtain the estimate $\hat{\beta}$, and then compute the sufficient predictors

$$\hat{\beta}^{\mathsf{T}} X_1^{(2)}, \ldots, \hat{\beta}^{\mathsf{T}} X_m^{(2)}.$$

In this example, the response variable is categorical, taking values at labels 0, 6, 9, which do not have numerical meanings. Naturally, we take three slices according to these labels. Note that, for SIR, the matrix $\mathrm{var}_n[E_h(X|\tilde{Y})]$ is of rank 2. So it can extract at most 2 sufficient predictors, but there is no such limit for the second-order methods such as SAVE and SIR-II.

The upper panel of Figure 5.9 shows the scatter plot for the two sufficient predictors of SIR; the two lower left panels show the perspective plots for the first three sufficient predictors of SAVE (left) and SIR-II (right). Plotting symbols "+", "x", and "." represent digits 0, 6, and 9, respectively. What is interesting is that SIR separates the locations of the three groups almost perfectly, but does not give much information about the variances of these three groups. On the other hand, both SAVE and SIR-II give very good separation of the variances of the three groups: the variances for the sufficient predictors for digit 9 are much greater than those of the sufficient predictors for 6 and 0. However, they failed almost completely to separate their locations. Intuitively, this is because SAVE and SIR-II give too much weight on the conditional variance $\mathrm{var}(X|Y)$, but too little weight on the conditional mean $E(X|Y)$, even though — as shown in Theorem 5.2 — SAVE does contain the information of $E(X|Y)$ in theory.

Another thing that we learned from this example is that the maximum number of sufficient predictors SIR can extract from a response variable with k categories is $k-1$, which limits its application particularly when the number of classes is small, such as binary response variables. Thus, the first-order method such as SIR, and the second-order methods such as SAVE and SIR-II, have their advantages and disadvantages.

This raises the question: is it possible to develop second-order methods that combine the information of the first two conditional moments more effectively than SAVE and SIR-II, so that we can separate both the locations and the variances of the sufficient predictors for data sets like this? This will be the goal of the next chapter.

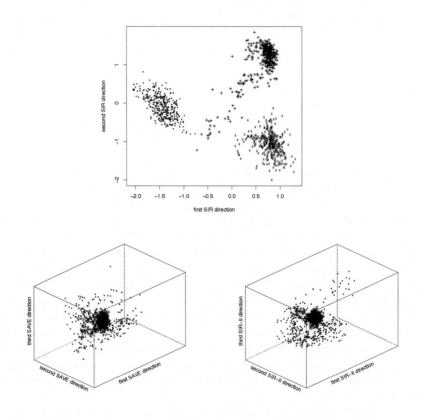

Figure 5.9 *First three predictors from SIR (upper panel), SAVE (lower-left panel), and SIR-II (lower-right panel).*

Chapter 6

Contour Regression and Directional Regression

6.1	Contour Directions and Central Subspace	63
6.2	Contour Regression at the Population Level	65
6.3	Algorithm and R Codes for CR	67
6.4	Exhaustiveness of Contour Regression	69
6.5	Directional Regression	70
6.6	Representation of Λ_{DR} Using Moments	74
6.7	Algorithm and R Codes for DR	76
6.8	Exhaustiveness Relation with SIR and SAVE	77
6.9	Pen Digit Case Study Continued	79

6.1 Contour Directions and Central Subspace

Intuitively, for a surface S defined on $\mathbb{R}^p$, the central subspace represents the directions in $\mathbb{R}^p$ in which the height of S changes the most. That is, the central subspace is aligned with the gradients of S. But gradient directions are orthogonal to the contour directions — those directions along which the height of S remains constant. Thus, if we can find the subspace spanned by the contour directions, then its orthogonal complement can be used to estimate the central subspace. This is the idea behind Contour Regression (Li et al. (2005)).

To illustrate the idea, consider the following model

$$Y = 2\sin X_1 + 0.2\varepsilon, \tag{6.1}$$

where $X = (X^1, X^2)^\top \sim N(0, I_2)$, $X \perp\!\!\!\perp \varepsilon$, and $\varepsilon \sim N(0,1)$. We generate an i.i.d. sample $(X_1, Y_1), \ldots, (X_{50}, Y_{50})$ from the distribution of (X,Y). The central subspace for this model is $\mathrm{span}\{(1,0)^\top\}$. For a small constant $\varepsilon > 0$, consider the collection of vectors

$$C = \{X_i - X_j : |Y_i - Y_j| < \varepsilon\}.$$

Figure 6.1 shows all these vectors with $\varepsilon = 0.3$. We can see that most of them are

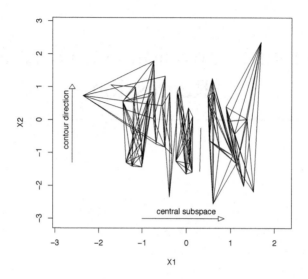

Figure 6.1 *Illustration of contour directions and central subspace.*

aligned roughly with the subspace $\mathrm{span}\{(0,1)^{\mathsf{T}}\}$, which is orthogonal to the central subspace. Indeed, we performed a spectral decomposition of

$$\textstyle\sum_{i<j}(X_i - X_j)(X_i - X_j)^{\mathsf{T}}I(|Y_i - Y_j| > \varepsilon).$$

The two eigenvectors are

$$v_1 = (-0.29, -0.96)^{\mathsf{T}}, \quad v_2 = c(0.96, -0.29)^{\mathsf{T}}.$$

where v_1, which is the principal direction of the vectors in Figure 6.1, is roughly orthogonal to $\mathscr{S}_{Y|X}$; whereas v_2, the orthogonal complement of the principal contour direction, is roughly aligned with $\mathscr{S}_{Y|X}$.

Li et al. (2005) refers to the set of vectors

$$\{X_i - X_j : 1 = i < j = n\}$$

as empirical directions. Contour Regression selects empirical directions whose increment in Y are small. These directions are aligned with the contours of $f(y|x)$. Since, for each fixed y, $f(y|x)$ is a function of $\beta^{\mathsf{T}}x$, its contours are affine subspaces. The selected empirical directions therefore roughly form a linear subspace orthogonal to the central subspace. We use the orthogonal complement of the contour subspace to recover the central subspace.

6.2 Contour Regression at the Population Level

Let $(\tilde{X}, \tilde{Y})$ be an independent copy of (X, Y). Let Z and $\tilde{Z}$ be the standardized versions of X and $\tilde{X}$. That is,

$$Z = \Sigma^{-1/2}[X - E(X)], \quad \tilde{Z} = \Sigma^{-1/2}[\tilde{X} - E(\tilde{X})].$$

Let $\varepsilon > 0$ be a small constant. Contour Regression is based on the following matrix at the population level

$$A(\varepsilon) = E[(Z - \tilde{Z})(Z - \tilde{Z})^{\mathsf{T}} | |Y - \tilde{Y}| < \varepsilon].$$

The next theorem is the theoretical basis for Contour Regression.

Theorem 6.1 *If* LCM(β) *and* CCV(β) *are satisfied, where* β *is a basis matrix of* $\mathscr{S}_{Y|Z}$, *then*

$$\operatorname{span}(2I_p - A(\varepsilon)) \subseteq \mathscr{S}_{Y|Z}.$$

PROOF. By definition,

$$\begin{aligned} A(\varepsilon) = &\, E(ZZ^{\mathsf{T}} | |Y - \tilde{Y}| < \varepsilon) - E(Z\tilde{Z}^{\mathsf{T}} | |Y - \tilde{Y}| < \varepsilon) \\ &- E(\tilde{Z}Z^{\mathsf{T}} | |Y - \tilde{Y}| < \varepsilon) + E(\tilde{Z}\tilde{Z}^{\mathsf{T}} | |Y - \tilde{Y}| < \varepsilon). \end{aligned}$$

Because $(Z, Y, \tilde{Y})$ and $(\tilde{Z}, \tilde{Y}, Y)$ have the same distribution, the conditional distributions of $Z|(Y, \tilde{Y})$ and $\tilde{Z}|(Y, \tilde{Z})$ are also the same. Consequently,

$$\begin{aligned} E(Z\tilde{Z}^{\mathsf{T}} | |Y - \tilde{Y}| < \varepsilon) &= E(\tilde{Z}Z^{\mathsf{T}} | |\tilde{Y} - Y| < \varepsilon) = E(\tilde{Z}Z^{\mathsf{T}} | |Y - \tilde{Y}| < \varepsilon) \\ E(ZZ^{\mathsf{T}} | |Y - \tilde{Y}| < \varepsilon) &= E(\tilde{Z}\tilde{Z}^{\mathsf{T}} | |\tilde{Y} - Y| < \varepsilon) = E(\tilde{Z}\tilde{Z}^{\mathsf{T}} | |Y - \tilde{Y}| < \varepsilon), \end{aligned}$$

which implies

$$A(\varepsilon) = 2E(ZZ^{\mathsf{T}} | |Y - \tilde{Y}| < \varepsilon) - 2E(Z\tilde{Z}^{\mathsf{T}} | |Y - \tilde{Y}| < \varepsilon). \tag{6.2}$$

Furthermore, by Theorem 2.1, part 3, we have

$$(Z, Y) \perp\!\!\!\perp (\tilde{Z}, \tilde{Y}) \Rightarrow (Z, Y) \perp\!\!\!\perp \tilde{Z} | \tilde{Y} \Rightarrow Z \perp\!\!\!\perp \tilde{Z} | (Y, \tilde{Y}).$$

Therefore the first term on the right-hand side of (6.2) reduces to

$$\begin{aligned} E(Z\tilde{Z}^{\mathsf{T}} | |Y - \tilde{Y}| < \varepsilon) &= E(Z | |Y - \tilde{Y}| < \varepsilon)E(\tilde{Z}^{\mathsf{T}} | |Y - \tilde{Y}| < \varepsilon) \\ &= E(Z | |Y - \tilde{Y}| < \varepsilon)E(Z^{\mathsf{T}} | |Y - \tilde{Y}| < \varepsilon), \end{aligned}$$

where the second equality holds because the conditional distribution $Z|(Y, \tilde{Y})$ is the same as that of $\tilde{Z}|(Y, \tilde{Y})$. To summarize, we now have

$$A(\varepsilon) = 2E(ZZ^{\mathsf{T}} | |Y - \tilde{Y}| < \varepsilon) - 2E(Z | |Y - \tilde{Y}| < \varepsilon)E(Z^{\mathsf{T}} | |Y - \tilde{Y}| < \varepsilon). \tag{6.3}$$

Next, let us further reduce $E(Z| |Y - \tilde{Y}| < \varepsilon)$ using LCM(β). Again, by Theorem 2.1, part 3, we have

$$(Z,Y) \perp\!\!\!\perp (\tilde{Z},\tilde{Y}) \Rightarrow (Z,Y) \perp\!\!\!\perp \tilde{Y} \Rightarrow \tilde{Y} \perp\!\!\!\perp Z|Y.$$

Hence

$$\begin{aligned}
E(Z| |Y - \tilde{Y}| < \varepsilon) &= E[E(Z|Y,\tilde{Y})| |Y - \tilde{Y}| < \varepsilon] \\
&= E[E(Z|Y)| |Y - \tilde{Y}| < \varepsilon] \\
&= P_\beta E[E(Z|Y)| |Y - \tilde{Y}| < \varepsilon] \\
&= P_\beta E[E(Z|Y,\tilde{Y})| |Y - \tilde{Y}| < \varepsilon] \\
&= P_\beta E[Z| |Y - \tilde{Y}| < \varepsilon],
\end{aligned} \tag{6.4}$$

where the third equality follows from LCM(β) and Theorem 3.1.

Finally, we further reduce $E(ZZ^\mathsf{T}| |Y - \tilde{Y}| < \varepsilon)$ using LCM(β) and CCV(β). Again because of $Z \perp\!\!\!\perp \tilde{Y}|Y$, we have

$$\begin{aligned}
E(ZZ^\mathsf{T}| |Y - \tilde{Y}| < \varepsilon) &= E[E(ZZ^\mathsf{T}|Y,\tilde{Y})| |Y - \tilde{Y}| < \varepsilon] \\
&= E[E(ZZ^\mathsf{T}|Y)| |Y - \tilde{Y}| < \varepsilon].
\end{aligned} \tag{6.5}$$

By LCM(β), CCV(β), Theorem 3.1, and Theorem 5.1, we have

$$\begin{aligned}
E(ZZ^\mathsf{T}|Y) &= \mathrm{var}(Z|Y) + E(Z|Y)E(Z|Y) \\
&= Q_\beta + P_\beta \mathrm{var}(Z|Y)P_\beta + P_\beta E(Z|Y)E(Z|Y)P_\beta \\
&= Q_\beta + P_\beta E(ZZ^\mathsf{T}|Y)P_\beta \\
&= Q_\beta + P_\beta E(ZZ^\mathsf{T}|Y,\tilde{Y})P_\beta.
\end{aligned} \tag{6.6}$$

Substituting (6.6) into (6.5), we find

$$\begin{aligned}
E(ZZ^\mathsf{T}| |Y - \tilde{Y}| < \varepsilon) &= E[Q_\beta + P_\beta E(ZZ^\mathsf{T}|Y,\tilde{Y})P_\beta| |Y - \tilde{Y}| < \varepsilon] \\
&= Q_\beta + P_\beta E(ZZ^\mathsf{T}| |Y - \tilde{Y}| < \varepsilon)P_\beta.
\end{aligned} \tag{6.7}$$

Substituting (6.4) and (6.7) into (6.3), we have

$$\begin{aligned}
A(\varepsilon) &= 2\left[Q_\beta + P_\beta E(ZZ^\mathsf{T}| |Y - \tilde{Y}| < \varepsilon)P_\beta\right] \\
&\quad - 2P_\beta E(Z| |Y - \tilde{Y}| < \varepsilon)E(Z^\mathsf{T}| |Y - \tilde{Y}| < \varepsilon)P_\beta \\
&= 2Q_\beta + P_\beta A(\varepsilon)P_\beta,
\end{aligned} \tag{6.8}$$

where the second equality follows from (6.3). Subtract both sides of the above equality by $2I_p$ to obtain

$$A(\varepsilon) - 2I_p = P_\beta A(\varepsilon)P_\beta - 2P_\beta = P_\beta[A(\varepsilon) - 2I_p]P_\beta,$$

which implies the desired relation. $\qquad\square$

The assertion of Theorem 6.1 is equivalent to

$$\mathrm{span}\{[2I_p - A(\varepsilon)]^2\} \subseteq \mathscr{S}_{Y|Z}.$$

Hence, by Theorem 2.2, if we let Λ_{CR} represent the matrix $[2I_p - A(\varepsilon)]^2$, then we have

$$\Sigma^{-1/2}\text{span}(\Lambda_{CR}) \subseteq \mathscr{S}_{Y|X}.$$

Consequently, if $v_1, \ldots, v_r$ are the eigenvectors of Λ_{CR} corresponding to its nonzero eigenvalues, then

$$\text{span}(\Sigma^{-1/2}v_1, \ldots, \Sigma^{-1/2}v_r) \subseteq \mathscr{S}_{Y|X}.$$

In the next section we mimic this relation to develop an estimator of $\mathscr{S}_{Y|X}$, which we call Contour Regression.

6.3 Algorithm and R Codes for CR

Let $(X_1, Y_1), \ldots, (X_n, Y_n)$ be an i.i.d. sample of (X, Y). The Contour Regression is implemented at the sample level through the following steps.

Algorithm 6.1 Contour Regression

---

1. Standardize $X_1, \ldots, X_n$ to $Z_1, \ldots, Z_n$ as before.
2. For fixed $\varepsilon > 0$, compute index set

$$J = \{(i,j) : i < j, |Y_j - Y_i| < \varepsilon\}.$$

For example, ε can be taken to be the 5% percentile of the sample $\{|Y_j - Y_i| : 1 \leq i < j \leq n\}$.
3. Compute the ratio

$$\hat{A}(\varepsilon) = \frac{1}{\text{card}(J)} \Sigma_{(i,j) \in J}(Z_i - Z_j)(Z_i - Z_j)^\mathsf{T}, \tag{6.9}$$

where $\text{card}(\cdot)$ represents the cardinality of a set.
4. Let $v_1, \ldots, v_r$ be the first r eigenvectors of $[I_p - A(\varepsilon)]^2$, and let $u_\ell = \hat{\Sigma}^{-1/2}v_\ell$. The sufficient predictors are

$$\{u_\ell^\mathsf{T}(X_i - E_n X) : i = 1, \ldots, n\}, \ \ell = 1, \ldots, r.$$

Below is an R-code to implement Contour Regression. To achieve an appropriate scale, we reset the threshold ε to εM, where

$$M = \Sigma_{i<j}|Y_i - Y_j| / \binom{n}{2}.$$

The input `percent` refers to the percentage of M.

```
cr = function(x,y,percent){
tradeindex12 = function(k,n){
j = ceiling(k/n)
i = k - (j-1)*n
return(c(i,j))}
mu=apply(x,2,mean);signrt=matpower(var(x),-1/2)
z=t(t(x)-mu)%*%signrt
n=dim(x)[1];p = dim(x)[2]
ymat=matrix(y,n,n)
deltay=c(abs(ymat - t(ymat)))
singleindex=(1:n^2)[deltay < percent*mean(deltay)]
contourmat=matrix(0,p,p)
for(k in singleindex){
doubleindex=tradeindex12(k,n)
deltaz=z[doubleindex[1],]-z[doubleindex[2],]
contourmat=contourmat+deltaz %*% t(deltaz)}
signrt=matpower(var(x),-1/2)
return(signrt%*%eigen(contourmat)$vectors)
}
```

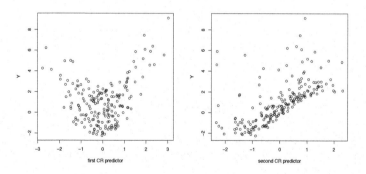

Figure 6.2 *Y versus first two CR predictors for the model in Example 5.1.*

Example 6.1 We now apply Contour Regression to the model in Example 5.1, where the central subspace contains two directions. In one direction the regression surface is roughly monotone; in the other it is symmetric about 0. Our goal is to see how Contour Regression picks up both the monotone trend and the symmetric trend. We take the threshold to be 20% of M. Figure 6.2 shows the Y versus the first two CR predictors.

We see that, like SAVE in Example 5.1, CR also recovers both predictors, but seems to have higher resolution than that achieved by SAVE in Figure 5.6. In fact, the Spearman's correlations between the first two CR predictors with the true predictors are 0.959 and 0.991, respectively; whereas Spearman's correlations between the first two SAVE predictors with the true predictors are 0.924 and 0.839, respectively. The number of slices used by SAVE is 6. □

6.4 Exhaustiveness of Contour Regression

In this section we investigate the problem of exhaustiveness. That is, whether $\mathrm{span}(\Lambda_{\mathrm{CR}})$ occupies the entire central subspace $\mathscr{S}_{Y|Z}$ under reasonable sufficient conditions.

Theorem 6.2 *Suppose LCM(β) and CCV(β) are satisfied with β being a basis matrix of $\mathscr{S}_{Y|Z}$. Suppose $(\tilde{Z}, \tilde{Y})$ is an independent copy of (Z, Y) and $\varepsilon > 0$. If, for any $v \in \mathscr{S}_{Y|Z}$, we have*

$$E\{[v^{\mathsf{T}}(Z - \tilde{Z})]^2 \,|\, |Y - \tilde{Y}| < \varepsilon\} < E\{[v^{\mathsf{T}}(Z - \tilde{Z})]^2\}, \tag{6.10}$$

then $\mathrm{span}(\Lambda_{\mathrm{CR}}) = \mathscr{S}_{Y|Z}$, *and consequently* $\mathrm{span}(\Sigma^{-1/2}\Lambda_{\mathrm{CR}}) = \mathscr{S}_{Y|X}$.

The condition (6.10) is a reasonable one: intuitively, since Y varies with $v^{\mathsf{T}}Z$ for $v \in \mathscr{S}_{Y|Z}$ but does not vary with $w^{\mathsf{T}}Z$ for $w \perp \mathscr{S}_{Y|Z}$, we expect that a small variation of Y entails a small variation in $v^{\mathsf{T}}Z$, but does not entail a small variation in $w^{\mathsf{T}}Z$. That is, we expect

$$E\{[v^{\mathsf{T}}(Z - \tilde{Z})]^2 \,|\, |Y - \tilde{Y}| < \varepsilon\} < E\{[w^{\mathsf{T}}(Z - \tilde{Z})]^2 \,|\, |Y - \tilde{Y}| < \varepsilon\}. \tag{6.11}$$

as long as $\|v\| = \|w\|$. The two conditions (6.11) and (6.11) turn out to be equivalent, as the following lemma shows.

Lemma 6.1 *Under LCM(β) and CCV(β), where β is a basis matrix $\mathscr{S}_{Y|Z}$, conditions (6.10) and (6.11) are equivalent.*

PROOF. It suffices to show that, for any $v \in \mathscr{S}_{Y|Z}$, $w \perp \mathscr{S}_{Y|Z}$, $\|v\| = \|w\|$,

$$E\{[v^{\mathsf{T}}(Z - \tilde{Z})]^2\} = E\{[w^{\mathsf{T}}(Z - \tilde{Z})]^2 \,|\, |Y - \tilde{Y}| < \varepsilon\}. \tag{6.12}$$

Without loss of generality, assume $\|v\| = \|w\| = 1$. The left-hand side is

$$v^{\mathsf{T}}E[(Z - \tilde{Z})(Z - \tilde{Z})^{\mathsf{T}}]v = v^{\mathsf{T}}[2E(ZZ^{\mathsf{T}})]v = 2.$$

By (6.8), under LCM(β) and CCV(β), the right-hand side of (6.12) is

$$w^{\mathsf{T}}E[(Z - \tilde{Z})(Z - \tilde{Z})^{\mathsf{T}} \,|\, |Y - \tilde{Y}| < \varepsilon]w = w^{\mathsf{T}}A(\varepsilon)w = w^{\mathsf{T}}[2Q_\beta + P_\beta A(\varepsilon)P_\beta]w = 2,$$

where the last equality holds because $w \perp \mathrm{span}(\beta)$. This completes the proof. □

PROOF OF THEOREM 6.2. By Lemma 5.1, it suffices to show that, for any $v \in \mathscr{S}_{Y|Z}$, we have $v^{\mathsf{T}}\Lambda_{\mathrm{CR}}v > 0$. Suppose $v \in \mathscr{S}_{Y|Z}$, $\|v\| = 1$, and $v\Lambda_{\mathrm{CR}}v = 0$. Then

$$2 - v^{\mathsf{T}}E[(Z - \tilde{Z})(Z - \tilde{Z})^{\mathsf{T}} \,|\, |Y - \tilde{Y}| < \varepsilon]v = 0,$$

which implies

$$E\{[v^{\mathsf{T}}(Z - \tilde{Z})]^2 \,|\, |Y - \tilde{Y}| < \varepsilon\} = 2,$$

contradicting (6.10).   □

Sufficient condition (6.10) indicates that Contour Regression is exhaustive under a reasonable condition. However, it does not give a method for actually checking whether Contour Regression is exhaustive given a specific model, because condition (6.10) involves infinitely many v. The next corollary, which follows easily from Lemma 5.1, gives a sufficient condition for the exhaustiveness that can be checked given a model.

Corollary 6.1 *Suppose LCM(β) and CCV(β) are satisfied with β being a basis matrix of $\mathscr{S}_{Y|Z}$. If there is an orthonormal basis of $\mathscr{S}_{Y|Z}$, say $\{v_1, \ldots, v_d\}$, such that*

$$\lambda_{\max}[V^\mathsf{T} A(\varepsilon)V] < 2,$$

where $V = (v_1, \ldots, v_d)$, then span(Λ_{CR}) = $\mathscr{S}_{Y|Z}$.

PROOF. By (6.10) and (6.12), we see that $2I_p - A(\varepsilon)$ is positive semidefinite. Hence, by Lemma 5.1, it suffices to show that $2I_d - V^\mathsf{T} A(\varepsilon)V > 0$, which is true because $V^\mathsf{T} A(\varepsilon)V < 2I_p$.   □

As an application of this criterion, let us check whether CR is exhaustive for the model in Example 5.1.

Example 6.2 In the model (5.8) in Example 5.1, take $v_1 = e_1$ and $v_2 = e_2$, where, recall that $\{e_i\}$ stands for the standard orthonormal basis in $\mathbb{R}^p$. We evaluate $\lambda_{\max}[V^\mathsf{T} A(\varepsilon)V]$ numerically for $\varepsilon = 0.1, \ldots, 0.5$, and they are $1.39, 1.41, 1.41, 1.41, 1.42$, none of which exceeds 2. This means Contour Regression is exhaustive for Model 5.8 for a wide range of contour thresholds. The conditional expectation $A(\varepsilon)$ is evaluated by Monte Carlo based on 1,000,000 draws of $(X,Y), (\tilde{X}, \tilde{Y})$ from the model. This explains why Contour Regression effectively recovers both directions in the central subspace, as shown by Figure 6.2.   □

For a more careful and elaborate investigation of the exhaustiveness of Contour Regression, see Li et al. (2005). The sufficient condition given here is somewhat simpler than that given in Li et al. (2005), in that the latter requires LCM(β) but not CCV(β); whereas we require both conditions. However, the ideas are similar, and the version presented here are more intuitive and requires a much simpler proof. Moreover, Li et al. (2005) did not give a further sufficient condition parallel to Corollary 6.1.

6.5 Directional Regression

Directional Regression (DR), introduced by Li and Wang (2007), is a second-order method somewhat similar to Contour Regression, but is much simpler to compute and in many ways more efficient than Contour Regression. If Contour Regression can be viewed as a method of *selecting* empirical directions to extract information about the central subspace, then Directional Regression can be viewed as a method of *reorganizing* empirical directions for extracting this information. Contour Regression relies on U-statistics of the form (6.9), which is a sum of approximately $\varepsilon\binom{n}{2}$

terms. Directional Regression, whose idea and performance is similar to Contour Regression, only involves sample moments. Instead of considering the matrix

$$E[(Z - \tilde{Z})(Z - \tilde{Z})^{\mathsf{T}} | |Y - \tilde{Y}| < \varepsilon],$$

DR is based on the following matrix

$$E[(Z - \tilde{Z})(Z - \tilde{Z})^{\mathsf{T}} | Y, \tilde{Y}].$$

Although, in appearance, the sample version of this matrix should also involve U-statistics, as we will see later, the above matrix appear as a term in an unconditional expectation, where $\tilde{Y}$ can be avoided by simplification. The next theorem shows how one can use the above matrix to recover vectors in $\mathscr{S}_{Y|Z}$.

Theorem 6.3 *Suppose LCM(β) and CCV(β) are satisfied with β being a basis matrix of $\mathscr{S}_{Y|Z}$, and that Z is standardized so that $E(Z) = 0$ and $\mathrm{var}(Z) = I_p$. Then*

$$\mathrm{span}\{2I_p - E[(Z - \tilde{Z})(Z - \tilde{Z})^{\mathsf{T}} | Y, \tilde{Y}]\} \subseteq \mathscr{S}_{Y|Z}.$$

Consequently, $\Sigma^{-1/2}\mathrm{span}\{2I_p - E[(Z - \tilde{Z})(Z - \tilde{Z})^{\mathsf{T}} | Y, \tilde{Y}]\} \subseteq \mathscr{S}_{Y|X}.$
PROOF. Let $A(Y, \tilde{Y}) = E[(Z - \tilde{Z})(Z - \tilde{Z})^{\mathsf{T}} | Y, \tilde{Y}]$. Then

$$A(Y, \tilde{Y}) = E(ZZ^{\mathsf{T}} | Y, \tilde{Y}) - E(Z\tilde{Z}^{\mathsf{T}} | Y, \tilde{Y}) - E(\tilde{Z}Z^{\mathsf{T}} | Y, \tilde{Y}) + E(\tilde{Z}\tilde{Z}^{\mathsf{T}} | Y, \tilde{Y}).$$

As we argued in the proof of Theorem 6.1, we have $Z \perp\!\!\!\perp \tilde{Z} | Y, \tilde{Y}$, $Z \perp\!\!\!\perp \tilde{Y} | Y$, and $\tilde{Z} \perp\!\!\!\perp Y | \tilde{Y}$. Hence the right-hand side above reduces to

$$
\begin{aligned}
A(Y, \tilde{Y}) &= E(ZZ^{\mathsf{T}} | Y, \tilde{Y}) - E(Z|Y, \tilde{Y})E(\tilde{Z}^{\mathsf{T}} | Y, \tilde{Y}) - E(\tilde{Z}|Y, \tilde{Y})E(Z^{\mathsf{T}} | Y, \tilde{Y}) + E(\tilde{Z}\tilde{Z}^{\mathsf{T}} | Y, \tilde{Y}) \\
&= E(ZZ^{\mathsf{T}} | Y) - E(Z|Y)E(\tilde{Z}^{\mathsf{T}} | \tilde{Y}) - E(\tilde{Z}|\tilde{Y})E(Z^{\mathsf{T}} | Y) + E(\tilde{Z}\tilde{Z}^{\mathsf{T}} | \tilde{Y}).
\end{aligned}
$$

By (3.1) and (6.6), the right-hand side can be rewritten as

$$
2Q_\beta + P_\beta[E(ZZ^{\mathsf{T}} | Y) - E(Z|Y)E(\tilde{Z}^{\mathsf{T}} | \tilde{Y}) - E(\tilde{Z}^{\mathsf{T}} | \tilde{Y})E(Z^{\mathsf{T}} | Y) + E(\tilde{Z}\tilde{Z}^{\mathsf{T}} | \tilde{Y})]P_\beta
$$
$$
= 2Q_\beta + P_\beta A(Y, \tilde{Y})P_\beta.
$$

Hence

$$A(Y, \tilde{Y}) - 2I_p = P_\beta[A(Y, \tilde{Y}) - 2I_p]P_\beta,$$

which implies the desired relation. $\square$

We now demonstrate the intuition behind the above mathematical development to show why Directional Regression is unbiased. Let $n = 40$. We generate $(X_1, Y_1), \ldots, (X_n, Y_n)$ as an i.i.d. sample from (X, Y), which follows the statistical model (6.1). The central subspace for this example is $\mathrm{span}\{(1, 0)\}$. The upper panel of Figure 6.3 shows the empirical directions $X_i - X_j$ plotted at (Y_i, Y_j), as represented by arrows. That is, the empirical direction $X_i - X_j$ is represented by an arrow begins at (Y_i, Y_j) and ends at $(Y_i, Y_j) + (X_i - X_j)$. From the plot we can see that, in the regions where Y_i and Y_j are far apart, which is the lower-right and upper-left corners of the panel, the vectors $X_k - X_\ell$ whose (Y_k, Y_ℓ) are near (Y_i, Y_j) tend to be longer and horizontal (i.e. in the direction of the central subspace), but in the region where Y_i and Y_j are close together, which is the diagonal region of the panel, the vectors $X_k - X_\ell$ whose (Y_k, Y_ℓ) are near (Y_i, Y_j) tend to be shorter and vertical (i.e. perpendicular to the central subspace).

This tendency becomes more pronounced in the plot shown in the lower-left panel of Figure 6.3. Here, at each (Y_i, Y_j), we construct the matrix

$$A(Y_i, Y_j) = \frac{\sum_{k=1}^{n}\sum_{\ell=1}^{n}(X_k - X_\ell)(X_k - X_\ell)^{\mathsf{T}}I(\|(Y_k, Y_\ell)^{\mathsf{T}} - (Y_i, Y_j)^{\mathsf{T}}\| \leq 0.2)}{\sum_{k=1}^{n}\sum_{\ell=1}^{n}I(\|(Y_k, Y_\ell)^{\mathsf{T}} - (Y_i, Y_j)^{\mathsf{T}}\| \leq 0.2)}, \qquad (6.13)$$

which is an approximation of

$$E[(X - \tilde{X})(X - \tilde{X})^{\mathsf{T}}|Y = Y_i, \tilde{Y} = Y_j].$$

We extract the first eigenvector $\hat{v}_{ij}$ and eigenvalue $\hat{\lambda}_{ij}$ from the matrix (6.13), and plot the vector $\hat{\lambda}_{ij}^{1/2}\hat{v}_{ij}$ at (Y_i, Y_j). The vector $\hat{\lambda}_{ij}^{1/2}\hat{v}_{ij}$ represents the principal direction of vectors $X_k - X_\ell$ whose (Y_k, Y_ℓ) are near (Y_i, Y_j), with length roughly the same length as the average lengths of empirical direction $X_k - X_\ell$ near $X_i - X_j$. We can see that this plot strengthens the tendency shown in the upper panel: the vectors are either aligned with or perpendicular to the central subspace, with longer arrows aligned with it and shorter ones perpendicular to it. This gives us a hint as to how to extract information about the central subspace from these re-organized and filtered empirical directions: if the arrow is short, we extract its orthogonal direction; if the arrow is long, we extract itself.

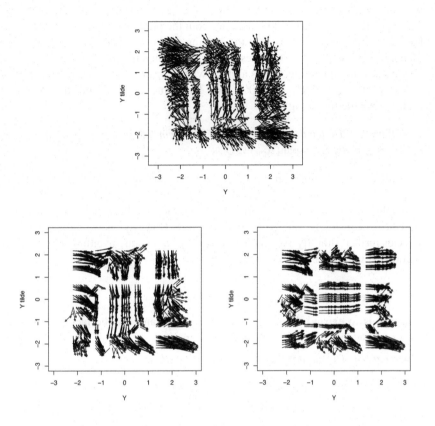

Figure 6.3 *Reorganization of empirical directions by Directional Regression.*

This automatic selection of a vector or its orthogonal complement is implemented systematically by spectral decomposition, as follows: we simply extract the first eigenvector of the matrix

$$[2I_p - A(Y_i, Y_j)]^2.$$

To see how this works, when Y_i and Y_j are near each other, a typical matrix $A(Y_i, Y_j)$ is similar to

$$\begin{pmatrix} 1 & 0 \\ 0 & 2 \end{pmatrix}$$

whose principal direction is $(0, 1)$. In this case the principal direction of $[2I_2 - A(Y_i, Y_j)]^2$ is $(1, 0)$, which is orthogonal to $(0, 1)$ but aligned with the central subspace. When Y_i and Y_j are far apart, a typical $A(Y_i, Y_j)$ is similar to

$$\begin{pmatrix} 5 & 0 \\ 0 & 2 \end{pmatrix}$$

whose principal direction is $(1,0)$. In this case the principal direction of $[2I_2 - A(Y_i, Y_j)]^2$ is $(1,0)$, which is aligned with both the principal direction of $A(Y_i, Y_i)$ and the central subspace. In this way the leading eigenvector of $[2I_2 - A(Y_i, Y_j)]^2$ automatically retains or orthogonalizes the principal direction of $A(Y_i, Y_j)$ according to whether this principal direction is aligned with or orthogonal to the central subspace.

Now let $\Lambda_{DR} = E[2I_p - A(Y, \tilde{Y})]^2$. By Theorem 6.3 and Proposition 5.2 we can easily prove the following result.

Corollary 6.2 *Under the assumptions of Theorem 6.3, we have* $\mathrm{span}(\Lambda_{DR}) \subseteq \mathscr{S}_{Y|Z}$, *and consequently* $\Sigma^{-1/2}\mathrm{span}(\Lambda_{DR}) \subseteq \mathscr{S}_{Y|X}$.

This corollary implies that the rank r of Λ_{DR} is no greater than d, and that if $v_1, \ldots, v_r$ be the eigenvectors of Λ_{DR}, then $\Sigma^{-1/2}v_\ell$, $\ell = 1, \ldots, r$ are vectors in $\mathscr{S}_{Y|X}$. In principle, we can mimic this matrix directly at the sample level, perhaps using a U-statistic, to construct a sample estimate of $\mathscr{S}_{Y|X}$. However, as we mentioned before, we can further simplify Λ_{DR} as a functions of moments, avoiding $\tilde{Z}, \tilde{Y}$. The next section explains in detail how this is done.

6.6 Representation of Λ_{DR} Using Moments

The idea is to expand the square $[2I_p - A(Y, \tilde{Y})]^2$, and use the fact that (Z, Y) and $(\tilde{Z}, \tilde{Y})$ are independent having the same distribution.

Theorem 6.4 *The matrix Λ_{DR} can be re-expressed as*

$$\begin{aligned}
\Lambda_{DR} = {} & 2E[E(ZZ^\mathsf{T}|Y)]^2 + 2E^2[E(Z|Y)E(Z^\mathsf{T}|Y)] \\
& + 2E[E(Z^\mathsf{T}|Y)E(Z|Y)]E[E(Z|Y)E(Z^\mathsf{T}|Y)] - 2I_p.
\end{aligned} \tag{6.14}$$

PROOF. First expand Λ_{DR} as

$$\Lambda_{DR} = 4I_p - 4EA(Y, \tilde{Y}) + A^2(Y, \tilde{Y})]. \tag{6.15}$$

We simplify $E[A(Y, \tilde{Y})]$ as follows:

$$\begin{aligned}
E[A(Y, \tilde{Y})] = {} & E\{E[(Z - \tilde{Z})(Z - \tilde{Z})^\mathsf{T}|Y, \tilde{Y}]\} \\
= {} & E[E(ZZ^\mathsf{T}|Y)] + E[E(Z|Y)E(\tilde{Z}^\mathsf{T}|\tilde{Y})] + E[E(\tilde{Z}|\tilde{Y})E(Z^\mathsf{T}|Y)] + E[E(\tilde{Z}\tilde{Z}^\mathsf{T}|\tilde{Y})],
\end{aligned}$$

where $E[E(ZZ^\mathsf{T}|Y)] = E(ZZ^\mathsf{T}) = I_p$, and similarly, the fourth term on the right is I_p. Because $Y \perp\!\!\!\perp \tilde{Y}$, the third term is $E(E(Z|Y)]E(\tilde{Z}^\mathsf{T}|\tilde{Y})] = E(Z)E(\tilde{Z}^\mathsf{T}) = 0$. Similarly, the third term is also 0. Thus we have

$$EA(Y, \tilde{Y}) = 2I_p. \tag{6.16}$$

Next, we simplify $E[A^2(Y, \tilde{Y})]$:

$$\begin{aligned}
E[A^2(Y, \tilde{Y})] = {} & E\{E[(Z - \tilde{Z})(Z - \tilde{Z})^\mathsf{T}|Y, \tilde{Y}]E[(Z - \tilde{Z})(Z - \tilde{Z})^\mathsf{T}|Y, \tilde{Y}]\} \\
= {} & E\{[E(ZZ^\mathsf{T}|Y) - E(Z|Y)E(\tilde{Z}^\mathsf{T}|\tilde{Y}) - E(\tilde{Z}|\tilde{Y})E(Z^\mathsf{T}|Y) + E(\tilde{Z}\tilde{Z}^\mathsf{T}|\tilde{Y})] \\
& [E(ZZ^\mathsf{T}|Y) - E(Z|Y)E(\tilde{Z}^\mathsf{T}|\tilde{Y}) - E(\tilde{Z}|\tilde{Y})E(Z^\mathsf{T}|Y) + E(\tilde{Z}\tilde{Z}^\mathsf{T}|\tilde{Y})]^\mathsf{T}\}.
\end{aligned}$$

Let $B(Y,\tilde{Y}) = E(ZZ^{\mathsf{T}}|Y) - E(Z|Y)E(\tilde{Z}^{\mathsf{T}}|\tilde{Y})$. Then

$$
\begin{aligned}
E[A^2(Y,\tilde{Y})] &= E\{[B(Y,\tilde{Y}) + B(\tilde{Y},Y)][B(Y,\tilde{Y}) + B(\tilde{Y},Y)]\} \\
&= E\{[B^2(Y,\tilde{Y})] + E[B(Y,\tilde{Y})B(\tilde{Y},Y)] + E[B(\tilde{Y},Y)B(Y,\tilde{Y})] + E[B^2(\tilde{Y},Y)]\}.
\end{aligned}
$$

But because $(Y,\tilde{Y})$ and $(\tilde{Y},Y)$ have the same distribution, we have

$$
E[A^2(Y,\tilde{Y})] = 2E[B^2(Y,\tilde{Y})] + 2E[B(Y,\tilde{Y})B(\tilde{Y},Y)]. \tag{6.17}
$$

The term $E[B^2(Y,\tilde{Y})]$ is further simplified as

$$
\begin{aligned}
EB^2(Y,\tilde{Y}) &= E\{[E(ZZ^{\mathsf{T}}|Y) - E(Z|Y)E(\tilde{Z}^{\mathsf{T}}|\tilde{Y})][E(ZZ^{\mathsf{T}}|Y) - E(Z|Y)E(\tilde{Z}^{\mathsf{T}}|\tilde{Y})]\} \\
&= E[E(ZZ^{\mathsf{T}}|Y)E(ZZ^{\mathsf{T}}|Y)] - E[(ZZ^{\mathsf{T}}|Y)E(Z|Y)E(\tilde{Z}^{\mathsf{T}}|\tilde{Y})] \\
&\quad - E[E(Z|Y)E(\tilde{Z}^{\mathsf{T}}|\tilde{Y})E(ZZ^{\mathsf{T}}|Y)] + E[E(Z|Y)E(\tilde{Z}^{\mathsf{T}}|\tilde{Y})E(Z|Y)E(\tilde{Z}^{\mathsf{T}}|\tilde{Y})].
\end{aligned}
$$

Because $Y \perp\!\!\!\perp \tilde{Y}$, the second and third terms each involves the factor $E(Z)$ or $E(\tilde{Z})$. So they are 0, and $E[B^2(Y,\tilde{Y})]$ reduces to

$$
\begin{aligned}
E[B^2(Y,\tilde{Y})] &= E[E(ZZ^{\mathsf{T}}|Y)^2] + E[E(Z|Y)E(\tilde{Z}^{\mathsf{T}}|\tilde{Y})E(Z|Y)E(\tilde{Z}^{\mathsf{T}}|\tilde{Y})] \\
&= E[E(ZZ^{\mathsf{T}}|Y)^2] + E[E(Z|Y)E(Z^{\mathsf{T}}|Y)E(\tilde{Z}|\tilde{Y})E(\tilde{Z}^{\mathsf{T}}|\tilde{Y})] \tag{6.18} \\
&= E[E(ZZ^{\mathsf{T}}|Y)^2] + \{E[E(Z|Y)E(Z^{\mathsf{T}}|Y)]\}^2.
\end{aligned}
$$

Similarly, the term $E[B(Y,\tilde{Y})B(\tilde{Y},Y)]$ in (6.17) is further simplified as

$$
\begin{aligned}
&E[B(Y,\tilde{Y})B(\tilde{Y},Y)] \\
&= E[E(ZZ^{\mathsf{T}}|Y)E(\tilde{Z}\tilde{Z}^{\mathsf{T}}|\tilde{Y})] + E[E(ZZ^{\mathsf{T}}|Y)E(\tilde{Z}|\tilde{Y})E(Z^{\mathsf{T}}|Y)] \\
&\quad + E[E(Z|Y)E(\tilde{Z}^{\mathsf{T}}|\tilde{Y})E(\tilde{Z}\tilde{Z}^{\mathsf{T}}|\tilde{Y})] + E[E(Z|Y)E(\tilde{Z}^{\mathsf{T}}|\tilde{Y})E(\tilde{Z}|\tilde{Y})E(Z^{\mathsf{T}}|Y)] \\
&\equiv C_1 + C_2 + C_3 + C_4,
\end{aligned}
$$

where $C_2 = 0$, $C_3 = 0$, and

$$
\begin{aligned}
C_1 &= E[E(ZZ^{\mathsf{T}}|Y)]E[E(\tilde{Z}\tilde{Z}^{\mathsf{T}}|\tilde{Y})] = I_p, \\
C_4 &= E[E(Z|Y)E(E(\tilde{Z}^{\mathsf{T}}|\tilde{Y})E(\tilde{Z}|\tilde{Y}))E(Z^{\mathsf{T}}|Y)] \\
&= E[E(Z^{\mathsf{T}}|Y)E(Z|Y)]E[E(Z|Y)E(Z^{\mathsf{T}}|Y)].
\end{aligned}
$$

Hence

$$
E[B(Y,\tilde{Y})B(\tilde{Y},Y)] = I_p + E[E(Z^{\mathsf{T}}|Y)E(Z|Y)]E[E(Z|Y)E(Z^{\mathsf{T}}|Y)]. \tag{6.19}
$$

Substitute (6.18) and (6.19) into (6.17) to obtain

$$
\begin{aligned}
E[A^2(Y,\tilde{Y})] &= 2E[E(ZZ^{\mathsf{T}}|Y)^2] + 2\{E[E(Z|Y)E(Z^{\mathsf{T}}|Y)]\}^2 \\
&\quad + 2I_p + 2E[E(Z^{\mathsf{T}}|Y)E(Z|Y)]E[E(Z|Y)E(Z^{\mathsf{T}}|Y)].
\end{aligned} \tag{6.20}
$$

Finally, substitute (6.16) and (6.20) into (6.15) to prove the desired equality. □

Algorithm 6.2 Directional Regression

---

1. Standardize $X_1, \ldots, X_n$ to $Z_1, \ldots, Z_n$ as before.
2. Let $\{J_\ell : \ell = 1, \ldots, h\}$ be partition of $[Y_{(1)}, Y_{(n)}]$. Compute:

$$M_{1\ell} = E(Z|Y \in J_\ell), \quad M_{2\ell} = E_n(ZZ^\top|Y \in J_\ell).$$

3. Compute

$$\Lambda_1 = h^{-1}\sum_{\ell=1}^{h} P_n(Y \in J_\ell) M_{2\ell}^2,$$

where $P_n(Y \in J_\ell) = E_n I(Y \in J_\ell)$. Similarly, compute

$$\Lambda_2 = h^{-1}\sum_{\ell=1}^{h} P_n(Y \in J_\ell)(M_{1\ell}M_{1\ell}^\top)^2, \quad \Lambda_3 = h^{-1}\sum_{\ell=1}^{h}[M_{1\ell}^\top M_{1\ell}(\tilde{Y})][h^{-1}\sum_{\ell=1}^{h}(M_{1\ell}M_{1\ell}^\top)].$$

Form the matrix $\Lambda_{\mathrm{DR}} = 2\Lambda_1 + 2\Lambda_2 + 2\Lambda_3 - 2I_p$.
4. Compute the first r eigenvectors of Λ_{DR}, say $v_1, \ldots, v_r$, and compute $u_\ell = \hat{\Sigma}^{-1/2}v_\ell$ for $\ell = 1, \ldots, r$. The sufficient predictor is

$$u_\ell^\top(X_i - E_n X), \quad i = 1, \ldots, n, \ \ell = 1, \ldots, r.$$

6.7 Algorithm and R Codes for DR

Let $(X_1, Y_1), \ldots, (X_n, Y_n)$ be a sample of (X, Y). As before, estimation procedure for Directional Regression is implemented by performing spectral decomposition of a matrix that mimics Λ_{DR} at the sample level.

The following is an R-code to implement Directional Regression:

```
dr = function(x,y,h,r,ytype){
p=ncol(x);n=nrow(x)
signrt=matpower(var(x),-1/2)
xc=t(t(x)-apply(x,2,mean))
xst=xc%*%signrt
if(ytype=="continuous") ydis=discretize(y,h)
if(ytype=="categorical") ydis=y
yless=ydis;ylabel=numeric()
for(i in 1:n) {
if(var(yless)!=0) {ylabel=c(ylabel,yless[1])
yless=yless[yless!=yless[1]]}}
ylabel=c(ylabel,yless[1])
prob=numeric()
for(i in 1:h) prob=c(prob,length(ydis[ydis==ylabel[i]])/n)
vxy = array(0,c(p,p,h));exy=numeric()
for(i in 1:h) {
vxy[,,i]=var(xst[ydis==ylabel[i],])
exy=rbind(exy,apply(xst[ydis==ylabel[i],],2,mean))}
mat1 = matrix(0,p,p);mat2 = matrix(0,p,p)
```

```
for(i in 1:h){
mat1 = mat1+prob[i]*(vxy[,,i]+exy[i,]%*%t(exy[i,]))%*%
          (vxy[,,i]+exy[i,]%*%t(exy[i,]))
mat2 = mat2+prob[i]*exy[i,]%*%t(exy[i,])}
out = 2*mat1+2*mat2%*%mat2+2*sum(diag(mat2))*mat2-2*diag(p)
return(signrt%*%eigen(out)$vectors[,1:r])
}
```

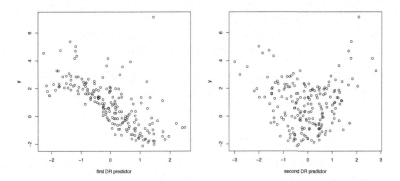

Figure 6.4 *Y versus first two DR predictors for the model in Example 5.1.*

Example 6.3 We again applied Directional Regression to the model in Example 5.1, with monotone and symmetric trend along the two directions in the central subspace. As before, we took the sample size n to be 200. We take 3 slices in the DR algorithm. Figure 6.4 shows the response versus the first two DR predictors.

We see that, like Contour Regression, Directional Regression successfully picks up both the monotone trend and the symmetric trend.   □

6.8   Exhaustiveness Relation with SIR and SAVE

In this section we investigate the sufficient conditions under which Directional Regression is an exhaustive estimator, as well as the relation among the subspaces spanned by Λ_{SIR}, Λ_{SAVE}, and Λ_{DR}.

Theorem 6.5 *Let $\tilde{Z}$ be an independent copy of Z, where Z is the standardized X. Suppose* $\text{LCM}(\beta)$ *and* $\text{CCV}(\beta)$ *hold for a spanning matrix β of $\mathscr{S}_{Y|Z}$. If, for each $v \in \mathscr{S}_{Y|Z}$,*

$$E\{[v^{\mathsf{T}}(Z-\tilde{Z})]^2|Y,\tilde{Y}\} \tag{6.21}$$

is nondegenerate. Then $\text{span}(\Lambda_{\text{DR}}) = \mathscr{S}_{Y|Z}$.

PROOF. By Theorem 6.3, under $\text{LCM}(\beta)$ and $\text{CCV}(\beta)$, $\text{span}(\Lambda_{\text{DR}}) \subseteq \mathscr{S}_{Y|Z}$. By Lemma 5.1, we need to show that, for all $v \in \mathscr{S}_{Y|Z}$, $\|v\| > 0$,

$$v^{\mathsf{T}}\Lambda_{\text{DR}}v > 0.$$

Because

$$v^{\mathrm{T}}\Lambda_{\mathrm{DR}}v = v^{\mathrm{T}}E[I_p - A(Y,\tilde{Y})]^2 v$$
$$= v^{\mathrm{T}}E\{[I_p - A(Y,\tilde{Y})](I_p - vv^{\mathrm{T}})[I_p - A(Y,\tilde{Y})]\}v + E\{v^{\mathrm{T}}[I_p - A(Y,\tilde{Y})]v\}^2$$
$$\geq E[v^{\mathrm{T}}(I_p - A(Y,\tilde{Y}))v]^2,$$

where the inequality holds because $I_p - vv^{\mathrm{T}}$, being a projection, is positive semidefinite. By assumption, $v^{\mathrm{T}}A(Y,\tilde{Y})v$ is non-degenerate, which implies that $v^{\mathrm{T}}[I_p - A(Y,\tilde{Y})]v$ is also non-degenerate. Therefore, by Jensen's inequality,

$$E(v^{\mathrm{T}}[[I_p - A(Y,\tilde{Y})]v)^2 > [E(v^{\mathrm{T}}([I_p - A(Y,\tilde{Y}))v]^2 \geq 0,$$

as to be proved. $\qquad\square$

Similar to Corollary 6.1, the next corollary gives a sufficient condition for (6.21) that we can check once we know a basis of the central subspace. It follows directly from Lemma 5.1.

Corollary 6.3 *Suppose LCM(β) and CCV(β) are satisfied with β being a basis matrix of $\mathscr{S}_{Y|Z}$. If there is a basis of $\mathscr{S}_{Y|Z}$, say $\{v_1,\ldots,v_d\}$, such that*

$$V^{\mathrm{T}}\Lambda_{\mathrm{DR}}V > 0,$$

where $V = (v_1,\ldots,v_d)$, then $\mathrm{span}(\Lambda_{\mathrm{DR}}) = \mathscr{S}_{Y|Z}$.

Applying the similar calculation of Example 6.2, it is easy to show that Directional Regression is exhaustive for Model (5.8) in Example 5.1.

We see that, under LCM(β), CCV(β), and additional mild conditions, DR and SAVE are both exhaustive, but SIR is not. In fact, more is true. One can show that, with essentially no condition, Λ_{SAVE} and Λ_{DR} in fact span the same subspace; whereas Λ_{SIR} spans a subspace of the common subspace spanned by Λ_{SAVE} and Λ_{DR}.

Theorem 6.6 *If $E(Z) = 0$ and $\mathrm{var}(Z) = I_p$, and all moments involved in the definition of Λ_{SIR}, Λ_{SAVE} and Λ_{DR} are finite, then $\mathscr{S}_{\mathrm{SIR}} \subseteq \mathscr{S}_{\mathrm{DR}} = \mathscr{S}_{\mathrm{SAVE}}$.*

PROOF. By Theorem 6.4,

$$\Lambda_{\mathrm{DR}} = 2E\{[E(ZZ^{\mathrm{T}}|Y)]^2\} + 2\{\mathrm{var}[E(Z|Y)]\}^2 + 2\mathrm{tr}\{\mathrm{var}[E(Z|Y)]\}\{\mathrm{var}[E(Z|Y)]\} - 2I_p$$
$$= 2E\{[E(ZZ^{\mathrm{T}}|Y)]^2\} + 2\Lambda_{\mathrm{SIR}}^2 + 2\mathrm{tr}(\Lambda_{\mathrm{SIR}})\Lambda_{\mathrm{SIR}} - 2I_p \qquad (6.22)$$

In the above, because $E(ZZ^{\mathrm{T}}) = I_p$, we have

$$E\{[E(ZZ^{\mathrm{T}}|Y)]^2\} = E\{[E(ZZ^{\mathrm{T}}|Y) - I_p + I_p]^2\}$$
$$= E\{[E(ZZ^{\mathrm{T}}|Y) - I_p]^2\} + 2E[E(ZZ^{\mathrm{T}}|Y) - I_p] + I_p \qquad (6.23)$$
$$= E\{[E(ZZ^{\mathrm{T}}|Y) - I_p]^2\} + I_p.$$

Because

$$E(ZZ^{\mathrm{T}}|Y) = \mathrm{var}(Z|Y) + E(Z|Y)E(Z^{\mathrm{T}}|Y),$$

we have

$$E\{[E(ZZ^{\mathrm{T}}|Y) - I_p]^2\}$$
$$= E\{[\mathrm{var}(Z|Y) + E(Z|Y)E(Z^{\mathrm{T}}|Y) - I_p]^2\}$$
$$= E\{[I_p - \mathrm{var}(Z|Y)]^2\} + E\{[\mathrm{var}(Z|Y) - I_p]E(Z|Y)E(Z^{\mathrm{T}}|Y)\} \qquad (6.24)$$
$$\quad + E\{E(Z|Y)E(Z^{\mathrm{T}}|Y)[\mathrm{var}(Z|Y) - I_p]\} + E\{[E(Z|Y)E(Z^{\mathrm{T}}|Y)]^2\}.$$

Substitute (6.24) into the right-hand side of (6.22) to obtain

$$\Lambda_{\mathrm{DR}} = 2\Lambda_{\mathrm{SAVE}} + 2E[(\mathrm{var}(Z|Y) - I_p)E(Z|Y)E(Z^{\mathsf{T}}|Y)]$$
$$+ 2E[E(Z|Y)E(Z^{\mathsf{T}}|Y)(\mathrm{var}(Z|Y) - I_p)] + 2E[E(Z|Y)E(Z^{\mathsf{T}}|Y)]^2 + 2\Lambda_{\mathrm{SIR}}^2 + 2\mathrm{tr}(\Lambda_{\mathrm{SIR}})\Lambda_{\mathrm{SIR}}$$
$$\equiv A_1 + \cdots + A_6.$$

Since, by Theorem 5.2, $E(Z|Y) \in \mathrm{span}(\Lambda_{\mathrm{SAVE}})$, if we let $v \perp \mathscr{S}_{\mathrm{SAVE}}$, then

$$v^{\mathsf{T}}A_1 v = 0, \ldots, v^{\mathsf{T}}A_6 v = 0,$$

which implies $v \perp \mathscr{S}_{\mathrm{DR}}$. So $\mathscr{S}_{\mathrm{DR}} \subseteq \mathscr{S}_{\mathrm{SAVE}}$. Also, by (6.22) and (6.23) we see that

$$\Lambda_{\mathrm{DR}} = 2E\{[E(ZZ^{\mathsf{T}}|Y) - I_p]^2\} + 2E\{[E(Z|Y)E(Z^{\mathsf{T}}|Y)]^2\}$$
$$+ 2\mathrm{tr}E[E(Z|Y)E(Z^{\mathsf{T}}|Y)]E[E(Z|Y)E(Z^{\mathsf{T}}|Y)] \equiv B_1 + B_2 + B_3.$$

Because B_1, B_2, and B_3 are all positive semidefinite, if we take $v^{\mathsf{T}}\Lambda_{\mathrm{DR}}v = 0$, then

$$v^{\mathsf{T}}B_1 v = 0, \ldots, v^{\mathsf{T}}B_3 v = 0.$$

Note that $v^{\mathsf{T}}B_3 v = 0$ implies $v^{\mathsf{T}}E(X|Y) = 0$. Hence, by (6.24), we have

$$0 = v^{\mathsf{T}}B_1 v = v^{\mathsf{T}}E\{[E(ZZ^{\mathsf{T}}|Y) - I_p]^2\}v = v^{\mathsf{T}}E[\mathrm{var}(Z|Y) - I_p]^2 v.$$

Therefore $v^{\mathsf{T}}\Lambda_{\mathrm{SAVE}}v = 0$, which implies $\mathscr{S}_{\mathrm{SAVE}} \subseteq \mathscr{S}_{\mathrm{DR}}$. Also, by Theorem 5.2, $\mathscr{S}_{\mathrm{SIR}} \subseteq \mathscr{S}_{\mathrm{SAVE}}$. $\square$

Even though, at the population level, DR and SAVE recovers the same subspace of $\mathscr{S}_{Y|Z}$, at the sample level they can behave quite differently, as can be seen from the case study in the next subsection.

6.9 Pen Digit Case Study Continued

We applied Contour Regression and Directional Regression to the pen digit data studied in Section 5.8. Again focusing on the three digits 0, 6, 9, we used CR and DR to develop the basis matrix $\hat{\beta}$ for the central subspace using the training set, and then applied it to both the training data and the test data to obtain the three sufficient predictors. For CR, we took all the empirical directions within each category to form the contour vectors; that is

$$\bigcup_{\alpha \in \{0,6,9\}} \{X_i - X_j : i < j, Y_i = Y_j = \alpha\},$$

where $\alpha = 0, 6, 9$. For DR, we took three slices corresponding to the three digits; that is,

$$\{X_i : Y_i = \alpha\}, \quad \alpha = 0, 6, 9.$$

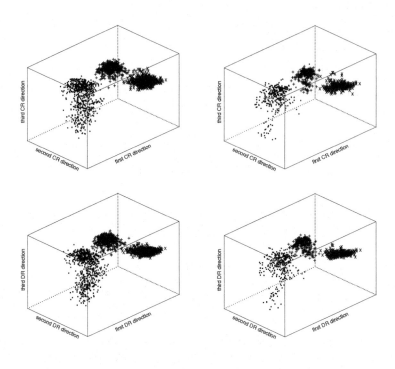

Figure 6.5 *Perspective plots for the pen digit case study.*

The upper-left plot in Figure 6.5 shows the first three CR predictors extracted from the training data and applied to the test data; the upper right plot shows the first three CR predictors using the test data. The two lower plots show the corresponding results for DR. Compared with the results obtained by SIR, SAVE, and SIR-II in Section 5.8, we see that CR and DR combines the advantages of SIR, SAVE, and SIR-II: CR and DR show separations of the three groups in both location and scale. Specifically, the third direction of CR and the third direction of DR separate the three groups by their variances (or scales): the digit 9 (as represented by ·) has much larger variance than the other two digits (+ for 0, x for 6). The first and second directions of CR and DR separate the locations of the three digits. In comparison, as we have seen in Section 5.8, SIR only provides separation in location: there is no obvious separation in variances in the three groups in the scatter plot (upper panel) in Figure 5.9. SAVE and SIR-II provide variance separation: the digit 9 has the largest variance than the other two digits, but provides no obvious location separation.

Scale separation provided by CR and DR in their third directions is obviously useful for classification. To illustrate this point, we produced in Figure 6.6 the scatter plots for the first two SIR predictors for the testing set (left panel), and for the first and third DR predictors for the testing set (right panel). We highlight two cases with plot symbol "o", case 823 (lower left "o" in the left panel) and case 894, both of which are digit 9. Based on the location separation provided by SIR on the left panel, it would be hard to identify case 823 and impossible to identify case 894 as digit 9. However, with the help of scale separation in the third DR direction, as shown on the right panel, we can reasonably confidently identify case 823 as the digit 9, and we have reasonable confidence to identify case 894 as the digit 9 — or at least we strongly suspect that case 894 does not belong to the digit 0 group.

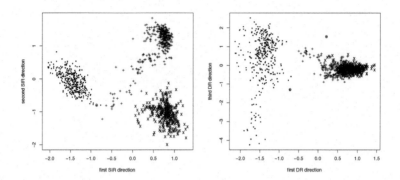

Figure 6.6 *Effect of scale separation by SIR and DR for the pen digit data.*

It turns out that the ambiguity of case 894 (it is placed somewhere between the 0 and 9 groups) is due to poor handwriting. In Figure 6.7 we show the rough shape of the handwritten digit produced by the 16 dimensional predictor, where the components 1, 3, ..., 15 are the x-coordinates and 2, 4, ..., 16 are the y-coordinates of the sampled points from the handwritten digit. In the middle of the five digit shapes is the case 894. To facilitate comparison, we also show typical shapes of handwritten 9 (case 892, first from left; case 893, second from left) and typical shapes of handwritten 0 (case 1, first from right; case 2, second from right). We see that case 894 is quite irregular, looking like neither 9 nor 0.

Another point worth noticing is that, while DR and CR produce rather similar results, DR requires much less computing time. In this data set, there are there are 363 digit 0, 336 digit 6, and 336 digit 9. Thus the number of empirical directions is

$$\binom{363}{2} + \binom{336}{2} + \binom{336}{2} = 178263,$$

which means the CR matrix Λ_{CR} is based on the sum of these many matrices $(X_i - X_j)(X_i - X_j)^T$. In comparison, the matrices in Λ_{DR} are sum of $n = 1035$ terms. The difference is already quite significant for this modest data set: in a desktop computer with Intel Core i5 processor, using R, CR took 31.42 seconds, whereas DR only took 0.06 seconds to complete the estimation of β.

Figure 6.7 *Five handwriting shapes in the pen digit data.*

Chapter 7

Elliptical Distribution and Predictor Transformation

7.1   Linear Conditional Mean and Elliptical Distribution   83
7.2   Box-Cox Transformation   88
7.3   Application to the Big Mac Data   92
7.4   Estimating Equations for Handling Non-Ellipticity   94

A crucial assumption for the methods developed in Chapters 3 through 6 is the linear conditional mean assumption (LCM(β)) and the constant conditional variance assumption (CCV(β)). In this chapter we study the theoretical relation between the linear conditional mean condition and elliptical distribution, and introduce a method to transform a set of vectors into roughly multivariate normal random vectors to satisfy the LCM(β) and CCV(β).

7.1   Linear Conditional Mean and Elliptical Distribution

In practice, the linear condition mean assumption may not be satisfied by the data and we need to first transform the predictor to meet this assumption. Since we do not know β, we cannot check LCM(β) directly. However, as was shown by Eaton (1986), if LCM(β) holds for all β, then the distribution of X is elliptical, and vice versa. This result links linear conditional mean with the shape of the distribution. We first define an elliptical distribution.

Definition 7.1 *A random vector $X \in \mathbb{R}^p$ is said to have an elliptical distribution if there is a positive definite matrix A such that the density of X is a function of $x^\mathsf{T} A x$; that is*

$$f_X(x) = h(x^\mathsf{T} A x).$$

The matrix A is called the shape parameter of the distribution of X. In the special case where $A = I_p$, X is said to have a spherical distribution.

In other words, X has an elliptical distribution if the contours $\{x : f_X(x) = c\}$ for any $c \in (\min[f_X(x)], \max[f_X(x)])$ is an ellipsoid $\{x : x^\mathsf{T} A x = c'\}$, or a sphere $\{x : x^\mathsf{T} x = c'\}$ when $A = I_p$. We now develop some technical properties of the elliptical distribution.

Lemma 7.1 *A random vector X has an elliptical distribution with shape parameter A if and only if $A^{1/2}X$ has a spherical distribution.*

PROOF. If $Z = A^{1/2}X$ has a spherical distribution, then $f_Z(z)$ is of the form $g_Z(z^{\mathsf{T}}z)$ for some function $g_Z : \mathbb{R}^+ \to \mathbb{R}^+$. The density of X is

$$f_X(x) = f_Z(A^{1/2}x)\det[\partial(A^{1/2}x)/\partial x^{\mathsf{T}}] = f_Z(A^{1/2}x)A^{1/2} = g_Z(x^{\mathsf{T}}Ax)\det(A^{1/2}),$$

where the right-hand side is a function of $x^{\mathsf{T}}Ax$. Thus X has an elliptical distribution.

If X has an elliptical distribution with shape parameter A, then $f_X(x) = g_X(x^{\mathsf{T}}Ax)$. The density of $Z = A^{1/2}X$ is

$$f_Z(z) = f_X(A^{-1/2}z)\det(A^{-1/2}) = g_X(z^{\mathsf{T}}z)\det(A^{-1/2}),$$

which is a function of $z^{\mathsf{T}}z$. Thus Z has a spherical distribution.   □

The next lemma shows that a function $f(x)$ depends only on $x^{\mathsf{T}}x$ if and only if this function is invariant under orthogonal transformations.

Lemma 7.2 Let $f : \mathbb{R}^p \to \mathbb{R}$ be a function. The following statements are equivalent:

1. f is a function of $x^{\mathsf{T}}x$;
2. $f(x) = f(\Gamma x)$ for any orthogonal matrix Γ.

PROOF. $1 \Rightarrow 2$. If Γ is an orthogonal matrix, then, by statement 1,

$$f(\Gamma x) = g(x^{\mathsf{T}}\Gamma^{\mathsf{T}}\Gamma x) = g(x^{\mathsf{T}}x) = f(x).$$

$2 \Rightarrow 1$. Let x_1, x_2 be members of $\mathbb{R}^p$ such that $\|x_1\| = \|x_2\|$. Then there is an orthogonal matrix Γ such that $x_2 = \Gamma x_1$. Hence $f(x_2) = f(\Gamma x_1) = f(x_1)$, which implies that f depends on x only through $\|x\|$.   □

Next, we show that if X has an elliptical distribution then its expectation is 0. In the following, the equality $\overset{\mathscr{D}}{=}$ means "equal in distribution"; that is, the random variables on the two sides of $\overset{\mathscr{D}}{=}$ have the same distribution.

Lemma 7.3 If X has an elliptical distribution and X is integrable, then $EX = 0$.

PROOF. First, assume X has a spherical distribution. Because $-I$ is an orthogonal matrix, we have $X \overset{\mathscr{D}}{=} -X$ and so $E(X) = -E(X)$, which implies $E(X) = 0$. If X has an elliptical distribution, then $X = A^{-1/2}Z$, where Z has a spherical distribution. Therefore $E(X) = A^{-1/2}E(Z) = 0$.   □

Let $\phi_X(t)$ denote the characteristic function of X; that is $\phi_X(t) = E(e^{it^{\mathsf{T}}X})$, where $i = \sqrt{-1}$ and $t \in \mathbb{R}^p$. The next lemma gives a sufficient and necessary condition for X to have an spherical distribution in terms of its characteristic function.

Lemma 7.4 A random vector X has a spherical distribution if and only if its characteristic function is a function of $t^{\mathsf{T}}t$; that is, $\phi_X(t) = g_0(t^{\mathsf{T}}t)$ for some function g_0.

PROOF. By Lemma 7.2, it suffices to show that, for any orthogonal matrix Γ, we have $\phi_X(t) = \phi_X(\Gamma t)$. Because Γ^{T} is also an orthogonal matrix, by Lemma 7.2 we have $X \overset{\mathscr{D}}{=} \Gamma^{\mathsf{T}}X$, which implies $\phi_X(t) = \phi_{\Gamma^{\mathsf{T}}X}(t)$. The right-hand side is

$$\phi_{\Gamma^{\mathsf{T}}X}(t) = E(e^{it^{\mathsf{T}}\Gamma^{\mathsf{T}}X}) = E(e^{i(\Gamma t)^{\mathsf{T}}X}) = \phi_X(\Gamma t),$$

as to be proved.   □

We now prove a technical lemma about the existence of a smooth curve between any two points on a sphere.

Lemma 7.5 *Let* $S = \{x : \|x\| = r\}$, *where* $r > 0$, *and let* $t_1, t_2 \in S$. *Then there is a differentiable mapping* $c : (0, 2\pi) \mapsto S$ *that passes through* t_1 *and* t_2; *that is, there exist* $\alpha_1, \alpha_2 \in (0, 2\pi)$ *such that* $c(\alpha_1) = t_1$ *and* $c(\alpha_2) = t_2$.

PROOF. Let H be a hyperplane that passes through $t_1, t_2, 0$ in $\mathbb{R}^p$. Then $C = H \cap S$ is a circle centered at the origin with radius r. Pick an arbitrary point t_3 in C that is not t_1 and t_2, and let t_4 be the point on C obtained by rotating t_3 counterclockwise $90°$ along C. Then any point t in C that is not t_3 can be written as $t = \cos(\alpha)t_3 + \sin(\alpha)t_4$ for some $\alpha \in (0, 2\pi)$. By construction, $\{c(\alpha) : \alpha \in (0, 2\pi)\}$ is a differentiable curve that passes through t_1 and t_2. □

We now establish the relation between spherical distribution and conditional expectation of the form $E(u^T X | v^T X)$ where $u \perp v$, which is an essential step toward linking LCM(β) with elliptical distributions.

Theorem 7.1 *Suppose X is integrable. Then the following statements are equivalent:*

1. for all $u, v \in \mathbb{R}^p$, $u \perp v$, *we have* $E(u^T X | v^T X) = 0$;

2. X has a spherical distribution.

PROOF. $1 \Rightarrow 2$. It suffices to show that, for any $t_1, t_2 \in \mathbb{R}^p$ such that $\|t_1\| = \|t_2\|$, we have $\phi_X(t_1) = \phi_X(t_2)$. Let $S = \{t : \|t\| = \|t_1\|\}$. By Lemma 7.5, there is a differential function $c : (0, 2\pi) \to S$ that passes through t_1 and t_2. It then suffices to show that ϕ_X is constant on C. A sufficient condition for this is

$$\partial \phi_X[c(\alpha)]/\partial \alpha = 0 \tag{7.1}$$

for all $\alpha \in (0, 2\pi)$. The left hand side is

$$\begin{aligned}
\partial \phi_X[c(\alpha)]/\partial \alpha &= \partial E(e^{ic(\alpha)^T X})/\partial \alpha \\
&= E[(e^{ic(\alpha)^T X} i X^T \dot{c}(\alpha)] \tag{7.2} \\
&= E\{(e^{ic(\alpha)^T X} i E[X^T \dot{c}(\alpha)|c(\alpha)^T X]\}.
\end{aligned}$$

Because $c(\alpha) \in S$ for all α, $c(\alpha)^T c(\alpha) = \|t_1\|$ for all α. Differentiate both sides of this equation to obtain

$$\dot{c}(\alpha)^T c(\alpha) = 0$$

for all $\alpha \in (0, 2\pi)$. Then, by statement *1*,

$$E[X^T \dot{c}(\alpha)|c(\alpha)^T X] = 0$$

for all $\alpha \in (0, 2\pi)$. Now substitute the above equality into the right-hand side of (7.2) to prove (7.1).

$2 \Rightarrow 1$. First we show

$$E(|u^T X||v^T X|) < \infty. \tag{7.3}$$

Because $E(|u^T X|) < \infty$, we have $E[E(|u^T X||v^T X)] < \infty$, which implies (7.3) because the integrand is nonnegative.

To prove *1*, we assume without loss of generality that $\|u\| = \|v\| = 1$. Let $Z_1 = u^T X$ and $Z_2 = v^T X$. Let $\Lambda \in \mathbb{R}^{2 \times 2}$ be an orthogonal matrix. Then

$$\Lambda \begin{pmatrix} Z_1 \\ Z_2 \end{pmatrix} = \Lambda \begin{pmatrix} u^T X \\ v^T X \end{pmatrix} = [(u, v)\Lambda]^T X.$$

Let $w_1, \ldots, w_{p-2}$ be vectors in $\mathbb{R}^p$ such that $\{u, v, w_1, \ldots, w_{p-2}\}$ forms an orthonormal basis in $\mathbb{R}^p$. Because $(u, v)\Lambda$ is just a rotation of (u, v) in the subspace spanned by u, v, the matrix

$$((u, v)\Lambda, w_1, \ldots, w_{p-2})$$

is an orthogonal matrix in $\mathbb{R}^{p \times p}$, which implies

$$(u, v)^\top X \stackrel{\mathscr{D}}{=} [(u, v)\Lambda]^\top X.$$

Hence $(Z_1, Z_2)^\top$ has a spherical distribution. In particular $(Z_1, Z_2) \stackrel{\mathscr{D}}{=} (-Z_1, Z_2)$, from which it follows that

$$E(Z_1 | Z_2) = -E(Z_1 | Z_2).$$

Therefore $E(Z_1 | Z_2) = 0.$ □

From Theorem 7.1 we can deduce that $E(X | v^\top X)$ is linear in $v^\top X$ for any $v \in \mathbb{R}^p$, which is already quite close to the LCM(β) assumption.

Corollary 7.1 *Suppose X is integrable. Then the following statements are equivalent.*

1. X has a spherical distribution;

2. for all $v \in \mathbb{R}^p$, $E(X | v^\top X) = P_v X$, where P_v is the projection $v(v^\top v)^{-1} v^\top$;

3. for all $v \in \mathbb{R}^p$, $E(X | v^\top X)$ is linear in $v^\top X$.

PROOF. *1* ⟺ *2*. By Theorem 7.1 it suffices to show that statement *2* is equivalent to statement *1* in Theorem 7.1. Let $Q_v = I_p - P_v$. If statement *2* of this corollary holds, then,

$$E(X | v^\top X) = P_v X \Leftrightarrow E(Q_v X | v^\top X) = 0$$
$$\Leftrightarrow a^\top E(Q_v X | v^\top X) = 0 \quad \text{for all } a \in \mathbb{R}^p, a \neq 0.$$

Since Q_v is the projection on to span$(v)^\perp$, any $u \perp v$ can be written as $Q_v a$ for some $a \in \mathbb{R}^p$. Hence

$$a^\top E(Q_v X | v^\top X) = 0 \quad \text{for all } a \in \mathbb{R}^p, a \neq 0$$
$$\Leftrightarrow E(u^\top X | v^\top X) = 0 \quad \text{for all } u \perp v.$$

2 ⟹ *3*. If *2* holds, then

$$E(X | v^\top X) = P_v X = v(v^\top v)^{-1} v^\top X,$$

which is a linear function of $v^\top X$.

3 ⟹ *2*. This is just a special case of Lemma 1.1. □

Corollary 7.1 can be easily extended from spherical distribution to elliptical distribution, which we do next.

Corollary 7.2 *Suppose X is integrable then the following statements are equivalent:*

1. X has elliptical distribution with shape parameter A;

2. for any $v \in \mathbb{R}^p$, $E(X | v^\top X) = P_v^\top (A^{-1}) X$;

3. for any $v \in \mathbb{R}^p$, $E(X | v^\top X)$ is linear in $v^\top X$.

PROOF. $1 \Leftrightarrow 2$. By Lemma 7.2, X has an elliptical distribution with shape parameter A if and only if $A^{1/2}X$ has a spherical distribution. By Corollary 7.1, this happens if and only if, for any $v \in \mathbb{R}^p$,

$$
\begin{aligned}
E(A^{1/2}X|v^{\mathsf{T}}X) &= E(A^{1/2}X|(A^{-1/2}v)^{\mathsf{T}}A^{1/2}X) \\
&= P_{A^{-1/2}v}(A^{1/2}X) \\
&= A^{-1/2}v(v^{\mathsf{T}}A^{-1}v)^{-1}v^{\mathsf{T}}X.
\end{aligned}
$$

Multiply both sides by $A^{-1/2}$ from the left to prove statement 2.
$2 \Leftrightarrow 3$. This is because $E(X|v^{\mathsf{T}}X)$ is linear in $v^{\mathsf{T}}X$ if and only if $E(A^{-1/2}X|v^{\mathsf{T}}X)$ is linear in $v^{\mathsf{T}}X$.
□

Finally, we further generalize Corollary 7.2 to the case where v is a matrix rather than a vector. In the following, if A and B are matrices, then $\mathrm{diag}(A,B)$ represents the block-diagonal matrix whose diagonal blocks are A and B.

Theorem 7.2 *Suppose X is a p-dimensional integrable random vector, and $k \in \{1,\ldots,p-1\}$. Then the following statements are equivalent.*

1. X has an elliptical distribution with shape parameter A;

2. for any matrix $\beta \in \mathbb{R}^{p \times k}$ and $\gamma \in \mathbb{R}^{p-k}$ of full column ranks such that $\gamma^{\mathsf{T}}A^{-1}\beta = 0$, we have

$$E(\gamma^{\mathsf{T}}X|\beta^{\mathsf{T}}X) = 0;$$

3. for any $\beta \in \mathbb{R}^{p \times k}$, LCM($\beta$) is satisfied;

4. for any $\beta \in \mathbb{R}^{p \times k}$, $E(X|\beta^{\mathsf{T}}X) = [P_{\beta}(A^{-1})]^{\mathsf{T}}X$.

PROOF. We first prove the equivalence of these statements when X has a spherical distribution.
$1 \Rightarrow 2$. Let $\mathscr{S} = \mathrm{span}(\beta)$. Then $\mathrm{span}(\gamma) = \mathscr{S}^{\perp}$, the orthogonal complement of $\mathscr{S}$ in $\mathbb{R}^p$. Let $A \in \mathbb{R}^{k \times k}$ and $B \in \mathbb{R}^{(p-k) \times (p-k)}$ be nonsingular matrices such that $\beta = \tilde{\beta}A$, $\gamma = \tilde{\gamma}B$ where the columns of $\tilde{\beta}$ and $\tilde{\gamma}$ are orthonormal sets. Then $(\tilde{\beta},\tilde{\gamma})$ is an orthogonal matrix. Let $\Gamma = (\tilde{\beta},\tilde{\gamma})$. Because X has a spherical distribution, by Lemma 7.2, we have $X \overset{\mathscr{D}}{=} \Gamma^{\mathsf{T}}X$, which implies $(\tilde{\beta}^{\mathsf{T}}X,\tilde{\gamma}^{\mathsf{T}}X)$ also has a spherical distribution. Because the matrix $\mathrm{diag}(I_k,-I_{p-k})$ is orthogonal, by Lemma 7.2 again, $(\tilde{\beta}^{\mathsf{T}}X,\tilde{\gamma}^{\mathsf{T}}X) \overset{\mathscr{D}}{=} (\tilde{\beta}^{\mathsf{T}}X,-\tilde{\gamma}^{\mathsf{T}}X)$. Consequently, $E(-\tilde{\gamma}^{\mathsf{T}}X|\tilde{\beta}^{\mathsf{T}}X) = E(\tilde{\gamma}^{\mathsf{T}}X|\tilde{\beta}^{\mathsf{T}}X)$, which implies $E(\tilde{\gamma}^{\mathsf{T}}X|\tilde{\beta}^{\mathsf{T}}X) = 0$. Hence

$$E(\gamma^{\mathsf{T}}X|\beta^{\mathsf{T}}X) = E(\gamma^{\mathsf{T}}X|\tilde{\beta}^{\mathsf{T}}X) = B^{\mathsf{T}}E(\tilde{\gamma}^{\mathsf{T}}X|\tilde{\beta}^{\mathsf{T}}X) = 0.$$

$2 \Rightarrow 1$. It suffices to prove the statement 1 of Theorem 7.1. Let $u \perp v$. Then there exist $\beta \in \mathbb{R}^{p \times k}$, $\gamma \in \mathbb{R}^{p \times (p-k)}$ with $\mathrm{span}(\beta) \perp \mathrm{span}(\gamma)$ such that

$$v \in \mathrm{span}(\beta), \quad u \in \mathrm{span}(\gamma).$$

Consequently, there exist vectors $v_1 \in \mathbb{R}^k$ and $u_1 \in \mathbb{R}^{p-k}$ such that $v = \beta v_1$, $u = \gamma u_1$, and we have

$$E(u^{\mathsf{T}}X|v^{\mathsf{T}}X) = u_1^{\mathsf{T}}E[E(\gamma^{\mathsf{T}}X|\beta^{\mathsf{T}}X)|v_1^{\mathsf{T}}\beta^{\mathsf{T}}X] = 0.$$

Thus the statement 1 of Theorem 7.1 holds.
$2 \Rightarrow 3$. Let $\Gamma = (\beta,\gamma)$. Then $X = \Gamma^{-\mathsf{T}}\Gamma^{\mathsf{T}}X$. So

$$E(X|\beta^{\mathsf{T}}X) = \Gamma^{-\mathsf{T}}\begin{pmatrix} E(\beta^{\mathsf{T}}X|\beta^{\mathsf{T}}X) \\ E(\gamma^{\mathsf{T}}X|\beta^{\mathsf{T}}X) \end{pmatrix} = \Gamma^{-\mathsf{T}}\begin{pmatrix} \beta^{\mathsf{T}}X \\ 0 \end{pmatrix},$$

where the right-hand side is a linear function of $\beta^{\mathsf{T}}X$.

$3 \Rightarrow 4$. The proof is essentially the same as that of Lemma 1.1.

$4 \Rightarrow 2$. Because $\gamma^{\mathsf{T}}\beta = 0$, we have

$$E(\gamma^{\mathsf{T}}X|\beta^{\mathsf{T}}X) = \gamma^{\mathsf{T}}\beta(\beta^{\mathsf{T}}\beta)^{-1}\beta^{\mathsf{T}}X = 0.$$

Thus statement 2 holds.

We now turn to the general case where X has an elliptical distribution with a shape parameter A. In the following, β and $\tilde{\beta}$ are matrices in $\mathbb{R}^{p\times k}$ with full column ranks; γ and $\tilde{\gamma}$ are matrices in $\mathbb{R}^{p\times(p-k)}$ with full column ranks.

$1 \Leftrightarrow 2$. This is true because

$$X \text{ has an elliptical distribution with shape parameter } A$$
$$\Leftrightarrow A^{1/2}X \text{ has a spherical distribution}$$
$$\Leftrightarrow E(\gamma^{\mathsf{T}}A^{1/2}X|\beta^{\mathsf{T}}A^{1/2}X) = 0 \text{ for all } \beta, \gamma \text{ with } \beta^{\mathsf{T}}\gamma = 0$$
$$\Leftrightarrow E(\tilde{\gamma}^{\mathsf{T}}X|\tilde{\beta}^{\mathsf{T}}X) = 0 \text{ for all } \tilde{\beta}, \tilde{\gamma} \text{ with } \tilde{\beta}^{\mathsf{T}}A^{-1}\tilde{\gamma} = 0$$
$$\Leftrightarrow E(\gamma^{\mathsf{T}}X|\beta^{\mathsf{T}}X) = 0 \text{ for all } \beta, \gamma \text{ with } \beta^{\mathsf{T}}A^{-1}\gamma = 0.$$

$2 \Leftrightarrow 3$. This is true because

$$E(\gamma^{\mathsf{T}}X|\beta^{\mathsf{T}}X) = 0 \text{ for all } \beta, \gamma \text{ with } \beta^{\mathsf{T}}A^{-1}\gamma = 0$$
$$\Leftrightarrow E((A^{-1/2}\gamma)^{\mathsf{T}}A^{1/2}X|(A^{-1/2}\beta)^{\mathsf{T}}A^{1/2}X) = 0 \text{ for all } \beta, \gamma \text{ with } \beta^{\mathsf{T}}A^{-1}\gamma = 0$$
$$\Leftrightarrow E(\gamma^{\mathsf{T}}A^{1/2}X|\beta^{\mathsf{T}}A^{1/2}X) = 0 \text{ for all } \beta, \gamma \text{ with } \beta^{\mathsf{T}}\gamma = 0$$
$$\Leftrightarrow A^{1/2}X \text{ satisfies LCM}(\beta) \text{ for all } \beta$$
$$\Leftrightarrow X \text{ satisfies LCM}(A^{1/2}\beta) \text{ for all } \beta$$
$$\Leftrightarrow X \text{ satisfies LCM}(\beta) \text{ for all } \beta.$$

$3 \Leftrightarrow 4$. This is true because

$$X \text{ satisfies LCM}(\beta) \text{ for all } \beta$$
$$\Leftrightarrow A^{1/2}X \text{ satisfies LCM}(\beta) \text{ for all } \beta$$
$$\Leftrightarrow E[A^{1/2}X|\beta^{\mathsf{T}}(A^{1/2}X)] = P_{\beta}(I_p)(A^{1/2}X) \text{ for all } \beta$$
$$\Leftrightarrow E[X|\beta^{\mathsf{T}}(A^{-1/2}X)] = A^{-1/2}P_{A^{-1/2}\beta}(I_p)(A^{1/2}X) \text{ for all } \beta$$
$$\Leftrightarrow E[X|\beta^{\mathsf{T}}(A^{-1/2}X)] = [P_{\beta}(A^{-1})]^{\mathsf{T}}X \text{ for all } \beta.$$

The proof is completed.   □

In summary, the main message of this section is that a random vector X satisfies LCM(β) for all $\beta \in \mathbb{R}^{p\times k}$ if and only if X has an elliptical distribution, where k can be any integer between 1 and $p - 1$. Thus, if our exploratory data analysis (such as the scatter plot matrix) indicates that the joint distribution of X is elliptical, then the LCM(β) condition is satisfied.

7.2   Box-Cox Transformation

As we can see from Figure 3.3, the scatter plot matrix for the variables in the Big Mac data set, the elliptical distribution assumption is seriously violated. For example, the distribution of (X^1, X^2) is not elliptical at all. The same can be said of the distributions X^1 versus the other predictors. A pragmatic solution is to transform each X^i marginally to Gaussian distribution, and hope that, after these marginal transformations, the random vector X has a multivariate Gaussian distribution. Underlying this heuristic approach is a *Gaussian Copula Assumption*, which is stated as follows.

Assumption 7.1 *There exist injective functions $f_1, \ldots, f_p : \mathbb{R} \to \mathbb{R}$ such that $f_1(X^1), \ldots, f_p(X^p)$ has a multivariate Gaussian distribution.*

A commonly used set of injective transformations is called the Box-Cox transformation (Box and Cox, 1964), which is defined as

$$f_\lambda(u) = \begin{cases} (u^\lambda - 1)/\lambda & \lambda \neq 0 \\ \log u & \lambda = 0 \end{cases}$$

for $u > 0$. Let $X = (X^1, \ldots, X^p)$ and, for the time being, assume its components are positive random variables. We assume there exist $\lambda_1, \ldots, \lambda_p$ such that

$$\begin{pmatrix} f_{\lambda_1}(X^1) \\ \vdots \\ f_{\lambda_p}(X^p) \end{pmatrix} \sim N(\mu, \Sigma)$$

for some $\mu \in \mathbb{R}^p$ and positive definite matrix $\Sigma \in \mathbb{R}^{p \times p}$. We then use maximum likelihood estimation to estimate the transformation parameters $(\lambda_1, \ldots, \lambda_p)$. We denote this vector as λ, and denote vector-valued function on the left-hand side above as F_λ; that is,

$$F_\lambda : \mathbb{R}^p \to \mathbb{R}^p, \ (x^1, \ldots, x^p) \mapsto (f_{\lambda_1}(x^1), \ldots, f_{\lambda_p}(x^p))^\mathsf{T}.$$

The joint density of X is derived as follows. Let $Z = F_\lambda(X)$. Then $X = F_\lambda^{-1}(Z)$, where $Z \sim N(\mu, \Sigma)$. The joint density of Z is

$$f_Z(z) = \frac{1}{(2\pi)^{p/2}[\det(\Sigma)]^{1/2}} \exp[-\tfrac{1}{2}(z - \mu)^\mathsf{T} \Sigma^{-1}(z - \mu)],$$

and the joint density of X is

$$f_X(x) \propto [\det(\Sigma)]^{-1/2} \exp\{-\tfrac{1}{2}[F_\lambda^{-1}(x) - \mu]^\mathsf{T} \Sigma^{-1}[F_\lambda^{-1}(x) - \mu]\} \left| \det\left(\frac{\partial F_\lambda^{-1}(x)}{\partial x^\mathsf{T}} \right) \right|.$$

The matrix $\partial F_\lambda^{-1}(x)/\partial x^\mathsf{T}$ is a diagonal matrix in $\mathbb{R}^{p \times p}$ with its ith diagonal entry

$$\frac{\partial f_{\lambda_i}^{-1}(x)}{\partial x^\mathsf{T}} = \frac{\partial}{\partial x^i}\left[\frac{(x^i)^{\lambda_i} - 1}{\lambda_i} \right] = \frac{1}{\lambda_i} \lambda_i (x^i)^{\lambda_i - 1} = (x^i)^{\lambda_i - 1}.$$

So the joint density of X becomes

$$f_X(x) \propto [\det(\Sigma)]^{-1/2} \exp\{-\tfrac{1}{2}[F_\lambda^{-1}(x) - \mu]^\mathsf{T} \Sigma^{-1}[F_\lambda^{-1}(x) - \mu]\} \prod_{i=1}^p (x^i)^{\lambda_i - 1}.$$

If $X_1, \ldots, X_n$ are an i.i.d. sample of X, then the joint distribution of $(X_1, \ldots, X_n)$ is proportional to

$$\prod_{a=1}^n \{[\det(\Sigma)]^{-1/2} \exp\left[-\tfrac{1}{2}(F_\lambda^{-1}(x_a) - \mu)^\mathsf{T} \Sigma^{-1}(F_\lambda^{-1}(x_a) - \mu) \right] \prod_{i=1}^p (x_a^i)^{\lambda_i}\},$$

where

$$F_{\lambda}^{-1}(x_a) = \begin{pmatrix} \frac{(x_a^1)^{\lambda_1}-1}{\lambda_1} \\ \vdots \\ \frac{(x_a^p)^{\lambda_p}-1}{\lambda_p} \end{pmatrix}, \quad a = 1,\ldots,n.$$

The log likelihood of $X_1,\ldots,X_n$, ignoring the constant term $-\frac{pn}{2}\log(2\pi)$, is

$$\ell(\mu,\Sigma,\lambda)$$
$$= -\frac{n}{2}\log[\det(\Sigma)] - \frac{1}{2}\sum_{a=1}^n [F_{\lambda}^{-1}(X_a)-\mu]^{\mathsf{T}}\Sigma^{-1}[F_{\lambda}^{-1}(X_a)-\mu] + \sum_{a=1}^n\sum_{i=1}^a \log(X_a^i).$$

To maximize the log likelihood $\ell(\mu,\Sigma,\lambda)$ over (μ,Σ,λ), we first fix λ and maximize over μ and Σ. This is simply the maximum likelihood of joint Gaussian likelihood pretending $F_{\lambda}^{-1}(X_1),\ldots,F_{\lambda}^{-1}(X_n)$ to be the original observations. The maximizer is just the sample mean and sample variance:

$$\hat{\mu}_{\lambda} = E_n[F_{\lambda}^{-1}(X)], \quad \hat{\Sigma}_{\lambda} = \mathrm{var}_n[F_{\lambda}^{-1}(X)].$$

We then maximize the profile likelihood $\tilde{\ell}(\lambda) = \ell(\hat{\mu}_{\lambda},\hat{\Sigma}_{\lambda},\lambda)$.

The profile likelihood $\tilde{\ell}(\lambda)$ does not have an explicit maximizer. We use the coordinate descent (or Gauss-Seidel) algorithm (Bertsekas (1999)) to maximize this function, which proceeds as follows.

1. Choose an interval $[a,b]$ and a grid in $[a,b]$, say $\mathsf{G} = \{a,a_1,\ldots,a_m,b\}$;

2. (outer loop) Start with an initial value of λ, say $(\lambda_1^{(0)},\ldots,\lambda_p^{(0)})$;

3. (inner loop) Maximize $\tilde{\ell}(\lambda_1,\lambda_2^{(0)},\ldots,\lambda_p^{(0)})$ over $\lambda_1 \in \mathsf{G}$, denote the maximizer as $\lambda_1^{(1)}$.

4. At the kth step, maximize $\tilde{\ell}(\lambda_1^{(1)},\ldots,\lambda_{k-1}^{(1)},\lambda_k,\lambda_{k+1}^{(0)},\ldots,\lambda_p^{(0)})$. Repeat this process until we reach $k = p$. (end of inner loop)

5. Reset $(\lambda_1^{(0)},\ldots,\lambda_p^{(0)})$ to $(\lambda_1^{(1)},\ldots,\lambda_p^{(1)})$, and return to step 2. (end of outer loop)

6. Repeat outer loop until $\|(\lambda_1^{(1)},\ldots,\lambda_p^{(1)}) - (\lambda_1^{(0)},\ldots,\lambda_p^{(0)})\|$ is smaller than a preassigned value ε, or until a maximum number of iterations is reached.

Usually, the Box-Cox transformation does not require the grid to be very fine, the shape of distribution will significantly improve (i.e. mimicking a multivariate Gaussian distribution, at least marginally) with only a roughly accurate maximizer of $\tilde{\ell}(\lambda)$.

Before the Box-Cox transformation, we standardize $X_1,\ldots,X_n$ marginally, so that each component has the same scale. That is, we let

$$Z_a^i = [X_a^i - E_n(X^i)]/\sqrt{\mathrm{var}_n(X^i)}, \quad i = 1,\ldots,p, \ a = 1,\ldots,n.$$

We then make them positive by subtracting each sample by its minimum and then adding a small constant, such as $\varepsilon = 0.5$, as follows:

$$U_a^i = Z_a^i - \min\{Z_1^i,\ldots,Z_n^i\} + \varepsilon, \quad i = 1,\ldots,p, \ a = 1,\ldots,n. \tag{7.4}$$

Then, $\{U_1,\ldots,U_n\}$ are treated as the sample of observations $\{X_1,\ldots,X_n\}$ when evaluating the profile likelihood $\tilde{\ell}(\lambda)$.

Below is the R-code to compute the optimal λ using the coordinate descent algorithm, with all the functions embedded in the main code.

R-code for computing optimal λ in the Box-Cox transformation

```
gauss=function(x,lam,eps,mlam){
matpower=function(a,alpha){
a=(a+t(a))/2;tmp=eigen(a)
return(tmp$vectors%*%diag((tmp$values)^alpha)%*%t(tmp$vectors))}
standvec=function(x) return((x - mean(x))/sd(x))
standmar=function(x,eps){
mu=apply(x,2,mean);sig=diag(diag(var(x)));signrt=matpower(sig,-1/2)
x1=t(t(x)-mu)%*%signrt;x2=t(t(x1)-apply(x1,2,min))+eps
return(x2)}
bocotranvec=function(x,lam,eps){
n=length(x);x1=standvec(x);x2=x1-min(x1)+eps
if(abs(lam)<10^(-10)) x3=log(x2)
if(abs(lam)>=10^(-10)) x3=(x2^lam - 1)/lam
return(x3)}
bocotranmat=function(x,lam,eps){
n=dim(x)[1];p=dim(x)[2];xlam=numeric()
for(i in 1:p) xlam=cbind(xlam,bocotranvec(x[,i],lam[i],eps))
return(xlam)}
likelihood=function(x,lam,eps){
xlam=bocotranmat(x,lam,eps);xmar=standmar(x,eps)
mu=apply(xlam,2,mean);sig=var(xlam);n=dim(x)[1]
loglik=0
for(i in 1:n){
loglik = loglik+(-1/2)*(xlam[i,]-mu)%*%solve(sig)%*%(xlam[i,]-mu)+
         (-1/2)*log(det(sig))+sum((lam-1)*log(xmar[i,]))}
return(c(loglik))}
argmax = function(x,y)  return(x[order(y)[length(x)]])
gaussonestep=function(x,lam,eps,ilam,mlam){
onelam=seq(from=-2,to=2,length=mlam)
loglik=numeric()
for(k in 1:mlam){lam[ilam]=onelam[k];loglik=
                   c(loglik,likelihood(x,lam,eps))}
lamopt=argmax(onelam,loglik);lamout=lam;lamout[ilam]=lamopt
return(lamout) }
p = length(lam)
for(igauss in 1:5) for(i in 1:p){
xlam=bocotranmat(x,lam,eps)
pairs(xlam)
lam1=gaussonestep(x,lam,eps,i,mlam)
lam=lam1}
return(lam)}
```

In the above R-code, the input variables are

1. x: $n \times p$ matrix of original predictor;

2. lam: p-dimensional vector containing the initial values of λ, which is taken to be $(1, \ldots, 1)$;

3. eps: a small positive number in (7.4), which we take to be 0.5;

4. mlam: the number of points in the equally spaced grid G, which we take to be 20.

The R-code is written in such a way that one can observe a dynamic scatter plot matrix in which the data clouds become increasingly elliptical as the iteration progresses. For the coordinate descent algorithm we choose the option of maximum number of iterations, and we take it to be 5. The dynamic scatter plot matrix usually looks rather stable after a few iterations, long before the 5th outerloop is finished. The interval $[a, b]$ of the Grid G is chosen to be $[-2, 2]$.

Embedded in the R-code are the following functions.

1. `matpower(a,alpha)`: compute the power (`alpha`) of a matrix a;

2. `standvec(x)`: standardize a vector (`x`) so that it has mean 0 and variance 1;

3. `standmar(x,eps)`: standardize an $n \times p$ matrix (`x`) so that each of its columns has mean 0 and variance 1, and then perform the calculation (7.4) on each column to make each entry of the matrix positive;

4. `bocotranvec(x,lam,eps)`: apply the Box-Cox transformation to a column vector (`x`), for a given number λ (`lam`); the standardization and transformation to positivity (with margin eps) is done within the function;

5. `bocotranmat(x,lam,eps)`: apply the Box-Cox transformation to each column of a matrix (`x`), for a given vector λ (`lam`); the standardization and transformation to positivity (with margin eps) is done within the function;

6. `likelihood(x,lam,eps)`: evaluate the profile likelihood $\tilde{\ell}(\lambda)$ based on $U_1, \ldots, U_n$ in (7.4) for a given λ (`lam`); the transformation to U_a^i is done within the function, with margin of positive eps; x is the original predictor matrix;

7. `argmax(x,y)`: find the maximizer of a response vector y among the independent vector x;

8. `gaussonestep(x,lam,eps,i,mlam)`: perform one iteration of the Gauss-Seidel algorithm (that is; steps 2 and 3 in the Gauss-Seidel algorithm). x is the matrix of original predictor; `lam` is the initial value of the value obtained from the last iteration; i is the index of λ_i for the ith component; `mlam` is as defined before.

7.3   Application to the Big Mac Data

We applied the optimal Box-Cox transformation to the 9-dimensional predictor of the Big Mac data. As Figure 3.3 shows, the joint distribution of X^1 with other predictors are severely non-elliptical. Figure 7.1 shows the scatter plot matrix of the 9 predictors after the optimal Box-Cox transformation. The ellipticity of the pairwise joint distributions (X^i, X^j) are significantly improved, particularly for the pairs (X^1, X^i), $i = 2, \ldots, 9$. The optimal λ for the Box-Cox transformation is computed to be

$$-1.16, \ 0.32, \ 1.37, \ 0.11, \ 0.74, \ -0.11, \ 0.74, \ 0.11, \ 0.53.$$

Here, we should make a cautionary note. As a marginal approach, the Box-Cox transformation does not guarantee the ellipticity of the joint distribution of the random vector $(X^1, \ldots, X^p)$, even though it transforms each component to a Gaussian variable. Furthermore, the scatter plot matrix can only display elliptical shapes in pairwise joint distributions (X^i, X^j), and non-ellipticity may be hidden in the joint distribution of $(X^1, \ldots, X^p)$, even though the pairwise joint distributions seem elliptical.

We then applied SIR to the transformed data using 8 slices of roughly equal sample sizes. Figure 7.2 shows the scatter plot of the response versus first SIR predictor. The Spearman's

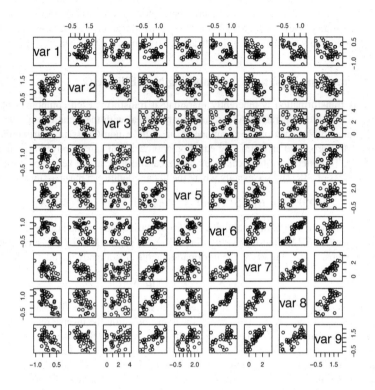

Figure 7.1 *Scatter plot matrix for Box-Cox transformed of the predictor of the Big Mac data.*

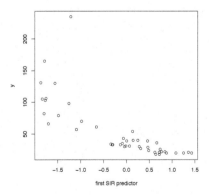

Figure 7.2 *First SIR predictor for the Box-Cox transformed Big Mac data.*

correlation between the response and the first SIR predictor is -0.900, slightly higher than -0.894, the corresponding Spearman's correlation for the untransformed data. Note that the response variable does not seem to have strong dependence on the the most skewed variable X_1, which probably explains the insignificant improvement for this data.

7.4   Estimating Equations for Handling Non-Ellipticity

Box-Cox transformation is inherently marginal: it transforms each component of X to the Gaussian variables separately. As a result it cannot deal with the situations where some hidden linear combinations of the components of X have non-Gaussian distributions. Li and Dong (2009) proposed an estimating equation approach that can adaptively account for non-ellipticity. To understand the idea, let us take a careful look at the key steps in the proof of unbiasedness of SIR, and see exactly where the linear conditional mean assumption was used:

$$E(X|Y) = E[E(X|\beta^\mathsf{T}X, Y)|Y] = E[E(X|\beta^\mathsf{T}X)|Y] = P_\beta^\mathsf{T}(\Sigma)E(X|Y). \tag{7.5}$$

We see that the linear conditional mean condition is used only in the last step, and all it does is to render $E(X|\beta^\mathsf{T}X)$ into the explicit form $P_\beta^\mathsf{T}(\Sigma)X$, so that $\Sigma^{-1}E(X|Y)$ is a member of the central subspace $\mathscr{S}_{Y|X}$. However, the information about β is already included in the equation

$$E(X|Y) = E[E(X|\beta^\mathsf{T}X)|Y], \tag{7.6}$$

and the last step in (7.5) can be dispensed of if we directly solve (the sample version of) this equation. Li and Dong (2009) called the above equation the inverse regression equation, and called the intersection of the collection all subspaces span(β) where β satisfies the above equation the *Central Solution Subspace*, which is denoted by $\mathscr{S}_{\text{SSC}}$. Let $\mathscr{S}_{\text{SIR}}$ be the subspace spanned by the SIR matrix $\Sigma^{-1}\text{cov}[E(X|Y)]\Sigma^{-1}$. The next theorem shows that the central solution subspace is contained in the central subspace without the linear conditional mean assumption, but when the linear conditional mean assumption does hold, it coincides with $\mathscr{S}_{\text{SIR}}$.

Theorem 7.3 *If X and Y are square integrable and $E(X) = 0$, then $\mathscr{S}_{\text{SSC}} \subseteq \mathscr{S}_{Y|X}$. Furthermore, if $\text{LCM}(\beta)$ holds where $\text{span}(\beta) = \mathscr{S}_{Y|X}$, then $\mathscr{S}_{\text{SSC}} = \mathscr{S}_{\text{SIR}}$.*

PROOF. Let β be a $p \times d$ matrix such that span(β) $= \mathscr{S}_{Y|X}$. Then $Y \perp\!\!\!\perp X|\beta^\mathsf{T}X$. Hence β satisfies the equation (7.6), and must contain the smallest subspace $\mathscr{S}_{\text{SSC}}$ that satisfies (7.6).

Next, suppose the linear conditional mean assumption holds and let η be a matrix such that span(η) $= \mathscr{S}_{\text{SSC}}$. Then

$$E(X|Y) = E[E(X|\eta^\mathsf{T}X)|Y] = P_\eta^\mathsf{T}(\Sigma)E(X|Y) = \Sigma P_\eta(\Sigma)\Sigma^{-1}E(X|Y).$$

Multiply both sides of the equations by Σ^{-1} and then take variance, to obtain

$$\Sigma^{-1}\text{var}[E(X|Y)]\Sigma^{-1} = P_\eta(\Sigma)\Sigma^{-1}\text{var}[E(X|Y)]\Sigma^{-1}P_\eta^\mathsf{T}(\Sigma),$$

which implies $\mathscr{S}_{\text{SIR}} \subseteq \mathscr{S}_{\text{SSC}}$. Conversely, if ξ is a matrix such that span(ξ) $= \mathscr{S}_{\text{SIR}}$, then $\Sigma^{-1}E(X|Y) = P_\xi(\Sigma)\Sigma^{-1}E(X|Y)$. By the linear conditional mean assumption we have $P_\xi^\mathsf{T}(\Sigma)E(X|Y) = E(X|\xi^\mathsf{T}X)$. Hence

$$\Sigma^{-1}E(X|Y) = \Sigma^{-1}E[E(X|\xi^\mathsf{T}X)|Y].$$

Canceling out the Σ^{-1} on both sides of the equation, we see that ξ satisfies the equation (7.6). Hence $\mathscr{S}_{\text{SIR}} \subseteq \mathscr{S}_{\text{SSC}}$. □

Li and Dong (2009) assumed that $E(X|\beta^T X)$ was a member of the functional space spanned by a set of known functions $f_1(\beta^T X), \ldots, f_k(\beta^T)$, such as a set of polynomials, so that $E(X|\beta^T X)$ can be written as the form

$$E[XG^T(\beta^T X)]\{E[G(\beta^T X)G^T(\beta^T X)]\}^{-1}G(\beta^T X),$$

where $G(u) = (f_1(u), \ldots, f_k(u))^T$. We can then use the slice means to approximate the equation (7.6) and then combine the equations obtained from different slices to form an objective function, whose minimizer is used to estimate the central subspace. Li and Dong (2009) also developed the similar estimating equations for Ordinary Least Squares, Kernel Inverse Regression, and Parametric Inverse Regression, as discussed in the previous chapters.

Dong and Li (2010) further extended the central solution subspace to second-order methods, such as SAVE and DR. For example, the estimating equation for SAVE can be developed along the following lines:

$$\begin{aligned}
\operatorname{var}(X|Y) &= E[\operatorname{var}(X|Y, \beta^T X)|Y] + \operatorname{var}[E(X|Y, \beta^T X)|Y] \\
&= E[\operatorname{var}(X|\beta^T X)|Y] + \operatorname{var}[E(X|\beta^T X)|Y].
\end{aligned}$$

In principle, we can model both $\operatorname{var}(X|\beta^T X)$ and $E(X|\beta^T X)$ using basis expansions, but for simplicity we assume the constant variance assumption as we did for SAVE and DR, so that the modeling of $\operatorname{var}(X|\beta^T X)$ is avoided. Under the constant variance assumption we have

$$E[\operatorname{var}(X|\beta^T)] = \operatorname{var}(X) - \operatorname{var}[E(X|\beta^T)],$$

which leads to the equation

$$\operatorname{var}(X) - \operatorname{var}(X|Y) = \operatorname{var}[E(X|\beta^T X)] - \operatorname{var}[E(X|\beta^T X)|Y].$$

We then use slice means and slice variances to approximate this equation, and combine the results from different slices to form an objective function as we did in the first-order methods. Dong and Li (2010) also extended Directional Regression in the similar fashion.

Ma and Zhu (2012) and Ma and Zhu (2013a) developed the estimating equation approach from the angle of semiparametric estimation. They derived a class of estimating equations by computing the orthogonal complement of the tangent space of a semeparametric family. We will further discuss their approach in Chapter 15.

Chapter 8

Sufficient Dimension Reduction for Conditional Mean

8.1   Central Mean Subspace   97
8.2   Ordinary Least Squares   100
8.3   Principal Hessian Direction   101
8.4   Iterative Hessian Transformation   104

In the development of Chapters 2 through 6, the Sufficient Dimension Reduction problem is posed as reducing the dimension of x in the conditional distribution of $Y|X$. However, in many applications, such as nonparametric regression and single index models (Ichimura (1993)), our primary interest of estimation is the conditional mean $E(Y|X)$ rather than entire conditional distribution of $F_{Y|X}$. In this case, it is beneficial to focus on the conditional mean when conducting dimension reduction, which can result in further reduction of dimension. In fact, some earlier Sufficient Dimension Reduction methods, such as Ordinary Least Squares and Principal Hessian Directions (Li and Duan (1989), Li (1992)) targeted the conditional mean, even though this was not realized or formulated explicitly. In this chapter we lay out the theoretical framework and some techniques for estimating Sufficient Dimension Reduction for conditional mean.

8.1   Central Mean Subspace

As proposed by Cook and Li (2002), instead of posing Sufficient Dimension Reduction through the conditional independence, $Y \perp\!\!\!\perp X | \beta^\mathsf{T} X$, in the applications where conditional mean is of main interest, we can formulate SDR through conditional mean, as follows

$$E(Y|X) = E(Y|\beta^\mathsf{T}X). \tag{8.1}$$

Like the original SDR problem, the β in the above relation is not uniquely defined, because $E(Y|\beta^\mathsf{T}X) = E(Y|A\beta^\mathsf{T}X)$ for any nonsingular matrix A. So the identifiable parameter here is the subspace span(β) rather than β itself. This leads to the following definition.

Definition 8.1 *If there is a $p \times r$ matrix β that satisfies (8.1), then* span(β) *is called a Sufficient Dimension Reduction subspace for conditional mean. The intersection of all such subspaces, if it is still sufficient in the sense of (8.1), is called the central mean subspace, written as $\mathscr{S}_{E(Y|X)}$.*

Because there is a one-to-one correspondence between $\beta^{\mathsf{T}}X$ and $P_{\text{span}(\beta)}X = \beta(\beta^{\mathsf{T}}\beta)^{-1}\beta^{\mathsf{T}}X$, we can write (8.1) equivalently as

$$E(Y|X) = E[Y|P_{\text{span}(\beta)}(X)].$$

Thus, $\mathscr{S}$ is a Sufficient Dimension Reduction subspace for conditional mean if and only if

$$E(Y|X) = E[Y|P_{\mathscr{S}}(X)]. \tag{8.2}$$

This representation is coordinate free; no coordinate matrix (which is not unique) has to be explicitly mentioned in this alternative definition.

It is possible that even when $\mathscr{S}_1$ and $\mathscr{S}_2$ both satisfy (8.2), $\mathscr{S}_1 \cap \mathscr{S}_2$ does not satisfy (8.2). However, such cases are quite rare and pathological. Similar to Assumption 2.1 we make the following assumption to rule out such cases.

Assumption 8.1 *If $\mathfrak{A}$ be the collection of all subspaces that satisfies (8.2), then*

$$\cap\{\mathscr{S} : \mathscr{S} \in \mathfrak{A}\} \in \mathfrak{A}.$$

There are two other equivalent conditions to (8.1), each of which can be used to define the central mean subspace.

Proposition 8.1 *Assuming all moments involved are finite, then the following statements are equivalent.*

1. *$E(Y|X) = E(Y|\beta^{\mathsf{T}}X)$;*
2. *$Y \perp\!\!\!\perp E(Y|X)|\beta^{\mathsf{T}}X$;*
3. *$\text{cov}[Y, E(Y|X)|\beta^{\mathsf{T}}X] = 0$.*

PROOF. Since the implications *1* ⇒*3*, *2* ⇒*3*, *2* ⇒*1*, *1* ⇒*2* are obvious, we only prove *3* ⇒*1*. By statement *3*,

$$E(YE(Y|X)|\beta^{\mathsf{T}}X) = E(Y|\beta^{\mathsf{T}}X)E(E(Y|X)|\beta^{\mathsf{T}}X) = [E(E(Y|X)|\beta^{\mathsf{T}}X)]^2.$$

On the other hand,

$$E[YE(Y|X)|\beta^{\mathsf{T}}X] = E[E(Y|X)E(Y|X)|\beta^{\mathsf{T}}X] = E\{[E(Y|X)]^2|\beta^{\mathsf{T}}X\}.$$

So, if we let $U = E(Y|X)$, then

$$(E(U|\beta^{\mathsf{T}}X))^2 = E(U^2|\beta^{\mathsf{T}}X),$$

which implies $\text{var}(U|\beta^{\mathsf{T}}X) = 0$. This means U is measurable with respect to $\sigma(\beta^{\mathsf{T}}X)$, and consequently,

$$E(U|\beta^{\mathsf{T}}X) = U \Rightarrow E(Y|X) = E(Y|\beta^{\mathsf{T}}X),$$

which is statement *1*.   □

Intuitively, the central mean subspace should be smaller than the central subspace, because there may be extra predictors beyond conditional mean that characterizes a conditional distribution. This is indeed the case, as the next theorem shows. In the following, if a set $\mathscr{S}$ is a linear subspace of a linear space $\mathscr{T}$, then we write $\mathscr{S} \leq \mathscr{T}$.

Theorem 8.1 *If Assumptions 2.1 and 8.1 hold, then $\mathscr{S}_{E(Y|X)} \subseteq \mathscr{S}_{Y|X}$.*

PROOF. Because $Y \perp\!\!\!\perp X | P_{\mathscr{S}} X \Rightarrow E(Y|X) = E(Y|P_{\mathscr{S}} X)$, we have

$$\{\mathscr{S} \leq \mathbb{R}^p : E(Y|X) = E(Y|P_{\mathscr{S}} X)\} \supseteq \{\mathscr{S} \leq \mathbb{R}^p : Y \perp\!\!\!\perp X | P_{\mathscr{S}} X\}.$$

So

$$\cap \{\mathscr{S} \leq \mathbb{R}^p : E(Y|X) = E(Y|P_{\mathscr{S}} X)\} \subseteq \cap \{\mathscr{S} \leq \mathbb{R}^p : Y \perp\!\!\!\perp X | P_{\mathscr{S}} X\},$$

which means $\mathscr{S}_{E(Y|X)} \subseteq \mathscr{S}_{Y|X}$.   □

The situations where $\mathscr{S}_{E(Y|X)}$ is a proper subspace of $\mathscr{S}_{Y|X}$ is illustrated by the following model:

$$Y = f(\beta_1^{\mathsf{T}} X) + g(\beta_2^{\mathsf{T}} X)\varepsilon,$$

where $X \perp\!\!\!\perp \varepsilon$, and β_1 and β_2 are vectors in $\mathbb{R}^p$ and $f : \mathbb{R} \to \mathbb{R}$, $g : \mathbb{R} \to \mathbb{R}$ are functions. In this case $\mathscr{S}_{Y|X} = \mathrm{span}(\beta_1, \beta_2)$, but $\mathscr{S}_{E(Y|X)} = \mathrm{span}(\beta_1)$.

The central mean subspace enjoys the similar affine equivariance property as the central subspace.

Theorem 8.2 *Suppose Assumption 8.1 is satisfied, $A \in \mathbb{R}^{p \times p}$ is a nonsingular matrix, and $b \in \mathbb{R}^p$. Then*

$$\mathscr{S}_{E(Y|X)} = A^{\mathsf{T}} \mathscr{S}_{E(Y|AX+b)}.$$

PROOF. Similar to the proof of Theorem 2.2, let

$$\begin{aligned}
\mathfrak{A}_{Y|X} &= \{\mathscr{S} \leq \mathbb{R}^p : E(Y|X) = E(Y|P_{\mathscr{S}} X)\}, \\
A^{-\mathsf{T}} \mathfrak{A}_{Y|X} &= \{A^{-\mathsf{T}} \mathscr{S} : \mathscr{S} \in \mathfrak{A}_{Y|X}\}, \\
\mathfrak{A}_{Y|AX+b} &= \{\mathscr{S} : E(Y|AX+b) = E(Y|P_{\mathscr{S}}(AX+b))\}.
\end{aligned}$$

We first show that

$$A^{-\mathsf{T}} \mathfrak{A}_{Y|X} = \mathfrak{A}_{Y|AX+b}. \tag{8.3}$$

Note that

$$\begin{aligned}
\mathscr{S} \in A^{-\mathsf{T}} \mathfrak{A}_{Y|X} &\Leftrightarrow A^{\mathsf{T}} \mathscr{S} \in \mathfrak{A}_{Y|X} \\
&\Leftrightarrow E(Y|X) = E(Y|P_{A^{\mathsf{T}} \mathscr{S}} X) \\
&\Leftrightarrow E(Y|X) = E[Y|P_{A^{\mathsf{T}} \mathscr{S}} A^{-1}(AX+b)].
\end{aligned}$$

Let $U = AX + b$, and let β be a basis matrix of $\mathscr{S}$. Then

$$\begin{aligned}
P_{A^{\mathsf{T}} \mathscr{S}} A^{-1} U &= A^{\mathsf{T}} \beta (\beta^{\mathsf{T}} AA^{\mathsf{T}} \beta)^{-1} \beta^{\mathsf{T}} AA^{-1} U \\
&= A^{\mathsf{T}} \beta (\beta^{\mathsf{T}} AA^{\mathsf{T}} \beta)^{-1} \beta^{\mathsf{T}} U.
\end{aligned}$$

This function of U has one-to-one correspondence with the following function of U:

$$P_{\mathscr{S}} U = \beta (\beta^{\mathsf{T}} \beta)^{-1} \beta^{\mathsf{T}} U.$$

Hence

$$\begin{aligned}
E(Y|X) = E[Y|P_{A^{\mathsf{T}} \mathscr{S}} A^{-1}(AX+b)] &\Leftrightarrow E(Y|X) = E[Y|P_{\mathscr{S}}(AX+b)] \\
&\Leftrightarrow \mathscr{S} \in \mathfrak{A}_{Y|AX+b}.
\end{aligned}$$

Thus we have proved (8.3).

The relation (8.3) can be equivalently written as

$$\{\mathscr{S} : \mathscr{S} \in \mathfrak{A}_{Y|X}\} = A^{\mathsf{T}}\{\mathscr{S}' : \mathscr{S}' \in \mathfrak{A}_{Y|AX+b}\} = \{A^{\mathsf{T}}\mathscr{S}' : \mathscr{S}' \in \mathfrak{A}_{Y|AX+b}\}.$$

Take intersection of the classes of sets on both sides to obtain

$$\begin{aligned}
\mathscr{S}_{Y|X} &= \cap\{\mathscr{S} : \mathscr{S} \in \mathfrak{A}_{Y|X}\} \\
&= \cap\{A^{\mathsf{T}}\mathscr{S}' : \mathscr{S}' \in \mathfrak{A}_{Y|AX+b}\} \\
&= A^{\mathsf{T}} \cap \{\mathscr{S}' : \mathscr{S}' \in \mathfrak{A}_{Y|AX+b}\} = A^{\mathsf{T}}\mathscr{S}_{Y|AX+b},
\end{aligned}$$

which is the desired equality.   □

Next, we introduce several estimators of the central mean subspace: the Ordinary Least Squares, the Principal Hessian Directions, the Iterative Hessian Transformations.

8.2   Ordinary Least Squares

This is the simplest dimension reduction method, which we have already discussed in Section 1.10 as a motivating example for Sufficient Dimension Reduction. It was first introduced by Li and Duan (1989) as a dimension reduction method. Let Σ_{XX} denote the variance matrix $\mathrm{var}(X)$, and let Σ_{XY} denote the covariance $\mathrm{cov}(X, Y)$, which is a p-dimensional vector. Note that $\mathrm{var}(X)$ was denoted simply by Σ in the previous chapters; we add subscript to distinguish it with Σ_{XY}.

Theorem 8.3 *Suppose X satisfies LCM(β) where β is a basis of $\mathscr{S}_{E(Y|X)}$. Assume all the moments involved are finite. Then*

$$\Sigma_{XX}^{-1}\Sigma_{XY} \in \mathscr{S}_{E(Y|X)}.$$

PROOF. Without loss of generality, assume $E(Y) = 0$ and $E(X) = 0$ (otherwise we can reset $X - E(X)$ as X and $Y - E(Y)$ as Y). Let $\beta \in \mathbb{R}^{p \times d}$ be a basis matrix of $\mathscr{S}_{E(Y|X)}$. Then

$$E(XY) = E[XE(Y|X)] = E[XE(Y|\beta^{\mathsf{T}}X)] = E[E(X|\beta^{\mathsf{T}}X)Y] = P_{\beta}^{\mathsf{T}}(\Sigma_{XX})E(XY),$$

where the second equality follows from the definition of $\mathscr{S}_{E(Y|X)}$, the third equality follows from Proposition 2.2, and the fourth equality follows from the LCM(β) assumption and Lemma 3.1. By the definition $P_{\beta}(\Sigma_{XX}) = \beta(\beta^{\mathsf{T}}\Sigma_{XX}\beta)^{-1}\beta^{\mathsf{T}}\Sigma_{XX}$, we have

$$\Sigma_{XX}^{-1}E(XY) = \Sigma_{XX}^{-1}\Sigma_{XY} = \beta(\beta^{\mathsf{T}}\Sigma_{XX}\beta)^{-1}\beta^{\mathsf{T}}E(XY) \equiv \beta u,$$

where $u \in \mathbb{R}^{d}$. The theorem follows because βu is a member of $\mathscr{S}_{E(Y|X)}$.   □

The sample estimate of $\Sigma_{XX}^{-1}\Sigma_{XY}$ is simply the usual Ordinary Least Squares estimate, $\hat{\Sigma}_{XX}^{-1}\hat{\Sigma}_{XY}$, where

$$\hat{\Sigma}_{XX} = E_n\{[X - E_n(X)][X - E_n(X)]^{\mathsf{T}}\}, \quad \hat{\Sigma}_{XY} = E_n\{[X - E_n(X)][Y - E_n(Y)]^{\mathsf{T}}\}.$$

The OLS predictor is, then,

$$\hat{\Sigma}_{YX}\hat{\Sigma}_{XX}^{-1}[X_a - E_n(X)], \quad a = 1, \dots, n,$$

where $\hat{\Sigma}_{YX} = \hat{\Sigma}_{XY}$. The R-code for OLS is only one line:

```
ols = function(x,y) return(solve(var(x))%*%cov(x,y))
```

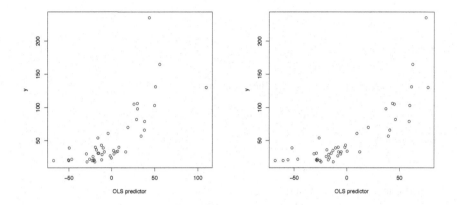

Figure 8.1 *OLS predictor for Big Mac data.*

The advantage of OLS is that it is very easy to use; the disadvantage is that it can only recover one direction in the central subspace, and may not be very accurate when the regression surface is complex.

We applied OLS to the Big Mac data, once with original predictor (left panel) and second time with the transformed predictor using the optimal Box-Cox transformation (right panel). The scatter plots are shown in Figure 8.1. The Spearman's correlation for the untransformed case is 0.817; for the transformed case it is 0.832. From the two plots we can see that the Box-Cox transformation result in a better predictor: the cloud is tighter around a curve for the transformed predictor.

8.3 Principal Hessian Direction

Principal Hessian Directions, or PHD, was proposed by Li (1992). Assuming, without loss of generality, that $E(Y) = 0$ and $E(X) = 0$, PHD is defined through the following generalized eigenvalue problem at the population level:

$$\text{GEV}(\Sigma_{XXY}, \Sigma_{XX})$$

where $\Sigma_{XXY} = E(XX^\mathsf{T}Y)$. The next theorem shows the unbiasedness of PHD.

Theorem 8.4 *Suppose* $\text{LCM}(\beta)$ *and* $\text{CCV}(\beta)$ *are satisfied with* β *being a basis matrix of* $\mathscr{S}_{E(Y|X)}$. *Assume, without loss of generality,* $E(Y) = 0$ *and* $E(X) = 0$. *Assume all the moments involved are finite. Then*

$$\text{span}(\Sigma_{XX}^{-1}\Sigma_{XXY}\Sigma_{XX}^{-1}) \subseteq \mathscr{S}_{E(Y|X)}. \qquad (8.4)$$

PROOF. Note that

$$E(XX^\mathsf{T}Y) = E[XX^\mathsf{T}E(Y|X)] = E[XX^\mathsf{T}E(Y|\beta^\mathsf{T}X)] = E[E(XX^\mathsf{T}|\beta^\mathsf{T}X)Y].$$

The term $E(XX^\mathsf{T}|\beta^\mathsf{T}X)$ on the right-hand side can be decomposed as

$$\text{var}(X|\beta^\mathsf{T}X) + E(X|\beta^\mathsf{T}X)E(X^\mathsf{T}|\beta^\mathsf{T}X) = \Sigma_{XX}Q\beta(\Sigma_{XX}) + P_\beta^\mathsf{T}(\Sigma_{XX})XX^\mathsf{T}P_\beta(\Sigma_{XX}),$$

where the second equality follows from Lemma 3.1 and Corollary 5.1. Hence

$$\begin{aligned}
\Sigma_{XXY} &= E\{[\Sigma_{XX}Q_\beta(\Sigma_{XX}) + P_\beta^\top(\Sigma_{XX})XX^\top P_\beta(\Sigma_{XX})]Y\} \\
&= \Sigma_{XX}Q_\beta(\Sigma_{XX})E(Y) + P_\beta^\top(\Sigma_{XX})E(XX^\top Y)P_\beta(\Sigma_{XX}) \\
&= P_\beta^\top(\Sigma_{XX})\Sigma_{XXY}P_\beta(\Sigma_{XX}).
\end{aligned}$$

Because $P_\beta^\top(\Sigma_{XX}) = \Sigma_{XX}P_\beta(\Sigma_{XX})\Sigma_{XX}^{-1}$, the above equality can be rewritten as

$$\Sigma_{XXY} = \Sigma_{XX}P_\beta(\Sigma_{XX})\Sigma_{XX}^{-1}\Sigma_{XXY}\Sigma_{XX}^{-1}P_\beta^\top(\Sigma_{XX})\Sigma_{XX}.$$

Consequently,

$$\Sigma_{XX}^{-1}\Sigma_{XXY}\Sigma_{XX}^{-1} = P_\beta(\Sigma_{XX})\Sigma_{XX}^{-1}\Sigma_{XXY}\Sigma_{XX}^{-1}P_\beta^\top(\Sigma_{XX}).$$

Hence $\mathrm{span}(\Sigma_{XX}^{-1}\Sigma_{XXY}\Sigma_{XX}^{-1}) \subseteq \mathrm{ran}\,[P_\beta(\Sigma_{XX})] = \mathscr{S}_{E(Y|X)}$.   □

The theorem tells us the eigenvectors of the generalized eigenvalue problem (8.4) corresponding to nonzero eigenvalues span at least a subspace in the central mean subspace. The estimation procedure, as summarized below, mimics this generalized eigenvalue problem at the sample level.

Algorithm 8.1 Principal Hessian Directions

---

1. Standardize $X_1,\ldots,X_n$ to $Z_1,\ldots,Z_n$ as before.
2. Compute

$$\hat{\Sigma}_{ZZY} = E_n(ZZ^\top Y), \quad \hat{\Sigma}_{XX} = \mathrm{var}_n(X).$$

3. Let $u_1,\ldots,u_r$ be the first r eigenvectors of $\hat{\Sigma}_{ZZY}$, the estimated vectors in $\mathscr{S}_{E(Y|X)}$ are $v_\ell = \hat{\Sigma}_{XX}^{-1/2}u_\ell$, $\ell = 1,\ldots,r$; the sufficient predictors are

$$\{v_\ell^\top[X_a - E_n(X)] : a = 1,\ldots,n\}, \quad \ell = 1,\ldots,r.$$

Below is an R-code for implementing PHD.

```
phd = function(x,y){
matpower=function(a,alpha){
a=(a+t(a))/2;tmp=eigen(a)
return(tmp$vectors%*%diag((tmp$values)^alpha)%*%t(tmp$vectors))}
n=length(y);signrt=matpower(var(x),-1/2)
z = center(x)%*%signrt
return(signrt%*%eigen(t(z*(y-mean(y)))%*%z/n)$vectors)}
```

The empirical behavior of PHD is somewhat similar to SAVE; that is, it is more sensitive to scale variation than location variation. We applied PHD to both the Big Mac data and pen digit data. Figure 8.2 shows the scatter plots of Y versus the first two PHD predictors, when the estimator is applied to the Big Mac data. We can see it looks somewhat similar to the scatter plots for SAVE in Figure 5.2. Figure 8.3 shows the 3-d perspective plot for the first three PHD predictors for the pen digit Data, where we used the same plotting symbos as in Figure 5.9.

Once again, it resembles the results by SAVE, with a clear separation in scale between the digit 9 group and the other two groups, but offers hardly any location separation among these three groups.

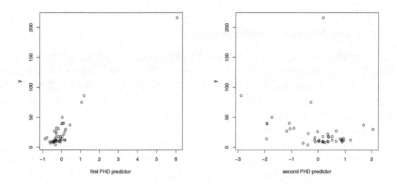

Figure 8.2 *PHD predictor for Big Mac data. Left panel: Y versus first predictor; right panel: Y versus the second predictor.*

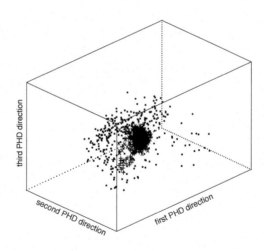

Figure 8.3 *Perspective plot for the first three PHD predictors for the pen digit data.*

Another thing worth noticing from the applications of both the OLS and PHD to the Big Mac data is that, in the SIR and SAVE scatter plots shown in Figure 3.4 (left panel) and

Figure 5.3 (left panel), the patterns are slightly "fan shaped"; that is, there appears to be heteroscedasticity in the response versus the sufficient predictor; whereas the fan-shaped pattern disappeared in the Figure 8.1 and the left panel of 8.2. This is precisely what the central mean subspace estimators tend to do: they focus on the mean pattern $E(Y|X)$ without paying attention to $\mathrm{var}(Y|X)$.

8.4   Iterative Hessian Transformation

This is a method introduced by Cook and Li (2002) and further developed in Cook and Li (2004). The motivations are two fold. First, to remove the reliance on the CCV(β) assumption. Second, to recover more vectors in the central mean subspace — unlike OLS, which can only recover 1, or SIR, which can only recover $k-1$ predictors, k being the number of classes in a categorical response variable.

The theoretical foundation of this method is that the matrix Σ_{XXY}, when considered as a linear transformation on $\mathbb{R}^p$, has the following property: the central mean subspace $\mathscr{S}_{E(Y|X)}$ is invariant subspace of this transformation, and this requires only the LCM(β) assumption. The following theorem, stated in terms of the standard random vector Z, describes this fact.

First, let us recall that if $T : \mathbb{R}^p \to \mathbb{R}^p$ is a linear transformation, and $\mathscr{S} \leq \mathbb{R}^p$ is a subspace of $\mathbb{R}^p$, then $\mathscr{S}$ is invariant subspace T if $T\mathscr{S} \subseteq \mathscr{S}$.

Theorem 8.5 *Suppose $E(Z) = 0$, $\mathrm{var}(Z) = I_p$, and LCM(β) is satisfied for a basis matrix β of $\mathscr{S}_{E(Y|Z)}$. Then $\mathscr{S}_{E(Y|Z)}$ is an invariant subspace of Σ_{ZZY}.*

PROOF. We need to show that if $v \in \mathscr{S}_{E(Y|Z)} \Rightarrow \Sigma_{ZZY}v \in \mathscr{S}_{E(Y|Z)}$. So let $v \in \mathscr{S}_{Y|Z}$. Then,

$$
\begin{aligned}
E(YZZ^\mathsf{T}v) &= E(E(Y|Z)ZZ^\mathsf{T}v) \\
&= E(E(Y|\beta^\mathsf{T}Z)ZZ^\mathsf{T}v) \\
&= E(YE(ZZ^\mathsf{T}v|\beta^\mathsf{T}Z)) \\
&= E(YZ^\mathsf{T}vE(Z|\beta^\mathsf{T}Z)) \\
&= E(YZ^\mathsf{T}vP_\beta Z) \\
&= P_\beta E(YZZ^\mathsf{T})v,
\end{aligned}
$$

where the second equality follows from the definition of $\mathscr{S}_{E(Y|Z)}$; the third from Proposition 2.2; the fourth follows from the fact that $v \in \mathscr{S}_{Y|Z}$, which implies that $v^\mathsf{T}Z$ is a function of $\beta^\mathsf{T}Z$; the fifth follows from Lemma 3.1. Hence $\Sigma_{ZZY}v \in \mathrm{ran}(P_\beta) = \mathscr{S}_{E(Y|Z)}$. □

This theorem implies that if we start with a seed vector in the central subspace $\mathscr{S}_{E(Y|Z)}$, then we can iteratively transform the seed vector to bring out more vectors in $\mathscr{S}_{E(Y|Z)}$. For example, we know that the OLS vector $\Sigma_{ZY} = \mathrm{cov}(Z,Y)$ is a member of $\mathscr{S}_{E(Y|Z)}$ under LCM(β); so Theorem 8.5 implies that

$$
\Sigma_{ZY}, \ \Sigma_{ZZY}\Sigma_{ZY}, \ \Sigma_{ZZY}^2\Sigma_{ZY}, \ldots \tag{8.5}
$$

are all in the central subspace. Since the unbiasedness of OLS only requires LCM(β), and the fact that $\mathscr{S}_{E(Y|Z)}$ is invariant subspace of Σ_{ZZY} also only requires LCM(β), this procedure of recovering vectors in the central mean subspace only requires LCM(β). Interestingly, this iterative transformation scheme has the flavor of partial least squares (Wold et al., 2001), and was one of the starting points in the development of the envelope models (Cook et al., 2007, 2010). The estimation procedure based on (8.5) is called Iterative Hessian Transformations (IHT), and its algorithm is summarized as follows.

Algorithm 8.2 Iterative Hessian Transformation

1. Standardize $X_1, \ldots, X_n$ to $Z_1, \ldots, Z_n$ as before.
2. Compute

$$\hat{\Sigma}_{ZY} = \mathrm{cov}_n(Z, Y), \ \hat{\Sigma}_{ZZY} = E_n(ZZ^\mathsf{T}Y)$$

3. Compute the $p \times p$ matrix

$$\hat{M} = (\hat{\Sigma}_{ZY}, \hat{\Sigma}_{ZZY}\hat{\Sigma}_{ZY}, \ldots, \hat{\Sigma}_{ZZY}^{p-1}\hat{\Sigma}_{ZY})$$

and form the matrix $\hat{\Lambda}_{\mathrm{IHT}} = \hat{M}\hat{M}^\mathsf{T}$. 4. Let $u_1, \ldots, u_r$ be the first r eigenvectors of $\hat{\Lambda}_{\mathrm{ITH}}$, and let $v_\ell = \hat{\Sigma}_{XX}^{-1/2}u_\ell$, $\ell = 1, \ldots, r$.
5. The sufficient predictors are

$$\{v_\ell^\mathsf{T}[X_a - E_n(X)] : a = 1, \ldots, n\}, \quad \ell = 1, \ldots, r.$$

Below is the R-code for implementing IHT.

```
iht=function(x,y,r){
matpower=function(a,alpha){
a=(a+t(a))/2;tmp=eigen(a)
return(tmp$vectors%*%diag((tmp$values)^alpha)%*%t(tmp$vectors))}
standmat=function(x){
mu=apply(x,2,mean);sig=var(x);signrt=matpower(sig,-1/2)
return(t(t(x)-mu)%*%signrt)}
z=standmat(x);szy=cov(z,y);szz=var(z);p=dim(z)[2];imat=szy
for(i in 1:(p-1)) imat=cbind(imat,matpower(szz,i)%*%szy)
return(eigen(imat%*%t(imat))$vectors[,1:r])}
```

We applied IHT to the pen digit data. Figure 8.4 shows the perspective plot for the first three predictors as well as their projections on the 1-2, 1-3, 2-3 predictor hyperplanes. It seems to have overcome the problem with SAVE, SIR-II, and PHD; that is, it achieves both location separation and scale separation. The digit 9 group clearly has larger variation than the other two groups; and the groups are clearly separated in location. One might argue that the performance of IHT is on a par with DR and CR. Note that, unlike DR and CR, IHT does not require the the CCV(β) assumption.

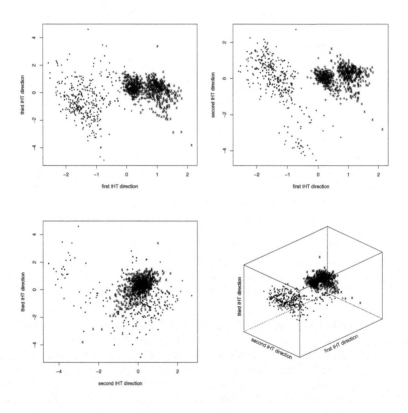

Figure 8.4 *Application of IHT to pen digit data.*

Chapter 9

Asymptotic Sequential Test for Order Determination

9.1   Stochastic Ordering and Von Mises Expansion   107
9.2   Von Mises Expansion and Influence Functions   109
9.3   Influence Functions of Some Statistical Functionals   110
9.4   Random Matrix with Affine Invariant Eigenvalues   112
9.5   Asymptotic Distribution of the Sum of Small Eigenvalues   115
9.6   General Form of the Sequential Tests   117
9.7   Sequential Test for SIR   118
9.8   Sequential Test for PHD   124
9.9   Sequential Test for SAVE   126
9.10 Sequential Test for DR   132
9.11 Applications   139

In the previous chapters we treated the dimension d of the central subspace as known, but in practice this number is unknown and must be estimated. In this chapter we develop the first method for estimating d — the asymptotic sequential test. We first review some asymptotic theories. Here we adopt the von Mises asymptotic expansion for statistical functionals, which covers all the matrix-valued statistics we used so far. The advantage of this method is that it is systematic and general, which allows us to derive the asymptotic distributions for the different Sufficient Dimension Reduction methods introduced in the previous chapters in a more or less unified manner. For more information on this topic, see, for example, Fernholz (1983) and von Mises (1947).

9.1   Stochastic Ordering and Von Mises Expansion

To introduce stochastic ordering we begin with the notion of bounded in probability of a sequence of random elements, say $\{X_n : n = 1, 2, \ldots\}$. We say that X_n is bounded in probability if, for all $\varepsilon > 0$, there exists a $K > 0$ such that

$$\limsup_{n \to \infty} P(\|X_n\| > K) < \varepsilon,$$

where $\| \cdot \|$ is the norm of the space where X_n takes values. If X_n is bounded in probability then we write $X_n = O_P(1)$. This notation is similar to the big O notation in calculus, where we write $x_n = O(1)$ if x_n is a bounded sequence. Let a_n be a sequence of positive numbers, such as $n^{1/2}, n, n^{-1/2}$. We say that X_n is of stochastic order a_n if X_n/a_n is bounded in probability; that is $X_n/a_n = O_P(1)$. In this case we write $X_n = O_P(a_n)$. If X_n converges in probability to 0; that is, if, for any $\varepsilon > 0$,

$$\lim_{n \to \infty} P(\|X_n\| > \varepsilon) = 0,$$

then we write $X_n = o_P(1)$. Again, there is a similar notation in calculus: if $x_n \to 0$, then $x_n = o(1)$. If a_n is a positive sequence and $X_n/a_n = o_P(1)$, then we write $X_n = o_P(a_n)$, and say that X_n is ignorable compared with a_n.

Suppose that $X_1, X_2, \ldots$ are i.i.d. copies of X. For each n, let F_n denote the empirical distribution based on the sample $X_1, \ldots, X_n$. Let F_0 be the true distribution of X. Let $\mathfrak{F}$ be the class of all distributions of X. As discussed in Chapter 2, a statistical functional T is a mapping from $\mathfrak{F}$ to some space $\mathbb{S}$. Usually, $T(F_n)$ corresponds to an estimator, and $T(F_0)$ corresponds to the parameter $T(F_n)$ is intended to estimate. Intuitively, since F_n is close to F_0, we can make an expansion of the statistics $T(F_n)$ around the true distribution F_0 — just like in calculus we can make an expansion of a function $f(x)$ around x_0 if x and x_0 are close and if f is differentiable to a certain order. There is a rigorous theory behind this kind of expansion, but in this book we will use the form of expansion without worrying too much about the regularity conditions that control the remainder of the expansion (which can be found in Fernholz (1983)). All the Sufficient Dimension Reduction estimators in the previous chapters, being essentially nice functions of sample moments, satisfy these regularity conditions.

Consider a line segment in $\mathfrak{F}$ from F_0 to F_n:

$$\{(1 - \varepsilon)F_0 + \varepsilon F_n : \varepsilon \in [0, 1]\}.$$

Let $G_\varepsilon = (1 - \varepsilon)F_0 + \varepsilon F_n$. Then $\varepsilon \mapsto T(G_\varepsilon)$ is a line in $\mathfrak{F}$, with $T(G_1) = T(F_n)$ being the statistic, and $T(G_0) = T(F_0)$ being the parameter we are interested in estimating using $T(F_n)$. Formally, we can take a Taylor expansion of $T(G_1)$ around $T(G_0)$ with ε being the variable of the expansion. Note that even though 1 and 0 are not close to each other, the quantities they represent, F_0 and F_n, are. The kth order Taylor expansion is

$$T(F_n) = T(F_0) + [dT(G_\varepsilon)/d\varepsilon]_{\varepsilon=0}(1 - 0) + \cdots + \frac{1}{k!}[d^k T(G_\varepsilon)/d\varepsilon^k]_{\varepsilon=0}(1 - 0)^k + R$$

$$= T(F_0) + [dT(G_\varepsilon)/d\varepsilon]_{\varepsilon=0} + \cdots + \frac{1}{k!}[d^k T(G_\varepsilon)/d\varepsilon^k]_{\varepsilon=0} + R.$$

The derivative $d^k T(G_\varepsilon)/d\varepsilon^k$ is called the kth Gateaux derivative at F_0 in the direction of $F_n - F_0$. Note that each term $[d^\ell T(G_\varepsilon)/d\varepsilon^\ell]_{\varepsilon=0}$ is a random element depending on n. Under some regularity conditions, such as Frechet differentiability, the remainder is of the order $o_P(n^{-k/2})$. Thus, the zeroth term has order $O_P(1)$, the first term has order $O_P(n^{-1/2})$, $\ldots$, and the kth term has order $O_P(n^{-k/2})$. Hence

$$T(F_n) = T(F_0) + [dT(G_\varepsilon)/d\varepsilon]_{\varepsilon=0} + \cdots + \frac{1}{k!}[d^k T(G_\varepsilon)/d\varepsilon^k]_{\varepsilon=0} + o_P(n^{-k/2}). \qquad (9.1)$$

Different sets of sufficient conditions for this approximation can be found in Boos and Serfling (1990), Reeds (1976), and Filippova (1962). This expansion is called the von Mises expansion, and its validity will be taken as an assumption throughout the rest of this chapter.

9.2 Von Mises Expansion and Influence Functions

When performing the von Mises expansion, it is convenient to use the notation of influence functions, which are defined as the Gateaux derivatives in the direction of $\delta_X - F_0$, where δ_X is the point mass — or Dirac measure — described in Section 1.1. In comparison, the von Mises expansion is based on the Gateaux derivative in the direction of $F_n - F_0$ as in the von Mises expansion. We first define influence functions formally.

Definition 9.1 *Let $T(F)$ be a statistical functional, and let δ_X be the Dirac measure at X. Then the derivative*

$$\partial^k T((1-\varepsilon)F_0 + \varepsilon\delta_X)/\partial\varepsilon^k|_{\varepsilon=0}$$

is called the kth-order influence function.

The first two influence functions will be written as T^* and T^{**}. The von Mises expansion can be readily expressed in terms of the influence functions. We now give the explicit relation between the first two terms in the von Mises expansion and the corresponding influence functions. All the statistical functionals encountered in this book are of the form

$$T(F) = g(T_1(F), \ldots, T_m(F)), \tag{9.2}$$

where g is either defined explicitly or implicitly through an equation, and $T_1(F), \ldots, T_m(F)$ are *linear statistical functionals* of the form $\int \phi_1 dF, \ldots, \int \phi_m dF$ for some functions $\phi_1, \ldots, \phi_m$ or x. Let $T_i(G_\varepsilon) = g_i(\varepsilon)$. Then

$$T(G_\varepsilon) = g(g_1(\varepsilon), \ldots, g_m(\varepsilon)). \tag{9.3}$$

Proposition 9.1 *Suppose $T(F)$ is of the form (9.2), g is differentiable, and the ϕ_i's has finite second moments. Then*

1. *$T^* = \sum_{i=1} b_i T_i^*$, where T_i^* are random variables with means 0 and finite variances, and b_i are the nonrandom constants*

$$\frac{\partial g(g_1(0), \ldots, g_m(0))}{\partial g_i};$$

2. *$\partial T(G_\varepsilon)/\partial\varepsilon|_{\varepsilon=0} = \sum_{i=1}^m E_n(T_i^*)$;*

3. *$T^{**} = \sum_{i=1}^k \sum_{j=1}^k c_{ij} T_i^* T_j^*$ where T_i^* are random variables with means 0 and finite variances, and c_{ij} are the nonrandom constants*

$$c_{ij} = \frac{\partial^2 g(g_1(0), \ldots, g_m(0))}{\partial g_i \partial g_j};$$

4. *$\partial^2 T(G_\varepsilon)/\partial\varepsilon^2|_{\varepsilon=0} = \sum_{i=1}^k \sum_{j=1}^k c_{ij} E_n(T_i^*) E_n(T_j^*)$.*

PROOF. Take the first two derivatives with respect to ε on both sides of (9.3), to obtain

$$\frac{\partial T(G_\varepsilon)}{\partial\varepsilon} = \sum_{i=1}^m \frac{\partial g(g_1(\varepsilon), \ldots, g_m(\varepsilon))}{\partial g_i} \int \phi_i d(F_n - F_0),$$

$$\frac{\partial^2 T(G_\varepsilon)}{\partial\varepsilon^2} = \sum_{i=1}^m \sum_{j=1}^m \frac{\partial^2 g(g_1(\varepsilon), \ldots, g_m(\varepsilon))}{\partial g_i \partial g_j} \int \phi_i d(F_n - F_0) \int \phi_j d(F_n - F_0),$$

where, for the second equality, we have used $\partial^2 \int \phi_i d((1-\varepsilon)F_0 + \varepsilon F_n))/\partial \varepsilon^2 = 0$. Evaluating these derivatives at $\varepsilon = 0$, we have

$$\left.\frac{\partial T(G_\varepsilon)}{\partial \varepsilon}\right|_{\varepsilon=0} = \sum_{i=1}^m b_i \int \phi_i d(F_n - F_0),$$

$$\left.\frac{\partial^2 T(G_\varepsilon)}{\partial \varepsilon^2}\right|_{\varepsilon=0} = \sum_{i=1}^m \sum_{j=1}^m c_{ij} \int \phi_i d(F_n - F_0) \int \phi_j d(F_n - F_0),$$

where b_i and c_{ij} are defined in the Proposition. Since

$$\int \phi_i d(F_n - F_0) = n \sum_{a=1}^n \phi_i d(\delta_{x_a} - F_0) = E_n T_i^*,$$

we have

$$\partial T(G_\varepsilon)/\partial \varepsilon|_{\varepsilon=0} = \sum_{i=1}^m b_i E_n(T_i^*), \quad \partial^2 T(G_\varepsilon)/\partial \varepsilon^2|_{\varepsilon=0} = \sum_{i=1}^m \sum_{j=1}^m c_{ij} E_n(T_i^*) E_n(T_j^*).$$

Meanwhile, by the similar procedures, it is easy to verify that

$$T^* = \sum_{i=1}^m b_i T_i^*, \quad T^{**} = \sum_{i=1}^m \sum_{j=1}^m c_{ij} T_i^* T_j^*.$$

Moreover, we note that $ET_i^* = E(\phi_i - E(\phi_i)) = 0$ and $\text{var}(T_i) = \text{var}[\phi_i(X)]$. Hence the asserted results hold.   □

Although we have proved the above proposition only for real-valued ϕ_i, the result also holds when ϕ_i are vector-valued or matrix-valued, with b_i and c_{ij} changing to corresponding gradient vector or matrix. The way we use this proposition is the following: for example, if we find the first two influence functions of T to be a function of $T_1^*, \ldots, T_m^*$, then the first two terms of the von Mises expansion are obtained by simply replacing T_i^* by $E_n(T_i^*)$. In this way, we can carry out the von Mises expansion entirely by manipulating the influence functions.

In the next proposition we develop the rules of calculating influence functions. Being derivatives, influence functions follow the same rules for differentiation.

Proposition 9.2 *Let $T(F)$ and $U(F)$ be statistical functionals.*

1. *(product rule) $(TU)^* = T^*U + TU^*$;*

2. *(addition rule) $(T+U)^* = T^* + U^*$;*

3. *(composite function) if $f[T(F)]$ is a function of $T(F)$ then $f^*(T) = f'(T)T^*$, where f' represents the usual derivative;*

4. *(constant) if $T(F)$ is a constant, then $T^* = 0$.*

9.3 Influence Functions of Some Statistical Functionals

We now derive the influence functions of some commonly used statistical functionals, such as the mean, the covariance matrix, the inverse of the covariance matrix, and the negative square root of a matrix. Before doing so we first review the notions of vectorization of a matrix and the Kronecker product between matrices, which we rely on heavily in the asymptotic development.

For a matrix A with columns $a_1, \ldots, a_k$, $\text{vec}(A)$ is the vector formed by stacking the columns of A. That is, $\text{vec}(A) = (a_1^\mathsf{T}, \ldots, a_k^\mathsf{T})^\mathsf{T}$. For two matrices $A_1 \in \mathbb{R}^{m_1 \times n_1}$ and $A_2 \in \mathbb{R}^{m_2 \times n_2}$, their Kronecker product is the $m_1 \times n_1$ block matrix, whose (i,j)th block is the matrix $a_{ij}A_2$, where a_{ij} is the (i,j)th entry of A_1. That is,

$$A \otimes B = \begin{pmatrix} a_{11}B & \cdots & a_{1n_1}B \\ \vdots & \ddots & \vdots \\ a_{m_1 1}B & \cdots & a_{m_1 n_1}B \end{pmatrix}.$$

Let $A_3 \in \mathbb{R}^{m_3 \times n_3}$ be a third matrix. If $n_1 = m_2$ and $n_2 = m_3$, then

$$\text{vec}(A_1 A_2 A_3) = (A_3^\mathsf{T} \otimes A_1)\text{vec}(A_2).$$

The most useful feature of this identity is that we can always single out any member of a string of matrices $A_1 \cdots A_r$ and put it on the right as a vector. For example,

$$\begin{aligned} \text{vec}(A_1 A_2 A_3) &= [(A_2 A_3)^\mathsf{T} \otimes I_{m_1}]\text{vec}(A_1) \\ &= (A_3^\mathsf{T} \otimes A_1)\text{vec}(A_2) \\ &= [I_{n_3} \otimes (A_1 A_2)]\text{vec}(A_3). \end{aligned} \tag{9.4}$$

Regarding the Kronecker product, these properties will also be useful:

$$\begin{aligned} (A_1 \otimes A_2)^{-1} &= A_1^{-1} \otimes A_2^{-1}; \\ (A_1 \otimes A_2)^\mathsf{T} &= A_1^\mathsf{T} \otimes A_2^\mathsf{T}; \\ (A_1 \otimes A_2)(A_3 \otimes A_4) &= (A_1 A_3) \otimes (A_2 A_4). \end{aligned} \tag{9.5}$$

We are now ready to derive some important influence functions.

Lemma 9.1 *Suppose* $f(X,Y) \equiv U$, $g(X,Y) = V$ *are functions of* (X,Y).

1. *If* U *is integrable, then* $(EU)^* = U - EU$;

2. *If* U *and* V *are square integrable, then*

$$[\text{cov}(U,V)]^* = (U - EU)(V - EV)^\mathsf{T} - E[(U - EU)(V - EV)^\mathsf{T}].$$

3. *If* $\Sigma = E[(X - EX)(X - EX)^\mathsf{T}]$, *then*

$$(\Sigma^{-1})^* = -\Sigma^{-1}\Sigma^*\Sigma^{-1};$$

4. $\text{vec}[(\Sigma^{-1/2})]^* = -(\Sigma^{1/2} \otimes \Sigma + \Sigma \otimes \Sigma^{1/2})^{-1}[\text{vec}(\Sigma)]^*$.

PROOF. *1.* This is because, by the definition of the influence function,

$$\begin{aligned} (EU)^* &= \int f(X,Y)d\{\partial[(1 - \varepsilon)F_0 + \varepsilon\delta_{XY}]/\partial\varepsilon\} \\ &= \int f(X,Y)d(\delta_{XY} - F_0) = f(X,Y) - Ef(X,Y). \end{aligned}$$

Here, δ_{XY} means the Dirac delta measure that puts the point mass 1 at (X,Y).
2. Rewrite $\text{cov}(U,V)$ as

$$\text{cov}(U,V) = E(UV^\mathsf{T}) - (EU)(EU^\mathsf{T}).$$

By the product rule in Proposition 9.2 and by part 1 of this proposition,

$$
\begin{aligned}
&[\mathrm{cov}(U,V)]^* \\
=& [E(UV^\mathsf{T})]^* - (EU)^*(EV^\mathsf{T}) - EU(EV^\mathsf{T})^* \\
=& UV^\mathsf{T} - E(UV^\mathsf{T}) - (U-EU)EV^\mathsf{T} - EU(V-EV)^\mathsf{T} \\
=& UV^\mathsf{T} - (EU)V^\mathsf{T} - UEV^\mathsf{T} - E[UV^\mathsf{T} - (EU)V^\mathsf{T} - U(EV^\mathsf{T})] \\
=& (U-EU)(V-EV)^\mathsf{T} - E[(U-EU)(V-EV)^\mathsf{T}].
\end{aligned}
$$

3. Since $\Sigma^{-1}\Sigma = I$, by the product rule and constant rule of Proposition 9.2, we have

$$
(\Sigma^{-1})^* \Sigma + \Sigma^{-1}(\Sigma)^* = 0.
$$

Solve this equation for $(\Sigma^{-1})^*$ to prove the desired relation.

4. Applying the product rule in Proposition 9.2 to the equation $\Sigma^{-1/2}\Sigma^{-1/2} = \Sigma^{-1}$, we obtain

$$
\left(\Sigma^{-1/2}\right)^* \Sigma^{-1/2} + \Sigma^{-1/2}\left(\Sigma^{-1/2}\right)^* = \left(\Sigma^{-1}\right)^*.
$$

Use the vec maneuver (9.4) to move $(\Sigma^{-1/2})^*$ to the right of each of the two terms on the left-hand side of this equation:

$$
\left(\Sigma^{-1/2} \otimes I_p\right) \mathrm{vec}\left(\Sigma^{-1/2}\right)^* + \left(I_p \otimes \Sigma^{-1/2}\right) \mathrm{vec}\left(\Sigma^{-1/2}\right)^* = \mathrm{vec}\left(\Sigma^{-1}\right)^*.
$$

Solving this equation for $\mathrm{vec}\left(\Sigma^{-1/2}\right)$, we find

$$
\mathrm{vec}\left(\Sigma^{-1/2}\right)^* = \left(\Sigma^{-1/2} \otimes I_p + I_p \otimes \Sigma^{-1/2}\right)^{-1} \mathrm{vec}\left(\Sigma^{-1}\right)^*.
$$

From part 3 of this proposition we see that $(\Sigma^{-1})^* = -(\Sigma^{-1} \otimes \Sigma^{-1})\mathrm{vec}(\Sigma)$. Substitute this relation into the right-hand side above, to obtain

$$
\mathrm{vec}\left(\Sigma^{-1/2}\right)^* = -\left(\Sigma^{-1/2} \otimes I_p + I_p \otimes \Sigma^{-1/2}\right)^{-1} \left(\Sigma^{-1} \otimes \Sigma^{-1}\right) \mathrm{vec}\left(\Sigma^*\right).
$$

By the first relation in (9.5), the right-hand side is

$$
\begin{aligned}
&- \left(\Sigma^{-1/2} \otimes I_p + I_p \otimes \Sigma^{-1/2}\right)^{-1} (\Sigma \otimes \Sigma)^{-1} \mathrm{vec}\left(\Sigma^*\right) \\
=& - \left(\Sigma^{1/2} \otimes \Sigma + \Sigma \otimes \Sigma^{1/2}\right)^{-1} \mathrm{vec}\left(\Sigma^*\right).
\end{aligned}
$$

This completes the proof. $\square$

9.4   Random Matrix with Affine Invariant Eigenvalues

All the statistics we are interested in in this chapter are matrix-valued statistical functionals $T : \mathfrak{F} \to \mathbb{R}^{p \times p}$. While they can be written as $T(F_n)$, in this section we will write them as $T(X_1, \ldots, X_n; Y_1, \ldots, Y_n)$, which is convenient for our development. We will abbreviate the sample $X_1, \ldots, X_n$ as $\mathbb{X}_n$, and the sample $Y_1, \ldots, Y_n$ as $\mathbb{Y}_n$. So we have several different symbols to describe $X_1, \ldots, X_n$: X is a random vector of which $X_1, \ldots, X_n$ is an i.i.d. sample, $\mathbb{X}_n$ is the sample, and X_a^i is the ith components of X_a. For a matrix $A \in \mathbb{R}^{p \times p}$, and a vector $b \in \mathbb{R}^p$, let

$$
A\mathbb{X}_n + b = (AX_1 + b, \ldots, AX_n + b).
$$

The asymptotic developments in the next few sections will be about the eigenvalues of the random matrix $T(\mathbb{X}_n, \mathbb{Y}_n)$, and will be greatly simplified if the eigenvalues enjoy an *affine invariance property*, as defined below.

Definition 9.2 *Suppose* $T(\mathbb{X}_n, \mathbb{Y}_n)$ *is symmetric, and let* $\lambda[T(\mathbb{X}_n, \mathbb{Y}_n)]$ *be any eigenvalue of* $T(\mathbb{X}_n, \mathbb{Y}_n)$. *Then we say that* $\lambda[T(\mathbb{X}_n, \mathbb{X}_n)]$ *is affine invariant if, for any nonsingular matrix* $A \in \mathbb{R}^{p \times p}$ *and* $b \in \mathbb{R}^p$, *we have*

$$\lambda[T(\mathbb{X}_n, \mathbb{Y}_n)] = \lambda[T(A\mathbb{X}_n + b, \mathbb{Y}_n)].$$

If the property holds for any eigenvalue of $T(\mathbb{X}_n, \mathbb{Y}_n)$, *then we say that* $T(\mathbb{X}_n, \mathbb{Y}_n)$ *is eigenvalue affine invariant, or e-affine invariant.*

This property simplifies the asymptotic development in the following way: if the above property holds, then

$$\lambda[T(\mathbb{X}_n, \mathbb{Y}_n)] \overset{\mathscr{D}}{=} \lambda[T(\mathbb{Z}_n, \mathbb{Y}_n)],$$

where $\mathbb{Z}_n = (Z_1, \ldots, Z_n)$, and $Z_i = \Sigma^{-1/2}[X_i - E(X_i)]$. That is, we can, without loss of generality, assume X_i has mean 0 and covariance matrix I_p. It turns out many of the candidate matrices introduced in the previous chapters, such as Λ_{SIR}, Λ_{SAVE}, and Λ_{DR}, enjoy this property. The next lemma gives a sufficient condition of eigenvalue affine invariance.

Lemma 9.2 *If, for any nonsingular matrix* $A \in \mathbb{R}^{p \times p}$ *and* $b \in \mathbb{R}^p$, $T(A\mathbb{X}_n + b, \mathbb{Y}_n)$ *satisfies*

$$\begin{aligned}
&T(A\mathbb{X}_n + b, \mathbb{Y}_n) \\
&= [A\Sigma(\mathbb{X}_n)A^\top]^{-1/2}A\Sigma^{1/2}(\mathbb{X}_n)T(\mathbb{X}_n, \mathbb{Y}_n)\Sigma^{1/2}(\mathbb{X}_n)A^\top[A\Sigma(\mathbb{X}_n)A^\top]^{-1/2},
\end{aligned} \tag{9.6}$$

where $\Sigma(\mathbb{X}_n)$ *is positive definite matrix depending on* $\mathbb{X}_n$, *then* $T(\mathbb{X}_n, \mathbb{Y}_n)$ *is e-affine invariant.*

PROOF. Suppose λ is an eigenvalue of $T(A\mathbb{X}_n + b, \mathbb{Y}_n)$. Then, and only then, there is a vector $v \in \mathbb{R}^p$ such that

$$[A\Sigma(\mathbb{X}_n)A^\top]^{-1/2}A\Sigma^{1/2}T(\mathbb{X}_n, \mathbb{Y}_n)\Sigma^{1/2}A^\top[A\Sigma(\mathbb{X}_n)A^\top]^{-1/2}v = \lambda v.$$

Multiplying both sides from the left by the matrix $A^\top[A\Sigma(\mathbb{X}_n)A^\top]^{-1/2}$, we see that the above equality is equivalent to

$$\Sigma^{-1}(\mathbb{X}_n)\Sigma^{1/2}(\mathbb{X}_n)T(\mathbb{X}_n, \mathbb{Y}_n)\Sigma^{1/2}(\mathbb{X}_n)A^\top[A\Sigma(\mathbb{X}_n)A^\top]^{-1/2}v = \lambda A^\top[A\Sigma(\mathbb{X}_n)A^\top]^{-1/2}v.$$

This is equivalent to

$$T(\mathbb{X}_n, \mathbb{Y}_n)\Sigma^{1/2}(\mathbb{X}_n)A^\top[A\Sigma(\mathbb{X}_n)A^\top]^{-1/2}v = \lambda\Sigma^{1/2}(\mathbb{X}_n)A^\top[A\Sigma(\mathbb{X}_n)A^\top]^{-1/2}v.$$

Letting $w = \Sigma^{1/2}(\mathbb{X}_n)A^\top[A\Sigma(\mathbb{X}_n)A^\top]^{-1/2}v$, the above is equivalent to

$$T(\mathbb{X}_n, \mathbb{Y}_n)w = \lambda w.$$

That is, λ is an eigenvalue of $T(\mathbb{X}_n, \mathbb{Y}_n)$. □

Consider the class of all statistical functionals satisfying (9.6). Then Lemma 9.2 says that every member of this class is e-affine invariant. The next lemma shows that this class is closed under addition and multiplication.

Lemma 9.3 *If* $T_1(\mathbb{X}_n, \mathbb{Y}_n)$, $T_2(\mathbb{X}_n, \mathbb{Y}_n)$ *satisfies (9.6), then so do*

$$T_1(\mathbb{X}_n, \mathbb{Y}_n) + T_2(\mathbb{X}_n, \mathbb{Y}_n) \quad and \quad T_1(\mathbb{X}_n, \mathbb{Y}_n)T_2(\mathbb{X}_n, \mathbb{Y}_n).$$

PROOF. That $T_1(\mathbb{X}_n, \mathbb{Y}_n) + T_2(\mathbb{X}_n, \mathbb{Y}_n)$ satisfies (9.6) is obvious. Note that

$$T_1(A\mathbb{X}_n + b, \mathbb{Y}_n)T_2(A\mathbb{X}_n + b, \mathbb{Y}_n)$$
$$= [A\Sigma(\mathbb{X}_n)A^\mathsf{T}]^{-1/2}A\Sigma^{1/2}(\mathbb{X}_n)T_1(\mathbb{X}_n, \mathbb{Y}_n)\Sigma^{1/2}(\mathbb{X}_n)A^\mathsf{T}[A\Sigma(\mathbb{X}_n)A^\mathsf{T}]^{-1/2}$$
$$[A\Sigma(\mathbb{X}_n)A^\mathsf{T}]^{-1/2}A\Sigma^{1/2}(\mathbb{X}_n)T_2(\mathbb{X}_n, \mathbb{Y}_n)\Sigma^{1/2}(\mathbb{X}_n)A^\mathsf{T}[A\Sigma(\mathbb{X}_n)A^\mathsf{T}]^{-1/2}.$$

Because

$$\Sigma^{1/2}(\mathbb{X}_n)A^\mathsf{T}[A\Sigma(\mathbb{X}_n)A^\mathsf{T}]^{-1/2}[A\Sigma(\mathbb{X}_n)A^\mathsf{T}]^{-1/2}A\Sigma^{1/2}(\mathbb{X}_n)$$
$$= \Sigma^{1/2}(\mathbb{X}_n)A^\mathsf{T}A^{-\mathsf{T}}\Sigma^{-1}(\mathbb{X}_n)A^{-1}A\Sigma^{1/2}(\mathbb{X}_n) = I_p,$$

we have

$$T_1(A\mathbb{X}_n + b, \mathbb{Y}_n)T_2(A\mathbb{X}_n + b, \mathbb{Y}_n)$$
$$= [A\Sigma(\mathbb{X}_n)A^\mathsf{T}]^{-1/2}A\Sigma^{1/2}(\mathbb{X}_n)T_1(\mathbb{X}_n, \mathbb{Y}_n)T_2(\mathbb{X}_n, \mathbb{Y}_n)\Sigma^{1/2}(\mathbb{X}_n)A^\mathsf{T}[A\Sigma(\mathbb{X}_n)A^\mathsf{T}]^{-1/2},$$

as desired.   □

We now use Lemma 9.2 to show that some of the candidate matrices developed in the previous chapters are, in fact, e-affine invariant. We first show that the candidate matrix for SIR is e-affine invariant.

Theorem 9.1 *Let $\Sigma(\mathbb{X}_n)$ denote the sample covariance matrix of $\mathbb{X}_n$. The candidate matrix for SIR, as defined by*

$$\Lambda_{\text{SIR}}(\mathbb{X}_n, \mathbb{Y}_n) = \Sigma^{-1/2}(\mathbb{X}_n)\left(\sum_{\ell=1}^h (E_n S_\ell)^{-1}E_n[(X - E_n X)S_\ell]E_n[(X - E_n X)^\mathsf{T} S_\ell]\right)\Sigma^{-1/2}(\mathbb{X}_n),$$

is e-affine invariant.

PROOF. By the definition of sample covariance matrix, it is easy to see that

$$\Sigma(A\mathbb{X}_n + b) = A\Sigma(\mathbb{X}_n)A^\mathsf{T}. \tag{9.7}$$

Moreover, note that

$$E_n[(AX + b) - E_n(AX + b)S_\ell] = AE_n(X - E_n X).$$

Hence

$$\Lambda_{\text{SIR}}(A\mathbb{X}_n + b, \mathbb{Y}_n)$$
$$= [A\Sigma(\mathbb{X}_n)A^\mathsf{T}]^{-1/2}A\left(\sum_{\ell=1}^h (E_n S_\ell)^{-1}E_n[(X - E_n X)S_\ell]E_n[(X - E_n X)^\mathsf{T} S_\ell]\right)A^\mathsf{T}[A\Sigma(\mathbb{X}_n)A^\mathsf{T}]^{-1/2}$$
$$= [A\Sigma(\mathbb{X}_n)A^\mathsf{T}]^{-1/2}A\Sigma^{1/2}(\mathbb{X}_n)\Lambda_{\text{SIR}}(A\mathbb{X}_n + b, \mathbb{Y}_n)\Sigma^{1/2}(\mathbb{X}_n)A^\mathsf{T}[A\Sigma(\mathbb{X}_n)A^\mathsf{T}]^{-1/2},$$

as desired.   □

Next, we show that the candidate matrix for SAVE is e-affine invariant.

Theorem 9.2 *The candidate matrix for SAVE,*

$$\Lambda_{SAVE}(\mathbb{X}_n, \mathbb{Y}_n) = \sum_{\ell=1}^{h} E_n(S_\ell)[I_p - \Sigma^{-1/2}(\mathbb{X}_n)\text{var}_n(X|Y \in J_\ell)\Sigma^{-1/2}(\mathbb{X}_n)]^2. \tag{9.8}$$

where the sample covariance matrix is defined as

$$\text{var}_n(X|Y \in J_\ell) = (E_n S_\ell)^{-1} E[(X - E_n X)(X - E_n X)^\mathsf{T} S_\ell],$$

is e-affine invariant.

PROOF. Let

$$M_\ell(\mathbb{X}_n, \mathbb{Y}_n) = I_p - \Sigma^{-1/2}(\mathbb{X}_n)\text{var}_n(X|Y \in J_\ell)\Sigma^{-1/2}(\mathbb{X}_n).$$

By Lemma 9.3, it suffices to show that $M_\ell(\mathbb{X}_n, \mathbb{Y}_n)$ satisfies (9.6). It is easy to check that

$$\text{var}_n(AX + b|Y \in J_\ell) = A\text{var}_n(X|Y \in J_\ell)A^\mathsf{T}.$$

Substituting this and (9.7) into the right-hand side of (9.9), we find

$$M_\ell(A\mathbb{X}_n + b, \mathbb{Y}_n) = I_p - [A\Sigma(\mathbb{X}_n)A^\mathsf{T}]^{-1/2}A\Sigma^{1/2}(\mathbb{X}_n)\Sigma^{-1/2}(\mathbb{X}_n)$$
$$\text{var}_n(X|Y \in J_\ell)\Sigma^{-1/2}(\mathbb{X}_n)\Sigma^{1/2}(\mathbb{X}_n)A^\mathsf{T}[A\Sigma(\mathbb{X}_n)A^\mathsf{T}]^{-1/2}. \tag{9.9}$$

Because

$$[A\Sigma(\mathbb{X}_n)A^\mathsf{T}]^{-1/2}A\Sigma^{1/2}(\mathbb{X}_n)\Sigma^{1/2}(\mathbb{X}_n)A^\mathsf{T}[A\Sigma(\mathbb{X}_n)A^\mathsf{T}]^{-1/2} = I_p,$$

we have

$$M_\ell(A\mathbb{X}_n + b, \mathbb{Y}_n)$$
$$= [A\Sigma(\mathbb{X}_n)A^\mathsf{T}]^{-1/2}A\Sigma^{1/2}(\mathbb{X}_n)[I_p - \Sigma^{-1/2}(\mathbb{X}_n)\text{var}_n(X|Y \in J_\ell)\Sigma^{-1/2}(\mathbb{X}_n)]\Sigma^{1/2}(\mathbb{X}_n)A^\mathsf{T}[A\Sigma(\mathbb{X}_n)A^\mathsf{T}]^{-1/2}$$
$$= [A\Sigma(\mathbb{X}_n)A^\mathsf{T}]^{-1/2}A\Sigma^{1/2}(\mathbb{X}_n)M_\ell(\mathbb{X}_n, \mathbb{Y}_n)\Sigma^{1/2}(\mathbb{X}_n)A^\mathsf{T}[A\Sigma(\mathbb{X}_n)A^\mathsf{T}]^{-1/2},$$

as desired. □

Using the same method, we can prove that the candidate matrices for Contour Regression, Directional Regression, PHD, IHT are all e-affine invariant.

9.5 Asymptotic Distribution of the Sum of Small Eigenvalues

In this section we give the asymptotic distribution of the sum of the small eigenvalues of a random matrix that converges a singular matrix, where the small eigenvalues of the former correspond to the 0 eigenvalues of the latter. The next lemma is derived from perturbation theory (Kato, 1980), which is also used in Li (1991) and Li (1992). All the sequential test statistics used in this book are in the form of this sum. As was done in Section 1.7, for a matrix A, we let $\ker(A)$ denote the kernel of A; that is, the linear subspace $\{x : Ax = 0\}$.

Lemma 9.4 *Suppose that a positive semidefinite random matrix $\Lambda(F_n) \in \mathbb{R}^{p \times p}$ has the following von Mises expansion*

$$\Lambda(F_n) = \Lambda(F_0) + n^{-1/2}\Lambda^{(1)}(F_n) + n^{-1}\Lambda^{(2)}(F_n) + o_P(n^{-1}),$$

where the $\Lambda(F_0)$ is a positive semi-definite matrix of rank $r < p$. Let Q be the projection on to the subspace $\ker(\Lambda(F_0))$. Then

$$\sum_{i=r+1}^{p} \lambda_i[\Lambda(F_n)] = n^{-1/2} L^{(1)}(F_n) + n^{-1} L^{(2)}(F_n) + o_P(n^{-1}),$$

where

$$L^{(1)}(F_n) = \text{tr}[Q\Lambda^{(1)}(F_n)Q]$$
$$L^{(2)}(F_n) = \text{tr}[Q\Lambda^{(2)}(F_n)Q - Q\Lambda^{(1)}(F_n)\Lambda^\dagger(F_0)\Lambda^{(1)}(F_n)Q],$$

with $(\cdot)^\dagger$ representing the Moore-Penrose inverse.

Essentially, this result simply says that the first two Gateaux derivatives of

$$\sum_{i=r+1}^{p} \lambda_i[\Lambda(G_\varepsilon)]$$

are given by $n^{-1/2} L^{(1)}(F_n)$ and $2n^{-1} L^{(2)}(F_n)$, respectively. Correspondingly, we can read off the first- and second-order influence functions of the statistical functional $L(F) = n\sum_{i=r+1}^{p} \lambda_i[\Lambda(F)]$:

$$L^* = \text{tr}(Q\Lambda^* Q), \quad L^{**} = 2\text{tr}[\tfrac{1}{2}Q\Lambda^{**}Q - Q\Lambda^*\Lambda^\dagger(F_0)\Lambda^* Q].$$

As we will see, all the candidate matrices for the SDR methods in the previous sections can be written in the form

$$\Lambda(F_n) = M(F_n)M^\mathsf{T}(F_n) \equiv \hat{M}\hat{M}^\mathsf{T}, \tag{9.10}$$

where $\hat{M}$ is a matrix of dimension $p \times s$, possessing the von Mises expansion

$$M(F_n) = M(F_0) + [\partial M(G_\varepsilon)/\partial\varepsilon]_{\varepsilon=0} + (1/2)[\partial^2 M(G_\varepsilon)/\partial\varepsilon^2]_{\varepsilon=0} + o_P(n^{-1}). \tag{9.11}$$

By Proposition 9.7, the first- and second-order influence functions of $\Lambda(F)$ are

$$\Lambda^* = M^* M^\mathsf{T} + M M^{*\mathsf{T}}, \quad \Lambda^{**} = M^{**}M^\mathsf{T} + 2M^* M^{*\mathsf{T}} + M(M^{**})^\mathsf{T},$$

where M is the population version of $\hat{M}$. Let Q be the projection on to $\text{span}(M)^\perp$. Then $QM = 0$, and hence

$$Q\Lambda^* Q = Q(M^* M^\mathsf{T} + M M^{*\mathsf{T}})Q = 0$$
$$Q\Lambda^* \Lambda^\dagger \Lambda^* Q = QM^* M^\mathsf{T}(MM^\mathsf{T})^\dagger MM^* Q$$
$$Q\Lambda^{**}Q = 2QM^* M^{*\mathsf{T}}Q.$$

Then, by Lemma 9.4,

$$L^* = 0, \quad L^{**} = 2\text{tr}[QM^* M^{*\mathsf{T}}Q - QM^* M^\mathsf{T}(MM^\mathsf{T})^\dagger MM^{*\mathsf{T}}Q] \equiv 2\text{tr}(QM^* Q_{M^\mathsf{T}} M^{*\mathsf{T}}Q),$$

where

$$Q_{M^\mathsf{T}} = I_s - M^\mathsf{T}(MM^\mathsf{T})^\dagger M$$

is the projection on to the orthogonal complement of the row space of M; that is, the subspace $\text{span}(M^\mathsf{T})^\perp$. Thus, we have the asymptotic expansion

$$n \sum_{i=r+1}^{p} \lambda_i[\Lambda(F_n)] = \text{vec}^\mathsf{T}[\sqrt{n}QE_n(M^*)Q_{M^\mathsf{T}}]\text{vec}[\sqrt{n}QE_n(M^*)Q_{M^\mathsf{T}}] + o_P(1).$$

By the central limit theorem

$$\text{vec}[\sqrt{n}QE_n(M^*)Q_{M^\mathsf{T}}] = \text{vec}[\sqrt{n}E_n(QM^*Q_{M^\mathsf{T}})] \xrightarrow{\mathscr{D}} N\left(0, \text{var}[\text{vec}(QM^*Q_{M^\mathsf{T}})]\right).$$

Since the rank of Q_{M^T} and Q are $s-r$ and $p-r$ respectively, the rank of the above matrix is at most $u = \min\{p-r, s-r, \text{rank}(M^*)\}$. Then

$$n \sum_{i=r+1}^{p} \lambda_i[\Lambda(F_n)] \xrightarrow{\mathscr{D}} \sum_{i=1}^{u} \omega_i K_i,$$

where $\omega_1, \ldots, \omega_u$ are the first u eigenvalues of $\text{var}[\text{vec}(QM^*Q_{M^\mathsf{T}})]$. We summarize this result as the following theorem. This theorem is equivalent to Theorem 1 of Bura and Yang (2011).

Theorem 9.3 *If $\Lambda(F_n)$ is of the form (9.10), where $M(F_n)$ has von Mises expansion (9.11), then*

$$L(F_n) \xrightarrow{\mathscr{D}} \sum_{i=1}^{u} \omega_i K_i, \tag{9.12}$$

where $u = \min\{(p-r)(s-r), \text{rank}(M^*)\}$, $K_1, \ldots, K_u$ are i.i.d. χ_1^2 variables, and $\omega_1, \ldots, \omega_u$ are the first u eigenvalues of the matrix $\text{var}[\text{vec}(QM^*Q_{M^\mathsf{T}})]$.

9.6 General Form of the Sequential Tests

All the sequential tests discussed in this book are of the following general form. Suppose a candidate matrix $\Lambda(F_n) \in \mathbb{R}^{p \times p}$ for a Sufficient Dimension Reduction satisfies (9.10) and (9.11), and we would like to determine the rank of $\Lambda(F_0)$, the leading term in the von Mises expansion (9.10). For any number $r = 0, 1, \ldots, p-1$, consider the test statistic

$$L_r(F_n) = n \sum_{i=r}^{p} \lambda_i[\Lambda(F_n)].$$

Intuitively, if the rank of $\Lambda(F_0)$ is r, then $L_r(F_n)$ would be relatively small; that is, it would not be extreme when compared against its asymptotic distribution under the hypothesis that the rank is r. Thus, we perform a sequence of hypothesis tests:

$$H_0^{(r)} : \text{rank}[\Lambda(F_0)] = r, \quad r = 0, \ldots, p-1. \tag{9.13}$$

We estimate the rank d of $\Lambda(F_0)$ is the first r for which this hypothesis is accepted. If this hypothesis is rejected for all $r = 0, \ldots, p-1$, then we estimate the rank d as p. We record this procedure in the following definition.

Definition 9.3 *We call the following estimate*

$$\hat{d} = \min\{r : H_0^{(r)} \text{ is accepted}, r = 0, \ldots, p-1\}$$

the sequential test procedure for order determination. Here, the minimum of an empty set is taken to be p.

To perform the sequential test for order determination, we need to derive the asymptotic null distribution of $L_r(F_n)$ under the hypothesis $H_0^{(r)}$. In the next few sections we will derive such asymptotic distributions for several Sufficient Dimension Reduction methods developed in the previous sections. While these asymptotic results are not exhaustive, the method is general enough to cover all the SDR procedures we have proposed so far. For a nice summary and further developments on this topic, see Bura and Yang (2011).

9.7   Sequential Test for SIR

In Section 9.4, we denoted the candidate matrices of a SDR method as $\Lambda_{\text{METHOD}}(\mathbb{X}_n, \mathbb{Y}_n)$. In this section we denote them by $\Lambda_{\text{METHOD}}(F_n)$ to emphasize that they are matrix-valued statistical functionals. Since $\Lambda_{\text{SIR}}(\mathbb{X}_n, \mathbb{Y}_n)$ is e-affine invariant, we assume $E(X) = 0$ and $\text{var}(X) = I_p$ for the rest of the section. As usual, we use β to denote a basis matrix of $\mathscr{S}_{Y|X}$.

We adopt the following notations:

$$p_\ell = ES_\ell, \quad \hat{p}_\ell = E_n S_\ell, \quad \mu_\ell = E(X|Y \in \ell), \quad \hat{\mu}_\ell = E_n(X|Y \in J_\ell) = \frac{E_n(XS_\ell)}{E_n S_\ell}, \quad \hat{\mu} = E_n(X).$$

Note that we have used $\hat{\mu}$ to denote $\bar{X}$. This is because, in our notational system $\hat{\mu}$ also means $\mu(F_n)$, which is clearer than $\bar{X}$ in the context of von Mises expansions. For the same reason, even though $\text{var}_n(X)$ now converges to I_p instead of a generic Σ, we still denote $\text{var}_n(X)$ by $\hat{\Sigma}$. As mentioned earlier, we denote the influence functions of $\hat{p}_\ell$, $\hat{\mu}_\ell$, $\hat{\Sigma}$ and $\hat{\mu}$ as $p_\ell^\star$, $\mu_\ell^\star$, $\Sigma^\star$, and $\mu^\star$.

The sample-level candidate matrix for SIR is

$$\Lambda(F_n) = \hat{\Sigma}^{-1/2} \left(\sum_{\ell=1}^h \hat{p}_\ell (\hat{\mu}_\ell - \hat{\mu})(\hat{\mu}_\ell - \hat{\mu})^\top \right) \hat{\Sigma}^{-1/2} \equiv \hat{M}\hat{M}^\top,$$

where

$$\hat{M} = \left(\hat{p}_1^{1/2} \hat{\Sigma}^{-1/2}(\hat{\mu}_1 - \hat{\mu}), \ldots, \hat{p}_h^{1/2} \hat{\Sigma}^{-1/2}(\hat{\mu}_h - \hat{\mu}) \right).$$

Correspondingly, the population-level candidate matrix is

$$\Lambda(F_0) = MM^\top, \quad M = (p_1^{1/2}\mu_1, \ldots, p_h^{1/2}\mu_h),$$

where we have used the fact that $E(X) = 0$ and $\Sigma = I_p$.

Recall from Chapter 3 that, under the LCM(β) assumption, $\text{span}(\Lambda_{\text{SIR}}) \subseteq \mathscr{S}_{Y|X}$. If we further assume that SIR is exhaustive, then estimating the dimension of the central subspace amounts to estimating the rank of Λ_{SIR}. The next lemma reveals a special structure of the SIR candidate matrix.

Lemma 9.5 *Let $\hat{\pi}$ denote the vector $(\hat{p}_1^{1/2}, \ldots, \hat{p}_h^{1/2})^\top$, and let $Q_{\hat{\pi}} = I_h - \hat{\pi}\hat{\pi}^\top$ be the projection on to the orthogonal complement of $\text{span}(\hat{\pi})$. Define π and Q_π as the population-level counterparts of $\hat{\pi}$ and $Q_{\hat{\pi}}$. Then*

$$\hat{M} = \hat{M}Q_{\hat{\pi}}, \quad M = MQ_\pi.$$

PROOF. To prove the first equality, it suffices to show that $\text{span}(\hat{M}^\top) \subseteq \text{span}(\hat{\pi})^\perp$, or equivalently, $\hat{M}\hat{\pi} = 0$. This is true because

$$\hat{M}\hat{\pi} = \hat{\Sigma}^{-1/2}(\hat{\pi}_1(\hat{\mu}_1 - \hat{\mu}), \ldots, \hat{\pi}_1(\hat{\mu}_1 - \hat{\mu}))\hat{\pi}$$

$$= \hat{\Sigma}^{-1/2} \sum_{\ell=1}^h \hat{p}_\ell(\hat{\mu}_\ell - \hat{\mu}) = \hat{\Sigma}^{-1/2}(\hat{\mu} - \hat{\mu}) = 0,$$

as desired. The second equality can be proved similarly. □

We now derive the null distribution of the sequential test statistic for SIR.

Theorem 9.4 *Suppose* LCM(β) *and* CCV(β) *hold, and that SIR is exhaustive. Then, for each* $r = 0, \ldots, p-1$, *under the null hypothesis*

$$H_0^{(r)} : \dim(\mathscr{S}_{Y|X}) = r,$$

we have

$$n \sum_{i=r+1}^{p} \lambda_i(\Lambda_{\text{SIR}}(F_n)) \xrightarrow{\mathscr{D}} \chi^2_{(p-r)(h-1-r)}.$$

PROOF. Since, by Lemma 9.2, $\Lambda_{\text{SIR}}(F_n)$ is e-affine invariant, we assume $E(X) = 0$ and $\text{var}(X) = I_p$ without loss of generality. By Lemma 9.5, we have $\hat{M} = \hat{M} Q_{\hat{\pi}}$. Hence

$$\Lambda_{\text{SIR}}(F_n) = \hat{M} Q_{\hat{\pi}} \hat{M}^{\mathsf{T}}.$$

The first two Frechet derivatives of $\Lambda_{\text{SIR}}(F)$ are:

$$\Lambda_{\text{SIR}}^* = M^* Q_\pi M^{\mathsf{T}} + M Q_\pi^* M^{\mathsf{T}} + M Q_\pi M^{*\mathsf{T}}$$
$$\Lambda_{\text{SIR}}^{**} = M^{**} Q_\pi M^{\mathsf{T}} + M^* Q_\pi^* M^{\mathsf{T}} + M^* Q_\pi M^{*\mathsf{T}}$$
$$+ M^* Q_\pi^* M^{\mathsf{T}} + M Q_\pi^{**} M^{\mathsf{T}} + M Q_\pi^* M^{*\mathsf{T}}$$
$$+ M^* Q_\pi M^{*\mathsf{T}} + M Q_\pi^* M^{*\mathsf{T}} + M Q_\pi M^{**\mathsf{T}}.$$

Let $Q = I_p - \beta(\beta^{\mathsf{T}}\beta)^{-1}\beta^{\mathsf{T}}$ be the projection on to $\text{span}(\beta)^\perp$. Because $QM = 0$, we have

$$Q\Lambda_{\text{SIR}}^* Q = 0,$$
$$Q\Lambda_{\text{SIR}}^* \Lambda_{\text{SIR}}^\dagger \Lambda_{\text{SIR}} Q = QM^* Q_\pi M^{\mathsf{T}} (M Q_\pi M^{\mathsf{T}})^\dagger M Q_\pi M^{*\mathsf{T}} Q,$$
$$Q\Lambda_{\text{SIR}}^{**} Q = 2QM^* Q_\pi M^{*\mathsf{T}} Q,$$

where, for the second equality, we used the relation $\Lambda_{\text{SIR}} = MM^{\mathsf{T}} = MQ_\pi M^{\mathsf{T}}$. By Lemma 9.4,

$$L_r^* = 0$$
$$L_r^{**} = \text{tr}\left(QM^* Q_\pi M^{*\mathsf{T}} Q - QM^* Q_\pi M^{\mathsf{T}} (M Q_\pi M^{\mathsf{T}})^\dagger M Q_\pi M^{*\mathsf{T}} Q \right)$$
$$= \text{tr}\left[QM^* \left(Q_\pi - Q_\pi M^{\mathsf{T}} (M Q_\pi M^{\mathsf{T}})^\dagger M Q_\pi \right) M^{*\mathsf{T}} Q \right].$$

Because Q_π is the projection on to $\text{span}(\pi)^\perp$, $Q_\pi M^{\mathsf{T}} (M Q_\pi M^{\mathsf{T}})^\dagger M Q_\pi$ is the projection on to $\text{span}(Q_\pi M^{\mathsf{T}})$, and $\text{span}(Q_\pi M^{\mathsf{T}}) \subseteq \text{span}(Q_\pi) = \text{span}(\pi)^\perp$, by Conway (1990), page 40, Problem 6,

$$Q_\pi - Q_\pi M^{\mathsf{T}} (M Q_\pi M^{\mathsf{T}})^\dagger M Q_\pi$$

is a projection, and we denote it by $P_{Q_\pi \ominus M^{\mathsf{T}}}$. By the same reference, the rank of this matrix is

$$\text{rank}(Q_\pi) - \text{rank}(Q_\pi M^{\mathsf{T}} (M Q_\pi M^{\mathsf{T}})^\dagger M Q_\pi).$$

The rank of the first matrix is $h-1$. Since the second matrix is simply $M^{\mathsf{T}}(MM^{\mathsf{T}})^\dagger M$, its rank is simply r. It follows that

$$\text{rank}(P_{Q_\pi \ominus M^{\mathsf{T}}}) = h - 1 - r.$$

By the relation between von Mises expansion and influence functions, we have

$$L_r(F_n) = n\text{tr}\left(E_n(QM^*P_{Q_\pi \ominus M^\mathsf{T}})E_n(QM^*P_{Q_\pi \ominus M^\mathsf{T}})^\mathsf{T}\right) + o_P(1),$$

where

$$\sqrt{n}E_n(QM^*P_{Q_\pi \ominus M^\mathsf{T}}) \xrightarrow{\mathscr{D}} N(0, \text{var}[\text{vec}(QM^*P_{Q_\pi \ominus M^\mathsf{T}})]).$$

This implies that

$$L_r(F_n) \xrightarrow{\mathscr{D}} \sum_{i=1}^{(p-r)(h-1-r)} \omega_i K_i,$$

where the ω_i's are eigenvalues of $\text{var}[\text{vec}(QM^*P_{Q_\pi \ominus M^\mathsf{T}})]$ and the K_i's are i.i.d. χ_1^2 variables.

It remains to show that $\omega_1 = \cdots = \omega_{(p-r)(h-r-1)} = 1$. To do so, we first derive the quantity

$$QM^*P_{Q_\pi \ominus M^\mathsf{T}}.$$

The influence functions $\hat{p}_\ell^{1/2}\hat{\Sigma}^{-1/2}(\hat{\mu}_\ell - \hat{\mu})$, $\ell = 1, \ldots, h$, are

$$(p_\ell^{1/2})^*\mu_\ell + p_\ell^{1/2}(\Sigma^{-1/2})^*\mu_\ell + p_\ell^{1/2}(\mu_\ell^* - \mu^*).$$

Assembling these influence function together, we obtain the influence function for $\hat{M}$, as follows:

$$\begin{aligned}M^* =& ((p_1^{1/2})^*\mu_1, \ldots, (p_h^{1/2})^*\mu_h) + (\Sigma^{-1/2})^*(p_1^{1/2}\mu_1, \ldots, p_h^{1/2}\mu_h) \\ &+ (p_1^{1/2}\mu_1^*, \ldots, p_h^{1/2}\mu_h^*) - (p_1^{1/2}\mu^*, \ldots, p_h^{1/2}\mu^*) \\ =& ((p_1^{1/2})^*\mu_1, \ldots, (p_h^{1/2})^*\mu_h) + (\Sigma^{-1/2})^*M + (p_1^{1/2}\mu_1^*, \ldots, p_h^{1/2}\mu_h^*) - \mu^*\pi^\mathsf{T}.\end{aligned}$$

Since $\hat{\mu} = \hat{p}_1\hat{\mu}_1 + \cdots + \hat{p}_h\hat{\mu}_h$, we have

$$\begin{aligned}\mu^* =& p_1^*\mu_1 + \cdots + p_h^*\mu_h + p_1\mu_1^* + \cdots + p_h\mu_h^* \\ =& p_1^*\mu_1 + \cdots + p_h^*\mu_h + (p_1^{1/2}\mu_1^*, \cdots, p_h^{1/2}\mu_h^*)\pi.\end{aligned}$$

So

$$\begin{aligned}M^* =& ((p_1^{1/2})^*\mu_1, \ldots, (p_h^{1/2})^*\mu_h) + (\Sigma^{-1/2})^*M + (p_1^{1/2}\mu_1^*, \ldots, p_h^{1/2}\mu_h^*)Q_\pi \\ &- (p_1^*\mu_1 + \cdots + p_h^*\mu_h)\pi^\mathsf{T}.\end{aligned}$$

Because $Q\mu_\ell = 0$ for any $\ell = 1, \ldots, h$, we have

$$QM^*P_{Q_\pi \ominus M^\mathsf{T}} = Q(\Sigma^{-1/2})^*MP_{Q_\pi \ominus M^\mathsf{T}} + Q(p_1^{1/2}\mu_1^*, \ldots, p_h^{1/2}\mu_h^*)Q_\pi P_{Q_\pi \ominus M^\mathsf{T}}.$$

Since $M = MQ_\pi$, we have

$$MP_{Q_\pi \ominus M^\mathsf{T}} = MQ_\pi P_{Q_\pi \ominus M^\mathsf{T}} = (P_{Q_\pi \ominus M^\mathsf{T}}Q_\pi M^\mathsf{T})^\mathsf{T} = 0.$$

Also note that $Q_\pi P_{Q_\pi \ominus M^\mathsf{T}} = P_{Q_\pi \ominus M^\mathsf{T}}$. Hence

$$QM^*P_{Q_\pi \ominus M^\mathsf{T}} = Q(p_1^{1/2}\mu_1^*, \ldots, p_h^{1/2}\mu_h^*)P_{Q_\pi \ominus M^\mathsf{T}}. \tag{9.14}$$

The influence function μ_ℓ^* is

$$\mu_\ell^* = [\hat{p}_\ell^{-1}E_n(XS_\ell)]^* = (\hat{p}_\ell^{-1})^*E(XS_\ell) + p_\ell^{-1}[XS_\ell - E(XS_\ell)] \tag{9.15}$$

Because

$$E(XS_\ell) = E[E(X|Y)S_\ell] = E\{E[E(X|\beta^\mathsf{T}X)|Y]S_\ell\} = PE(XS_\ell),$$

where $P = I_p - Q$, we have $Q\mu_\ell^* = p_\ell^{-1}XS_\ell$. Substitute this into (9.14) to obtain

$$QM^*P_{Q\pi\ominus M^\mathsf{T}} = Q(p_1^{-1/2}XS_1, \ldots, p_h^{-1/2}XS_h)P_{Q\pi\ominus M^\mathsf{T}}.$$

Next, we compute the variance of $\mathrm{vec}(QM^*P_{Q\pi\ominus M^\mathsf{T}})$, which can be rewritten as

$$(P_{Q\pi\ominus M^\mathsf{T}} \otimes Q) \begin{pmatrix} p_1^{-1/2}XS_1 \\ \vdots \\ p_h^{-1/2}XS_h \end{pmatrix}.$$

Let $R = (p_1^{-1/2}S_1, \ldots, p_h^{-1/2}S_h)^\mathsf{T}$. Then the above can be further rewritten as

$$\mathrm{vec}(QM^*P_{Q\pi\ominus M^\mathsf{T}}) = (P_{Q\pi\ominus M^\mathsf{T}} \otimes Q)(R \otimes X).$$

So

$$\begin{aligned}
&\mathrm{var}[\mathrm{vec}(QM^*P_{Q\pi\ominus M^\mathsf{T}})] \\
&= (P_{Q\pi\ominus M^\mathsf{T}} \otimes Q)\mathrm{var}(R \otimes X)(P_{Q\pi\ominus M^\mathsf{T}} \otimes Q) \\
&= (P_{Q\pi\ominus M^\mathsf{T}} \otimes Q)E(RR^\mathsf{T} \otimes XX^\mathsf{T})(P_{Q\pi\ominus M^\mathsf{T}} \otimes Q) \\
&\quad - (P_{Q\pi\ominus M^\mathsf{T}} \otimes Q)E(R \otimes X)E(R \otimes X)^\mathsf{T}(P_{Q\pi\ominus M^\mathsf{T}} \otimes Q).
\end{aligned} \tag{9.16}$$

We now compute the two terms on the right-hand side. Note that

$$E(RR^\mathsf{T} \otimes XX^\mathsf{T}) = E[RR^\mathsf{T} \otimes E(XX^\mathsf{T}|Y)]. \tag{9.17}$$

By $Y \perp\!\!\!\perp X|\beta^\mathsf{T}X$, LCM($\beta$), CCV($\beta$), and expression (6.6), we have

$$E(XX^\mathsf{T}|Y) = Q + PE(XX^\mathsf{T}|Y)P,$$

Substituting this into (9.17), we find

$$E(RR^\mathsf{T} \otimes XX^\mathsf{T}) = E[RR^\mathsf{T} \otimes (Q + PE(XX^\mathsf{T}|Y)P)].$$

Hence the first term on the right-hand side of (9.16) becomes

$$\begin{aligned}
P_{Q\pi\ominus M^\mathsf{T}} \otimes QE(RR^\mathsf{T} \otimes XX^\mathsf{T})P_{Q\pi\ominus M^\mathsf{T}} \otimes Q \\
= (P_{Q\pi\ominus M^\mathsf{T}} \otimes Q)[E(RR^\mathsf{T}) \otimes Q](P_{Q\pi\ominus M^\mathsf{T}} \otimes Q).
\end{aligned}$$

The (k, ℓ)th element of the matrix $E(RR)$ is

$$p_k^{-1/2}p_\ell^{-1/2}E(S_kS_\ell) = p_k^{-1/2}p_\ell^{-1/2}P(Y \in J_k \cap J_\ell).$$

Because $P(Y \in J_k \cap J_\ell)$ is 0 if $k \neq \ell$, and is p_ℓ if $k = \ell$, the right-hand side above is simply the Kronecker delta $\delta_{k\ell}$; that is, $E(RR^{\mathsf{T}}) = I_h$. So the first term on the right-hand side of (9.16) reduces to

$$P_{Q\pi\ominus M^{\mathsf{T}}} \otimes QE(RR^{\mathsf{T}} \otimes XX^{\mathsf{T}})P_{Q\pi\ominus M^{\mathsf{T}}} \otimes Q = P_{Q\pi\ominus M^{\mathsf{T}}} \otimes Q.$$

To compute the second term on the right-hand side of (9.16), we note that

$$E(R \otimes X) = E(R \otimes E(X|Y)) = E[R \otimes PE(X|Y)],$$

where the second equality follows from $\mathrm{LCM}(\beta)$ and Theorem 3.1. Hence

$$(P_{Q\pi\ominus M^{\mathsf{T}}} \otimes Q)E(R \otimes X)E(R \otimes X)^{\mathsf{T}}(P_{Q\pi\ominus M^{\mathsf{T}}} \otimes Q) = 0.$$

In summary, we have

$$\mathrm{var}[\mathrm{vec}(QM^*P_{Q\pi\ominus M^{\mathsf{T}}})] = P_{Q\pi\ominus M^{\mathsf{T}}} \otimes Q.$$

This is a projection matrix of rank $(p-r)(h-1-r)$, whose first $(p-r)(h-1-r)$ are 1, which is what we need to prove.   □

We now summarize the procedure for the sequential test for SIR as the following algorithm. We will be brief on the part that involves computing the SIR candidate matrix, which was already described in Chapter 3.

Algorithm 9.1 Sequential test for SIR

---

1. Divide the range of $X_1, \ldots, X_n$ into h subintervals and compute $\hat{p}_\ell$, $\hat{\mu}_\ell$ for $\ell = 1, \ldots, h$, and the sample covariance matrix $\hat{\Sigma}$. Then compute the candidate matrix $\hat{\Lambda}_{\mathrm{SIR}} = \hat{M}\hat{M}^{\mathsf{T}}$, where

$$\hat{M} = \left(\hat{p}_1^{-1/2}\hat{\Sigma}^{-1/2}(\hat{\mu}_1 - \hat{\mu}), \ldots, \hat{p}_h^{-1/2}\hat{\Sigma}^{-1/2}(\hat{\mu}_h - \hat{\mu}) \right).$$

2. For each $r = 0, \ldots, p-1$, compute the sequential test statistic

$$L_r(F_n) = n \sum_{i=r+1}^{p} \hat{\lambda}_i[\Lambda_{\mathrm{SIR}}(F_n)].$$

3. For a significance level $\alpha > 0$, if $L_r(F_n) \le \chi^2_{(p-r)(h-1-r)}(1-\alpha)$, then set $\hat{d} = r$. Otherwise return to step 2 if $r < p-1$, or set $\hat{d} = p$ if $r = p-1$.

R-code to implement sequential test for SIR

```
seqtestsir=function(x,y,h,r,ytype){
p=ncol(x);n=nrow(x)
signrt=matpower(var(x),-1/2)
xc=t(t(x)-apply(x,2,mean))
xst=xc%*%signrt
```

```
if(ytype=="continuous") ydis=discretize(y,h)
if(ytype=="categorical") ydis=y
yless=ydis;ylabel=numeric()
for(i in 1:n) {if(var(yless)!=0) {ylabel=c(ylabel,yless[1]);
                                  yless=yless[yless!=yless[1]]}}
ylabel=c(ylabel,yless[1])
prob=numeric();exy=numeric()
for(i in 1:h) prob=c(prob,length(ydis[ydis==ylabel[i]])/n)
for(i in 1:h) exy=rbind(exy,apply(xst[ydis==ylabel[i],],2,mean))
sirmat=t(exy)%*%diag(prob)%*%exy
test=n*sum(eigen(sirmat)$values[(r+1):p])
pval=1-pchisq(test,(p-r)*(h-1-r))
return(pval)}
```

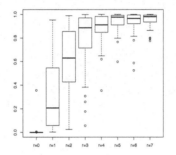

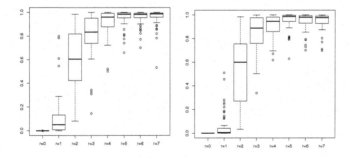

Figure 9.1 *Boxplots of p-values for sequential test based on SIR as applied to Model (9.18).*

Example 9.1 We applied the sequential test for SIR to the model

$$y = 2\sin(1.4X_1) + (X_2 + 1)^2 + .4\varepsilon, \tag{9.18}$$

where $X \sim N(0, I_p)$, $p = 10$, $\varepsilon \sim N(0, 1)$. Here, we have avoided models similar to Example

5.1, under which SIR is not exhaustive. This is to separate the issues of estimation accuracy and the order determination accuracy. We took $n = 100, 200, 300$, and chose the number of slices for SIR to be $h = 10$ for all three sample sizes. We generated 50 samples for each n, and produced the boxplots of the p-values of $n\sum_{i=r+1}^{p} \lambda_i(\Lambda_{\text{SIR}}(F_n))$ against the null asymptotic distribution $\chi^2_{(p-r)(h-1-r)}$ for each step of the sequential test: $r = 1, \ldots, 8$. The upper panel of Figure 9.1 shows that boxplots for $n = 100$; the lower left panel for $n = 200$; the lower right panel for $n = 300$.

In all three cases we can observe a substantial jump in p-values at $r = 3$, and this tendency is strengthened n increases, indicating that the estimation of d tends to be 2.   □

9.8   Sequential Test for PHD

Using the notations of Section 9.5 , the candidate matrix for PHD is

$$\Lambda_{\text{PHD}}(F_n) = \left(\hat{\Sigma}_{XX}^{-1/2} \hat{\Sigma}_{XXY} \hat{\Sigma}_{XX}^{-1/2} \right)^2 = \hat{M}\hat{M},$$

where

$$\hat{\Sigma}_{XX} = E_n[(X - \hat{\mu}_X)(X - \hat{\mu}_X)^{\mathsf{T}}], \quad \hat{\Sigma}_{XXY} = E_n[(Y - \hat{\mu}_Y)(X - \hat{\mu}_X)(X - \hat{\mu}_X)^{\mathsf{T}}],$$

where $\hat{\mu}_X$ and $\hat{\mu}_Y$ are the sample means of X and Y.

Theorem 9.5 *Suppose $X \sim N(0, I_p)$ and PHD is exhaustive. Then, under the null hypothesis*

$$H_0^{(r)} : \dim(\mathscr{S}_{Y|X}) = r, \quad r = 0, \ldots, p-1,$$

we have

$$\frac{n}{2\text{var}_n(Y)} \sum_{i=r+1}^{p} \lambda_i[\Lambda_{\text{PHD}}(F_n)] \xrightarrow{\mathscr{D}} \chi^2_{(p-r)(p-r+1)/2}. \tag{9.19}$$

PROOF. Using Lemma 9.2, it is easy to show that $\Lambda_{\text{PHD}}(F_n)$ is e-affine invariant. Hence we assume $E(X) = 0$ and $\text{var}(X) = I_p$ without loss of generality. Let P be the projection on to $\text{span}(M)$ and $Q = I_p - P$. Under the assumption that X is Gaussian and PHD is exhaustive, we have

$$\text{span}(M) = \text{span}(M^{\mathsf{T}}) = \mathscr{S}_{Y|X} = \text{span}(P).$$

We first derive the influence function of M. Because $\Sigma = I$, we have

$$M^* = (\Sigma_{XX}^{-1/2})^* \Sigma_{XXY} + \Sigma_{XXY}^* + \Sigma_{XXY} (\Sigma_{XX}^{-1/2})^*.$$

By the definition of $\hat{\Sigma}_{XXY}$,

$$\begin{aligned}
\hat{\Sigma}_{XXY} &= E_n(Y - \hat{\mu}_Y)(X - \hat{\mu}_X)(X - \hat{\mu}_X)^{\mathsf{T}} \\
&= E_n(XX^{\mathsf{T}}Y) - E_n(YX)\hat{\mu}_X^{\mathsf{T}} - \hat{\mu}_X E_n(YX^{\mathsf{T}}) - \hat{\mu}_Y E_n(XX^{\mathsf{T}}) + 2\hat{\mu}_Y \hat{\mu}_X \hat{\mu}_X^{\mathsf{T}}.
\end{aligned} \tag{9.20}$$

Note that

$$\begin{aligned}
&\mu_X = 0, \ E(XX^{\mathsf{T}}) = I_p, \ E[XX^{\mathsf{T}}(Y - \mu_Y)] = \Sigma_{XXY}, \\
&\mu_Y^* = Y - \mu_Y, \ \mu_X^* = X, \\
&[E_n(XXY)]^* = XX^{\mathsf{T}}Y - E(XX^{\mathsf{T}}Y), \ [E_n(XX^{\mathsf{T}})]^* = XX^{\mathsf{T}} - I_p.
\end{aligned} \tag{9.21}$$

By applying the rules in Proposition 9.2 to equation (9.20), and using (9.21), we find

$$\Sigma_{XXY}^* = XX^\top(Y - \mu_Y) - \Sigma_{XXY} - E(YX)X^\top - XE(YX^\top) - (Y - \mu_Y)I_p.$$

Since X is multivariate normal, both LCM(β) and CCV(β) are satisfied. By Theorem 8.3, under LCM(β), $E(XY) \in \mathscr{S}_{Y|X}$. By Theorem 8.4, under both LCM(β) and CCV(β), span(Σ_{XXY}) $\subseteq \mathscr{S}_{Y|X}$. Also, since M is symmetric, we have $Q = Q_{M^\top}$. It follows that

$$QM^*Q_{M^\top} = QM^*Q = Q\Sigma_{XXY}^*Q = Q[(Y - \mu_Y)(XX^\top - I_p)]Q \equiv UU^\top(Y - \mu_Y) - Q(Y - \mu_Y),$$

where $U = QX$. Hence,

$$\mathrm{var}[\mathrm{vec}(QM^*Q_{M^\top})] = \mathrm{var}\{[U \otimes U - \mathrm{vec}(Q)](Y - \mu_Y)\}.$$

The right-hand side is expanded as

$$\mathrm{var}\{[U \otimes U - \mathrm{vec}(Q)](Y - \mu_Y)\}$$
$$= E\left(\mathrm{var}[(U \otimes U)|Y](Y - \mu_Y)^2\right) + \mathrm{var}\left(E\{[U \otimes U - \mathrm{vec}(Q)]|Y\}(Y - \mu_Y)\right).$$

Because $X \perp\!\!\!\perp Y|\beta^\top X$, we have $U \perp\!\!\!\perp Y|\beta^\top X$. Moreover, by the multivariate normal assumption and the fact that $\mathrm{cov}(U, \beta^\top X) = Q\beta = 0$, we have $U \perp\!\!\!\perp \beta^\top X$. Hence, by Theorem 2.1, statement 4, we have $U \perp\!\!\!\perp (\beta^\top X, Y)$, and in particular, $U \perp\!\!\!\perp Y$. Consequently, the right-hand side of the above equality becomes

$$\mathrm{var}(U \otimes U)\mathrm{var}(Y) + E[U \otimes U - \mathrm{vec}(Q)]\mathrm{var}(Y) = \mathrm{var}(U \otimes U)\mathrm{var}(Y).$$

But by Theorem 7.10 of Schott (1997) (see also Shao et al. (2007)), $\frac{1}{2}\mathrm{var}(U \otimes U)$ is a projection matrix with rank $(p - r + 1)(p - r)/2$. Hence the eigenvalues of

$$\mathrm{var}[\mathrm{vec}(QM^*Q)] = \mathrm{var}\{[U \otimes U - \mathrm{vec}(Q)](Y - \mu_Y)]\}$$

are either $2\mathrm{var}(Y)$ or 0, and the rank of this matrix is $(p - r + 1)(p - r)/2$. By Theorem 9.3, we have the convergence

$$\frac{n}{2\mathrm{var}(Y)} \sum_{i=r+1}^{p} \lambda_i[\Lambda_{\mathrm{PHD}}(F_n)] \overset{\mathscr{D}}{\to} \chi^2_{(p-r)(p-r+1)/2}.$$

Because $\mathrm{var}_n(Y) \overset{P}{\to} \mathrm{var}(Y)$, we have the convergence (9.19) by Slutzky's theorem. $\qquad\square$

To perform the sequential test, we simply compute the sum of the eigenvalues of the last $p - r$ eigenvalues of the matrix $(\hat{\Sigma}_{XX}^{-1/2}\hat{\Sigma}_{XXY}\hat{\Sigma}_{XX}^{-1/2})^2$ and then compare it with the $(1 - \alpha) \times 100\%$ quantile of $\chi^2_{(p+1-r)(p-r)/2}$. Below is the R-code to implement the sequential test.

R-code to implement sequential test for PHD

```
seqtestphd=function(x,y,r){
matpower=function(a,alpha){
a=(a+t(a))/2;tmp=eigen(a)
return(tmp$vectors%*%diag((tmp$values)^alpha)%*%t(tmp$vectors))}
n=length(y);p=ncol(x);signrt=matpower(var(x),-1/2)
z = center(x)%*%signrt
phdmat=t(z*(y-mean(y)))%*%z/n
test=n*sum((eigen(phdmat)$values[r:p])^2)/(2*var(y))
r=r-1;pval=1-pchisq(test,(p-r+1)*(p-r)/2)
return(list(pval=pval,test=test))}
```

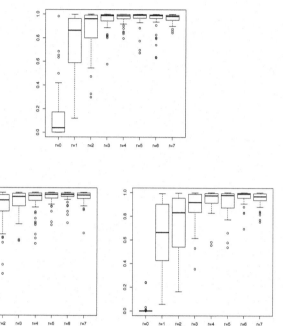

Figure 9.2 *Boxplots of p-values for sequential test based on PHD as applied to Model (9.18).*

Example 9.2 We applied the sequential test for PHD to model (9.18) with $n = 100, 200, 300$. Figure 9.2 shows the boxplots of the p-values of the PHD sequential test statistic (9.19) based on 50 simulated samples for each n. Inevitably, the performance of the sequential test is affected by the PHD estimator itself, which doesn't pick up the second direction in the central subspace. From the boxplots we see that the p-values for $H_0 : r = 0$ are small but those for $H_0 : r = 1$ are large. That means in most cases this test would estimate d to be 1. In general, PHD does not perform well for estimating monotone trends. □

9.9   Sequential Test for SAVE

To facilitate the von Mises expansion, we reexpress the candidate matrix for SAVE, as given in Section 5.4, as

$$\Lambda_{\text{SAVE}}(F_n) = \sum_{k=1}^{h} \hat{p}_k (I_p - \hat{\Sigma}^{-1/2} \hat{\Omega}_k \hat{\Sigma}^{-1/2})^2,$$

where

$$\hat{\Omega}_k = \text{var}_n(X|Y \in J_k) = \hat{H}_k - \hat{\mu}_k\hat{\mu}_k^\mathsf{T}, \quad \hat{H}_k = \hat{p}_k^{-1} E_n(XX^\mathsf{T}S_k) \tag{9.22}$$

and S_k, $\hat{p}_k$, $\hat{\mu}_k$, $\hat{\mu}$, and $\hat{\Sigma}$ are as defined in Section 9.7. The type of asymptotic distribution parallel to the chi-squared distributions for SIR and PHD turns out to be quite elusive and, to our knowledge, no such simple form has been developed in the literature. Because of this intractability, Shao et al. (2007) adopted a modified version of SAVE, whose asymptotic distribution for sum of small eigenvalues is a chi-squared distribution under the Gaussian assumption for the predictor X. The modified statistics is no longer in the form of sum of the small eigenvalues.

The asymptotic expansion for SAVE is not itself difficult; the difficulty arises when one tries to simplify it to a chi-squared distribution (or something comparatively simple), which involves demonstrating the weights ω_i's in (9.12) are either 0 or 1. To be consistent with the other methods, here we still use the sum of smallest eigenvalues as the test statistic and derive its asymptotic distribution as a weighted sum of independent χ_1^2 variables, where the weights are to be estimated from the sample. The advantage of this approach is it is sufficiently general to be applicable to all other methods introduced so far, such as DR and PIR.

Theorem 9.6 *Suppose that* $\text{LCM}(\beta)$ *and* $\text{CCV}(\beta)$ *are satisfied for a basis matrix* β *of* $\mathscr{S}_{Y|X}$, *and that* $E(X) = 0$ *and* $\text{var}(X) = I_p$. *Moreover, suppose that SAVE is exhaustive. Then, under the null hypothesis* $H_0^{(r)} : \dim(\mathscr{S}_{Y|X}) = r, r = 0, \ldots, p-1$,

$$n \sum_{i=r+1}^{p} \lambda_i[\Lambda_{\text{SAVE}}(F_n)] \overset{\mathscr{L}}{\to} \sum_{i=1}^{(p-r)(ph-r)} \omega_i K_i,$$

where $K_1, \ldots, K_{(p-r)(ph-r)}$ *are i.i.d.* χ_1^2 *random variables, and* $\omega_1, \ldots, \omega_{(p-r)(ph-r)}$ *are the eigenvalues of the matrix*

$$V = \text{var}\left(\text{vec}\{Q[U_1^\mathsf{T} \otimes (XX^\mathsf{T} - I_p) + XU_2^\mathsf{T}]Q_{M^\mathsf{T}}\} \right), \tag{9.23}$$

where

$$U_1 = (p_1^{-1/2}(p_1 - S_1), \ldots, p_h^{-1/2}(p_h - S_h))^\mathsf{T}, \quad U_2 = (p_1^{-1/2}S_1\mu_1^\mathsf{T}, \ldots, p_h^{-1/2}S_h\mu_h^\mathsf{T})^\mathsf{T}.$$

In practice, we replace ω_i by the eigenvalues of the sample estimate of the matrix V. It may well be that the last few eigenvalues of $\{\omega_i : i = 1, \ldots, (ph-r)(p-r)\}$ are 0. Instead of trying to figure out the exact rank of V exactly, we rely on the smallness of the sample estimates $\hat{\omega}_i$ corresponding to the null eigenvalues.

PROOF OF THEOREM 9.6. Let

$$\hat{M}_k = \hat{p}_k^{1/2}(I_p - \hat{\Sigma}^{-1/2}\hat{\Omega}_k\hat{\Sigma}^{-1/2}), \quad k = 1, \ldots, h,$$

and let $\hat{M}$ be the $p \times ph$ matrix $(\hat{M}_1, \ldots, \hat{M}_h)$. Then $\Lambda_{\text{SAVE}}(F_n)$ can be written as

$$\Lambda_{\text{SAVE}}(F_n) = \hat{M}\hat{M}^\mathsf{T}.$$

By Theorem 9.3, the asymptotic distribution of $\Lambda_{\text{SAVE}}(F_n)$ is of the form (9.12). We now derive the explicit form of ω_i with s being ph, the number of columns of M.

Note that

$$Q\hat{M}Q_{M^\top} = (Q\hat{M}_1, \ldots, Q\hat{M}_h)Q_{M^\top}.$$

By the rules in Proposition 9.2, we have

$$M_k^\star = (p_k^{1/2})^\star(I_p - \Omega_k) + p_k^{1/2}(I_p - \Sigma^{-1/2}\Omega_k\Sigma^{-1/2})^\star. \tag{9.24}$$

By Theorem 5.1, under LCM(β) and CCV(β), we have span$(I_p - \Omega_k) \subseteq \mathscr{S}_{Y|X}$. Hence $Q(I_p - \Omega_k) = 0$, which means the first term in $M_k^\star$ in (9.24) can be ignored after being multiplied by Q. That is,

$$QM_k^\star = p_k^{1/2}Q(I_p - \Sigma^{-1/2}\Omega_k\Sigma^{-1/2})^\star.$$

By Proposition 9.2, the influence function on the right-hand side is

$$(I_p - \Sigma^{-1/2}\Omega_k\Sigma^{-1/2})^\star = -(\Sigma^{-1/2})^\star\Omega_k - \Omega_k^\star - \Omega_k(\Sigma^{-1/2})^\star.$$

By Lemma 9.1, part 4, and the fact that $Q(I_p - \Omega_k) = 0$, we have

$$\begin{aligned}
Q(I_p - \Sigma^{-1/2}\Omega_k\Sigma^{-1/2})^\star &= -Q[-\tfrac{1}{2}(XX^\top - I_p)](\Omega_k - I_p) - Q(\Sigma^{-1/2})^\star \\
&\quad - Q\Omega_k^\star - Q(\Omega_k - I_p)(\Sigma^{-1/2})^\star - Q(\Sigma^{-1/2})^\star \\
&= \tfrac{1}{2}QXX^\top(\Omega_k - I_p) - Q\Omega_k^\star + Q(XX^\top - I_p).
\end{aligned}$$

Now let us compute $Q\Omega_k^\star$. By the definition of Ω_k in (9.22), we have

$$\Omega_k^\star = H_k^\star - \mu_k^\star\mu_k^\top - \mu_k\mu_k^{\star\top}. \tag{9.25}$$

By the LCM(β) assumption and Theorem 3.1, $\mu_k \in \mathscr{S}_{Y|X}$. Hence

$$Q\Omega_k^\star = QH_k^\star - Q\mu_k^\star\mu_k^\top. \tag{9.26}$$

By Proposition 9.2 and Lemma 9.1, part 1, we have

$$H_k^\star = (p_k^{-1})^\star E(XX^\top S_k) + p_k^{-1}[XX^\top S_k - E(XX^\top S_k)]. \tag{9.27}$$

The term $E(XX^\top S_k)$ is further computed as

$$E(XX^\top S_k) = E[E(XX^\top|Y)S_k] = E\{E[E(XX^\top|\beta^\top X)|Y]S_k\} = E[E(XX^\top|\beta^\top X)S_k],$$

where the third equality follows from Proposition 2.2. By LCM(β) and CCV(β), the term $E(XX^\top|\beta^\top X)$ is simplified as

$$\mathrm{var}(X|\beta^\top X) + E(X|\beta^\top X)E(X^\top|\beta^\top X) = Q + PXX^\top P,$$

where $P = I_p - Q$. Hence

$$E(XX^\top S_k) = p_k Q + PE(XX^\top S_k)P.$$

Substituting this into (9.27), we have

$$H_k^\star = (p_k^{-1})^\star[p_k Q + PE(XX^\top S_k)P] + p_k^{-1}\{XX^\top S_k - [p_k Q + PE(XX^\top S_k)P]\}.$$

Multiplying both sides of the above equation by Q from the left, and using the fact that $QP = 0$, we have

$$QH_k^* = (p_k^{-1})^* p_k Q + p_k^{-1}(QXX^\mathsf{T} S_k - p_k Q). \tag{9.28}$$

Moreover, by (9.15) and the fact that $E(XS_k) \in \mathscr{S}_{Y|X}$ under LCM(β),

$$Q\mu_k^* = p_k^{-1} QXS_k. \tag{9.29}$$

Substituting (9.28) and (9.29) into (9.26), we find

$$Q\Omega_k^* = (p_k^{-1})^* p_k Q + p_k^{-1}(QXX^\mathsf{T} S_k - p_k Q) - p_k^{-1} QXS_k \mu_k^\mathsf{T}.$$

It follows that

$$
\begin{aligned}
QM_k^* &= p_k^{1/2} Q(I_p - \Sigma^{-1/2}\Omega_k\Sigma^{-1/2})^* \\
&= p_k^{1/2}\left[\tfrac{1}{2}QXX^\mathsf{T}(\Omega_k - I_p) - Q\Omega_k^* + Q(XX^\mathsf{T} - I_p)\right] \\
&= p_k^{1/2}\left[\tfrac{1}{2}QXX^\mathsf{T}(\Omega_k - I_p) - (p_k^{-1})^* p_k Q - p_k^{-1}(QXX^\mathsf{T} S_k - p_k Q) \right. \\
&\qquad\left. + p_k^{-1} QXS_k\mu_k^\mathsf{T} + Q(XX^\mathsf{T} - I_p)\right] \\
&= \tfrac{1}{2}QXX^\mathsf{T} p_k^{1/2}(\Omega_k - I_p) + p_k^{-1/2}(S_k - p_k)Q - p_k^{-1/2}(QXX^\mathsf{T} S_k - p_k Q) \\
&\qquad + p_k^{-1/2} QXS_k\mu_k^\mathsf{T} + p_k^{1/2}Q(XX^\mathsf{T} - I_p) \\
&= \tfrac{1}{2}QXX^\mathsf{T} p_k^{1/2}(\Omega_k - I_p) + p_k^{-1/2}QXS_k\mu_k^\mathsf{T} + p_k^{-1/2}(p_k - S_k)Q(XX^\mathsf{T} - I_p).
\end{aligned}
$$

So QM^* is the sum of three terms, the first of which is of the form

$$\left(-\tfrac{1}{2}QXX^\mathsf{T} p_k^{1/2}(\Omega_k - I_p), \ldots, -\tfrac{1}{2}QXX^\mathsf{T} p_k^{1/2}(\Omega_k - I_p)\right) = -\tfrac{1}{2}QXX^\mathsf{T} M.$$

Because, by construction, $MQ_{M^\mathsf{T}} = 0$, this term vanishes after being multiplied from the right by the matrix Q_{M^T}. As a result, the matrix $QM^* Q_{M^\mathsf{T}}$ is of the form given in (9.23). □

At the sample level, we use moments to estimate the asymptotic variance

$$\mathrm{var}[\mathrm{vec}(QM^* Q_{M^\mathsf{T}})],$$

which are relatively straightforward. The p-values for the sequential test are derived from the distribution of linear combinations of independent χ_1^2 variables, which can be calculated using the approximation method of Bentler and Xie (2000) or simply using Monte Carlo simulation. We summarize the steps for sequential test for order determination for SAVE in the following algorithm.

Algorithm 9.2 Sequential test for SAVE

1. For each $k = 1, \ldots, h$, evaluate the $\hat{\Omega}_k$ matrix as defined in (9.22); evaluate $\hat{p}_k = E_n S_k$.

2. Compute the following matrices

$$U_{1i} = (\hat{p}_1^{-1/2}(\hat{p}_1 - S_{1i}), \ldots, \hat{p}_h^{-1/2}(\hat{p}_h - S_{hi}))^\mathsf{T}, \quad U_{2i} = (\hat{p}_1^{-1/2} S_{1i} \mu_1^\mathsf{T}, \ldots, \hat{p}_h^{-1/2} S_{hi} \mu_h^\mathsf{T})^\mathsf{T}.$$

3. Form the matrices

$$\hat{M} = \left(\hat{p}_1^{1/2}(I_p - \hat{\Sigma}^{-1/2} \hat{\Omega}_1 \hat{\Sigma}^{-1/2}), \ldots, \hat{p}_h^{1/2}(I_p - \hat{\Sigma}^{-1/2} \hat{\Omega}_h \hat{\Sigma}^{-1/2}) \right).$$

Form the test statistic $L(F_n) = n \sum_{i=r+1}^{p} \hat{\sigma}_i^2(\hat{M})$, where $\hat{\sigma}_i(\hat{M})$ are the singular values of $\hat{M}$.

4. For the current r, perform singular value decomposition on $\hat{M}$. Let $u_1, \ldots, u_r$ be the first r left singular vectors and $v_1, \ldots, v_r$ be the first r right singular vectors of $\hat{M}$. Form the projection matrices

$$\hat{Q} = I_p - (u_1, \ldots, u_r)(u_1, \ldots, u_r)^\mathsf{T}, \quad \hat{Q}_{M\mathsf{T}} = I_{ph+p+h} - (v_1, \ldots, v_r)(v_1, \ldots, v_r)^\mathsf{T}.$$

6. For each $i = 1, \ldots, n$, compute the $p^2 h$-dimensional vectors

$$S_i = \mathrm{vec}\left(\hat{Q}[U_{1i}^\mathsf{T} \otimes (X_i X_i^\mathsf{T} - I_p) + X_i U_{2i}^\mathsf{T}] \hat{Q}_{M\mathsf{T}} \right).$$

Then compute the $p^2 h \times p^2 h$ dimensional sample covariance matrix of $S_1, \ldots, S_n$. Evaluate the eigenvalues $\hat{\omega}_1, \ldots, \hat{\omega}_{p^2 h}$ of $\mathrm{var}_n(S)$.

7. Find the p-value of $L(F_n)$ against the distribution of $\sum_{i=1}^{(p-r)(ph-r)} \omega_i K_i$, where $K_1, \ldots, K_{(p-r)(ph-r)}$ are i.i.d. χ_1^2 variables.

8. If the p-value is smaller than a pre-assigned value (such as 0.10), then return to step 3. Continue until the $H_0^{(r)}$ is accepted. Use $r - 1$ as the estimate of the rank of Λ_{SAVE}.

For the evaluation of the p-value in step 7, we can use either one of the following two methods. The first is Monte Carlo. Given a large number N, say $N = 10000$, we generate N samples of $K_1, \ldots, K_{(p-r)(ph-r)}$ of i.i.d. χ_1^2 observations. This results in N observations on $\sum_{i=1}^{(p-r)(ph-r)} \omega_i K_i$. The actual percentage of its values exceeding the statistic $n \sum_i \lambda_i [\Lambda_{\mathrm{SAVE}}(F_n)]$ is then taken as the p-value. The second approach is proposed by Bentler and Xie (2000) originally in the context of principal Hessian directions. Let A_n denote the matrix $\mathrm{var}_n(S)$ in step 6. Let s be the nearest integer to $\mathrm{tr}[L(F_n)]^2 / \mathrm{tr}[L^2(F_n)]$. Then $L(F_n)$ has the approximate distribution $[\mathrm{tr}(A_n)/s]\chi_s^2$. Below is an R-code for implementing sequential SAVE test, where the p-values are evaluated by Monte Carlo with $N = 20000$.

R-code to implement sequential test for SAVE

```
seqtestsave=function(x,y,h,r,ytype){
p=ncol(x);n=nrow(x)
xst=t(t(x)-apply(x,2,mean))%*%matpower(var(x),-1/2)
if(ytype=="continuous") ydis=discretize(y,h)
```

```
if(ytype=="categorical") ydis=y
yless=ydis;ylabel=numeric()
for(i in 1:n) {
if(var(yless)!=0){
ylabel=c(ylabel,yless[1]);yless=yless[yless!=yless[1]]}}
ylabel=c(ylabel,yless[1])
sk=matrix(0,n,h)
for(i in 1:n) for(k in 1:h) {
if(ydis[i]==ylabel[k]) sk[i,k]=1
if(ydis[i]!=ylabel[k]) sk[i,k]=0}
prob=numeric()
for(i in 1:h) prob=c(prob,length(ydis[ydis==ylabel[i]])/n)
muk=numeric()
for(i in 1:h) muk = rbind(muk,apply(xst[ydis==ylabel[i],],2,mean))
vxy = array(0,c(p,p,h))
for(i in 1:h) vxy[,,i]=var(xst[ydis==ylabel[i],])
mmat=numeric();for(i in 1:h) mmat=cbind(mmat,prob[i]^(1/2)*(diag(p)-vxy[,,i]))
tmp=svd(mmat);uu=tmp$u[,1:r];vv=tmp$v[,1:r]
qmat=diag(p)-uu%*%t(uu);qmatmt=diag(p*h)-vv%*%t(vv)
qmstqmt=array(0,c(n,p,p*h))
for(i in 1:n){
u1=numeric();for(k in 1:h) u1=c(u1,prob[k]^(-1/2)*(prob[k]-sk[i,k]))
u2=numeric();for(k in 1:h) u2=c(u2,prob[k]^(-1/2)*sk[i,k]*muk[k,])
mat1=qmat%*%kronecker(t(u1),(xst[i,]%*%t(xst[i,])-diag(p)))%*%qmatmt
mat2=qmat%*%xst[i,]%*%t(u2)%*%qmatmt
qmstqmt[i,,]=mat1+mat2}
vecqmstqmt=numeric()
for(i in 1:n) vecqmstqmt=rbind(vecqmstqmt,c(qmstqmt[i,,]))
vmat=var(vecqmstqmt);omega=eigen(vmat)$values[1:((p-r)*(p*h-r))]
teststat=n*sum((svd(mmat)$d[(r+1):p])^2)
obs=numeric();nsim=20000
for(isim in 1:nsim) {
ki=rnorm(length(omega))^2;obs=c(obs,sum(omega*ki))}
pval=length((1:nsim)[obs>teststat])/nsim;return(pval)
}
```

Description of the R-code   The R-code includes the following input variables.

1. x: an $n \times p$ matrix containing the predictors.

2. y: an n-dimensional vector containing the observations of the response.

3. h: the number of slices. A relatively small h tends to perform better for the sequential test for SAVE. For example, when $n = 200$ and $p = 10$, we take h to be 3.

4. r: this represents the rth test $H_0^{(r)}$ in the sequence of tests.

The output of the R-code is the p-value for the rth sequential test. It is computed by Monte Carlo; that is, we generate nsum=10000 observations from the distribution of $\sum_{i=1}^{p^2h} \hat{\omega}_i K_i$. The p-value is simply the percentage of the observations exceeding the test statistic $n \sum_{i=r}^{p} \lambda_i [\Lambda_{SAVE}(F_n)]$.

Example 9.3 We applied the sequential test for SAVE to the model in Example 5.1, with $n = 100, 200, 300$, $p = 10$. For $n = 100, 200$, we choose the number of slices to be $h = 3$ in all three cases. The boxplots for p-values for sequential test with $r = 0, 1, \ldots, 7$ based on 50 simulated samples for each n are shown in Figure 9.3. We see a substantial jump in the p-values at $r = 2$. And this tendency becomes stronger as n increases. This indicates that the sequential test would select the correct dimension $d = 2$ when the sample size is large. □

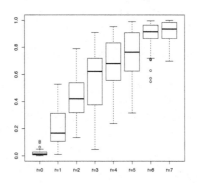

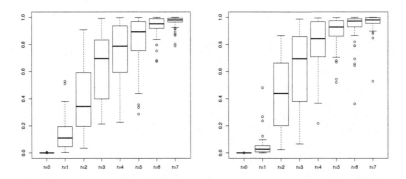

Figure 9.3 *Boxplots of p-values for sequential test based on SAVE as applied to the model in Example 5.1.*

9.10 Sequential Test for DR

Recall from Chapter 6 that the candidate matrix for Directional Regression is defined as

$$\Lambda_{\mathrm{DR}}(F) = E[2I_p - A(Y, \tilde{Y})]^2,$$

where $A(Y, \tilde{Y}) = E[(Z - \tilde{Z})(Z - \tilde{Z})^{\mathsf{T}} | Y, \tilde{Y}]$. Since the asymptotic expansion of the estimator of Λ_{DR} is rather tedious, we adopt the strategy of estimating the rank of a matrix whose rank is the same as Λ_{DR}. This is the strategy used in Li and Wang (2007). We first prove the following lemma, which was stated in Li and Wang (2007) without a proof.

Lemma 9.6 *Suppose Σ is nonsingular and all the moments involved are finite. Then the Λ_{DR} has the same rank as*

$$E\{2\Sigma - A_0(Y, \tilde{Y})\}^2, \tag{9.30}$$

where $A_0(Y, \tilde{Y}) = E[(X - \tilde{X})(X - \tilde{X})^{\mathsf{T}}|Y, \tilde{Y}]$.

PROOF. By definition,

$$E[2I_p - A(Y, \tilde{Y})]^2 = E\{\Sigma^{-1/2}[2\Sigma_p - A_0(Y, \tilde{Y})]\Sigma^{-1/2}\}^2.$$

Let $U = 2I_p - A(Y, \tilde{Y})$, $V = 2\Sigma_p - A_0(Y, \tilde{Y})$, and $C = \Sigma^{-1/2}$. Then the above equality can be rewritten as

$$E(U^2) = E[(CVC)^2],$$

and we need to show that $E(U^2)$ and $E(V^2)$ have the same rank. Let $\mathcal{S}$ be the subspace spanned by $E(V^2)$. Then, by the proof of Proposition 5.2,

$$\mathrm{span}(V) = \mathcal{S} \ a.s. \ P$$

in the sense that $v \perp \mathcal{S}$ if and only if $v^{\mathsf{T}}U = 0$ a.s. P. Since C is nonsingular, we have $\mathrm{span}(V) = \mathrm{span}(VC)$. Hence

$$\mathrm{span}(CVC) = C\mathrm{span}(VC) = C\mathrm{span}(V) = C\mathcal{S} \ a.s. \ P.$$

It follows that

$$\mathrm{span}[E(U^2)] = \mathrm{span}[E(CVC)^2] = C\mathcal{S} = C\mathrm{span}[E(V^2)].$$

Since C is nonsingular, $\mathrm{span}[E(U^2)]$ and $\mathrm{span}[E(V^2)]$ have the same dimension; that is, $E(U^2)$ and $E(V^2)$ have the same rank. $\qquad\square$

By focusing on (9.30), we exclude the term $\Sigma^{-1/2}$ from consideration, which greatly simplifies the asymptotic expansion. By straightforward computation it can be shown that (9.30) can be rewritten as

$$2E\left[E^2(XX^{\mathsf{T}} - I_p|Y)\right] + 2\left\{E[E(X|Y)E(X^{\mathsf{T}}|Y)]\right\}^2$$
$$+ 2\left\{E[E(X^{\mathsf{T}}|Y)E(X|Y)]\right\}\left\{E[E(X|Y)E(X|Y^{\mathsf{T}})]\right\}.$$

The discretized version of this matrix, ignoring the proportionality constant 2, is

$$\sum_{k=1}^{h} E^2(XX^{\mathsf{T}} - I_p|Y \in J_k)p_k + \left[\sum_{k=1}^{h} E(X|Y \in J_k)E(X^{\mathsf{T}}|Y \in J_k)p_k\right]^2$$
$$+ \left[\sum_{k=1}^{h} E(X^{\mathsf{T}}|Y \in J_k)E(X|Y \in J_k)p_k\right]\left[\sum_{k=1}^{h} E(X|Y \in J_k)E(X^{\mathsf{T}}|Y \in J_k)p_k\right].$$

Let us denote this matrix by $\Upsilon_{\mathrm{DR}}(F_0)$. The corresponding statistical functional is the above matrix with E replaced by E_F, X replaced by $X - E_F X$, and p_k by $p_k = E_F S_k$. That is,

$$\Upsilon_{\mathrm{DR}}(F) = \sum_{k=1}^{h} E_F^2[(X - E_F X)(X - E_F X)^{\mathsf{T}} - I_p | Y \in J_k] p_k(F)$$

$$+ \left\{ \sum_{k=1}^{h} E_F[(X - E_F X) | Y \in J_k] E_F[(X - E_F X)^{\mathsf{T}} | Y \in J_k] p_k(F) \right\}^2$$

$$+ \left\{ \sum_{k=1}^{h} E_F[(X - E_F X)^{\mathsf{T}} | Y \in J_k] E_F[(X - E_F X) | Y \in J_k] p_k(F) \right\}$$

$$\times \left\{ \sum_{k=1}^{h} E_F[(X - E_F X) | Y \in J_k] E_F[(X - E_F X)^{\mathsf{T}} | Y \in J_k] p_k(F) \right\}.$$

Let

$$U_k(F) = E_F[(X - E_F X) | Y \in J_k], \quad V_k = E_F[(X - E_F X)(X - E_F X)^{\mathsf{T}} | Y \in J_k] - \mathrm{var}_F(X). \quad (9.31)$$

Then $\Upsilon_{\mathrm{DR}}(F)$ can be rewritten as

$$\Upsilon_{\mathrm{DR}}(F) = 2 \sum_{k=1}^{h} V_k(F) V_k(F) p_k(F) + 2 \left(\sum_{k=1}^{h} U_k(F) U_k(F)^{\mathsf{T}} p_k(F) \right)^2$$

$$+ 2 \left(\sum_{k=1}^{h} U_k(F)^{\mathsf{T}} U_k(F) p_k(F) \right) \left(\sum_{k=1}^{h} U_k(F) U_k(F)^{\mathsf{T}} p_k(F) \right).$$

Further define

$$M_{1k}(F) = p_k^{1/2}(F) V_k(F), \quad k = 1, \ldots, h,$$

$$M_2(F) = \sum_{\ell=1}^{h} p_\ell(F) U(F)_\ell U_\ell(F)^{\mathsf{T}},$$

$$M_{3k}(F) = p_k^{1/2}(F) \left(\sum_{\ell=1}^{h} U_\ell(F)^{\mathsf{T}} U_\ell(F) p_\ell(F) \right)^{1/2} U_k(F), \quad k = 1, \ldots, h,$$

$$M(F) = (M_{11}(F), \ldots, M_{1h}(F), M_2(F), M_{31}(F), \ldots, M_{3h}(F)).$$

Then $\Upsilon_{\mathrm{DR}}(F)$ can be written as the quadratic form $M(F)M(F)^{\mathsf{T}}$. Correspondingly, let

$$p_k(F_n), \quad U_k(F_n), \quad V_k(F_n), \quad M_{1k}(F_n), \quad M_2(F_n), \quad M_{3k}(F_n), \quad M(F_n)$$

be the sample estimates with E_F replaced by E_n wherever possible. Let $\Upsilon_{\mathrm{DR}}(F_n) = M(F_n)M^{\mathsf{T}}(F_n)$. As before, we denote quantities such as $M(F_n)$ by $\hat{M}$, and quantities such as $M(F_0)$ by M. Let $L_r(F) = n \sum_{k=r+1}^{p} \lambda_k[\Upsilon_{\mathrm{DR}}(F)]$. We first derive the influence functions for $U_k(F)$ and $V_k(F)$.

Lemma 9.7 *Suppose that* $(X_1, Y_1), \ldots, (X_n, Y_n)$ *are independent copies of* (X, Y) *with* $E(X) = 0$, *and that all the moments involved are finite. Then*

$$U_k^* = (X - U_k)S_k / p_k - X$$
$$V_k^* = (XX^{\mathsf{T}} - \Sigma - V_k)S_k / p_k - U_k X^{\mathsf{T}} - X U_k^{\mathsf{T}} - XX^{\mathsf{T}} + \Sigma. \quad (9.32)$$

PROOF. Rewrite the definition of $U_k(F)$ in (9.31) as follow,

$$U_k(F) = \mathrm{cov}_F(X, S_k) / E_F(S_k).$$

Applying the rules for influence functions in Proposition 9.2, we have

$$U_k^\star = \text{cov}_F^\star(X,S_k)/p_k + \text{cov}(X,S_k)(-p_k^{-2})(S_k - p_k). \tag{9.33}$$

By part 2 of Lemma 9.1, together with the assumption $E(X) = 0$, the first term on the right-hand side is

$$X(S_k - p_k)/p_k - \text{cov}(X,S_k)/p_k.$$

The second term on the right-hand side of (9.33) is $-U_k(S_k - p_k)/p_k$. Substitute both results into (9.33) to prove the first equality in (9.32).

To prove the second equality in (9.32), first rewrite $V_k(F)$ in (9.31) as

$$V_k(F) = E_F[(X - E_F X)(X - E_F X)_T S_k]/E_F S_k - E_F[(X - E_F X)(X - E_F X)].$$

Applying the rules for influence functions to the above equation, we have

$$\begin{aligned} V_k^\star &= E_F^\star[(X - E_F X)(X - E_F X)^\mathsf{T} S_k]/p_k \\ &\quad + E[(X - EX)(X - EX)^\mathsf{T} S_k](1/E_F S_k)^\star - \text{var}_F^\star(X). \end{aligned} \tag{9.34}$$

The first term on the right-hand side of (9.34), without the factor $(1/p_k)$, is

$$\begin{aligned} &E_F^\star[(X - E_F X)(X - E_F X)^\mathsf{T} S_k] \\ &= E_F^\star(XX^\mathsf{T} S_k) - [E_F(XS_k)E_F(X^\mathsf{T})]^\star - [E_F(X)E_F(X^\mathsf{T} S_k)]^\star + [E_F(X)E_F(X^\mathsf{T})E_F(S_k)]^\star. \end{aligned} \tag{9.35}$$

Because $E(X) = 0$, the last term on the right-hand side of (9.35) is

$$E_F^\star(X)E(X^\mathsf{T})E(S_k) + E(X)E_F^\star(X^\mathsf{T})E(S_k) + E(X)E(X^\mathsf{T})E_F^\star(S_k) = 0.$$

The second term on the right-hand side of (9.35) is

$$[E_F(XS_k)E_F(X^\mathsf{T})]^\star = [XS_k - E(XS_k)]E(X^\mathsf{T}) - E(XS_k)(X - EX)^\mathsf{T} = -E(XS_k)X^\mathsf{T}.$$

Similarly, the third term on the right-hand side of (9.35) is $-XE(X^\mathsf{T} S_k)$. Meanwhile, the first term on the right-hand side of (9.35) is $XX^\mathsf{T} S_k - E(XX^\mathsf{T} S_k)$. In summary, we have

$$E_F^\star[(X - E_F X)(X - E_F X)^\mathsf{T} S_k] = XX^\mathsf{T} S_k - E(XX^\mathsf{T} S_k) - E(XS_k)X^\mathsf{T} - XE(X^\mathsf{T} S_k). \tag{9.36}$$

The second term on the right-hand side of (9.34) is

$$p_k(V_k + \Sigma)(-p_k^{-2})(S_k - p_k) = -(V_k + \Sigma)(S_k - p_k)/p_k. \tag{9.37}$$

By Lemma 9.1 and $E(X) = 0$, the third term on the right-hand side of (9.34), without the sign, is

$$E^\star[(X - EX)(X - EX)^\mathsf{T}] = XX^\mathsf{T} - \Sigma. \tag{9.38}$$

Now substitute (9.36), (9.37), and (9.38) into (9.34) to prove the second equality in (9.32). $\square$

Next, we derive the influence function for $M(F)$. Recall that $p_k^\star = S_k - p_k$.

Lemma 9.8 *Under the assumptions of Lemma 9.7,*

$$M^\star = (M_{11}^\star, \ldots, M_{1h}^\star, M_2^\star, M_{31}^\star, \ldots, M_{3h}^\star), \tag{9.39}$$

where

$$\begin{aligned} M_{1k}^\star &= (1/2)p_k^{-1/2}p_k^\star V_k + p_k^{1/2}V_k^\star, \quad M_2^\star = \sum(p_\ell^\star U_\ell U_\ell^\mathsf{T} + p_\ell U_\ell^\star U_\ell^\mathsf{T} + p_\ell U_\ell U_\ell^{\star\mathsf{T}}), \\ M_{3k}^\star &= (1/2)p_k^{-1/2}p_k^\star[\text{tr}(M_2)]^{1/2}U_k + (1/2)p_k^{1/2}[\text{tr}(M_2)]^{-1/2}\text{tr}(M_2^\star)U_k + p_k^{1/2}[\text{tr}(M_2)]^{1/2}U_k^\star. \end{aligned}$$

PROOF. Applying Proposition 9.2, it is easy to show that influence functions of M_{1k} and M_{2k} are as given in the lemma. Reexpress M_{3k} in terms of M_2 as

$$M_{3k} = p_k^{1/2} \left(\sum_\ell U_\ell^T U_\ell p_\ell \right)^{1/2} U_k = p_k^{1/2} [\text{tr}(M_2)]^{1/2} U_k.$$

Now apply the rules in Proposition 9.2 to obtain M_{3k}^*.   □

The asymptotic distribution of $L(F_n)$ follows directly from Theorem 9.3.

Theorem 9.7 *Suppose that LCM(β) and CCV(β) are satisfied for a basis matrix β of $\mathscr{S}_{Y|X}$, $E(X) = 0$, and that all the moments involved are finite. Moreover, suppose that DR is exhaustive. Then, under the null hypothesis $H_0^{(r)} : \dim(\mathscr{S}_{Y|X}) = r, \ r = 0, \dots, r-1$,*

$$L_r(F_n) \equiv n \sum_{r+1}^p \lambda_i[\Upsilon_{\text{DR}}(F_n)] \xrightarrow{\mathscr{L}} \sum_{i=1}^{(p-r)(ph+p+h-r)} \omega_i K_i, \tag{9.40}$$

where the K_i's are independent χ_1^2 variables, and ω_i's are the first $(p-r)(ph+h+p-r)$ eigenvalues of the matrix $\text{var}[\text{vec}(QM^*Q_{M^T})]$, with M^* given by (9.39).

The numerical procedure to implement the sequential DR test is presented in Algorithm 9.3. Let us first see an example.

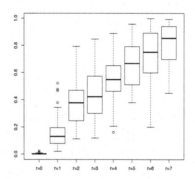

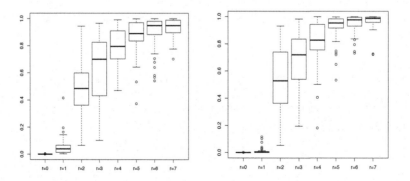

Figure 9.4 *Boxplots of p-values for sequential test based on DR for the model in Example 5.1.*

Example 9.4 We applied the sequential test for DR to the model in Example 5.1, with $n = 100, 200, 300$, $p = 10$ and $h = 4$ in all three scenarios. As before, we show in Figure 9.4 the boxplots for p-values for sequential test with $r = 0, 1, \ldots, 7$ based on 50 simulated samples for each n. Compared with the sequential test for SAVE, the jump in p-values at $r = 2$ is more clear cut, implying a greater chance of estimating the dimension d correctly. □

Algorithm 9.3 Sequential test for DR

---

1. Center $X_1, \ldots, X_n$ by $X_1 - E_n(X), \ldots, X_n - E_n(X)$. For ease of presentation of the rest of the algorithm, reset X_i to $X_i - E_n(X)$, and reset the generic X to $X - E_n(X)$.
2. For each $k = 1, \ldots, h$, compute

$$\hat{\Sigma} = E_n(XX^\mathsf{T}), \quad \hat{p}_k = E_n S_k, \quad \hat{U}_k = E_n(X S_k)/\hat{p}_k, \quad \hat{V}_k = E(XX^\mathsf{T} S_k)/\hat{p}_k - \hat{\Sigma}.$$

3. Compute the sample-level influence functions

$$\hat{p}_k^\star = S_k - \hat{p}_k, \quad \hat{U}_k^\star = (X - \hat{U}_k)S_k/\hat{p}_k - X,$$
$$\hat{V}_k^\star = (XX^\mathsf{T} - \hat{\Sigma} - \hat{V}_k)S_k/\hat{p}_k - \hat{U}_k X^\mathsf{T} - X\hat{U}_k^\mathsf{T} - XX^\mathsf{T} + \hat{\Sigma}.$$

4. Compute the sample-level quantities

$$\hat{M}_{1k} = \hat{p}_k^{1/2} \hat{V}_k, \quad \hat{M}_2 = \sum_{\ell=1}^h \hat{p}_\ell \hat{U}_\ell \hat{U}_\ell^\mathsf{T},$$
$$\hat{M}_{3k} = \hat{p}_k^{1/2} [\mathrm{tr}(\hat{M}_2)]^{1/2} \hat{U}_k, \quad \hat{M} = (\hat{M}_{11}, \ldots, \hat{M}_{1h}, \hat{M}_2, \hat{M}_{31}, \ldots, \hat{M}_{3h}).$$

5. Compute the sample-level influence functions

$$\hat{M}_{1k}^\star = (1/2)\hat{p}_k^{-1/2} \hat{p}_k^\star \hat{V}_k + \hat{p}_k^{1/2} \hat{V}_k^\star,$$
$$\hat{M}_2^\star = \sum_{\ell=1}^h \left(\hat{p}_\ell^\star \hat{U}_\ell \hat{U}_\ell^\mathsf{T} + \hat{p}_\ell U_\ell^\star \hat{U}_\ell^\mathsf{T} + \hat{p}_\ell \hat{U}_\ell \hat{U}_\ell^{\star\mathsf{T}} \right),$$
$$\hat{M}_{3k}^\star = 2^{-1} p_k^{-1/2} \hat{p}_k^\star [\mathrm{tr}(\hat{M}_2)]^{1/2} \hat{U}_k + 2^{-1} \hat{p}_k^{1/2} [\mathrm{tr}(\hat{M}_2)]^{-1/2} \mathrm{tr}(\hat{M}_2^\star) \hat{U}_k + \hat{p}_k^{1/2} [\mathrm{tr}(\hat{M}_2)]^{1/2} \hat{U}_k^\star.$$
$$\hat{M}^\star = \left(\hat{M}_{11}^\star, \ldots, \hat{M}_{1h}^\star, \hat{M}_2^\star, \hat{M}_{31}^\star, \ldots, \hat{M}_{3h}^\star \right).$$

6. At the current r $(r = 0, 1, 2, \ldots)$, compute the first r left and right singular vectors of $\hat{M}$, say $u_1, \ldots, u_r$ and $v_1, \ldots, v_r$. Let

$$P = \sum_{\ell=1}^r u_\ell u_\ell^\mathsf{T}, \quad P_{\hat{M}^\mathsf{T}} = \sum_{\ell=1}^r v_\ell v_\ell^\mathsf{T}, \quad Q = I_p - P, \quad Q_{\hat{M}^\mathsf{T}} = I_{ph+p+h} - P_{\hat{M}^\mathsf{T}}.$$

7. Compute the sample covariance matrix $\mathrm{var}_n[\mathrm{vec}(QM^\star Q_{\hat{M}^\mathsf{T}})]$, and let $\hat{\omega}_1, \ldots, \hat{\omega}_{(p-r)(ph+p+h-r)}$ be the first $(p - r)(ph + p + h - r)$ eigenvalues of the above matrix.
8. By either the Monte Carlo method or the approximation in Bentler and Xie (2000), find the p-value of $L(F_n)$ against the distribution of $\sum_{i=1}^{(p-r)(ph+p+h-r)} \omega_i K_i$, where the K_i's are i.i.d. χ_1^2 variables. If the p-value is smaller than a pre-assigned value (such as 0.10), then return to step 6. Continue until the $H_0^{(r)}$ is accepted. Use $r - 1$ as the estimate of the rank of Υ_{DR}.

R-code to implement sequential test for DR

```
seqtestdr=function(x,y,h,r,ytype){
p=ncol(x);n=nrow(x)
xc=t(t(x)-apply(x,2,mean))
if(ytype=="continuous") ydis=discretize(y,h)
if(ytype=="categorical") ydis=y
yless=ydis;ylabel=numeric()
for(i in 1:n) {
if(var(yless)!=0){
ylabel=c(ylabel,yless[1]);yless=yless[yless!=yless[1]]}}
ylabel=c(ylabel,yless[1])
sk=matrix(0,n,h); for(i in 1:n) for(k in 1:h)  if(ydis[i]==ylabel[k]){
                                                sk[i,k]=1}
pk=apply(sk,2,mean)
sig=var(xc)
uk=numeric()
for(i in 1:h) uk=rbind(uk,apply(xc[ydis==ylabel[i],],2,mean))
vk=array(0,c(p,p,h)); for(i in 1:h) vk[,,i]=
                      var(xc[ydis==ylabel[i],])-sig
m1=numeric(); for(k in 1:h) m1=cbind(m1,pk[k]^(1/2)*vk[,,k])
m2=t(uk)%*%diag(pk)%*%uk
tr=function(a) return(sum(diag(a)))
m3=numeric(); for(k in 1:h) m3=
                cbind(m3,pk[k]^(1/2)*(tr(m2))^(1/2)*uk[k,])
m=cbind(m1,m2,m3)
pkst=numeric(); for(i in 1:n) pkst=rbind(pkst,sk[i,]-pk)
ukst=array(0,c(p,h,n)); for(i in 1:n) for(k in 1:h) {
ukst[,k,i]=(xc[i,]-uk[k,])*sk[i,k]/pk[k]-xc[i,]}
vkst=array(0,c(p,p,h,n)); for(k in 1:h) for(i in 1:n) {
vkst[,,k,i]=(xc[i,]%*%t(xc[i,])-sig-vk[,,k])*sk[i,k]/pk[k]-
uk[k,]%*%t(xc[i,])-xc[i,]%*%t(uk[k,])-xc[i,]%*%t(xc[i,])+sig}
m1kst=array(0,c(p,p,h,n)); for(k in 1:h) for(i in 1:n) {
m1kst[,,k,i]=
(1/2)*pk[k]^(-1/2)*pkst[i,k]*vk[,,k]+pk[k]^(1/2)*vkst[,,k,i]}
m2st=array(0,c(p,p,n)); for(i in 1:n){
for(k in 1:h) {
m2st[,,i]=m2st[,,i]+pkst[i,k]*uk[k,]%*%t(uk[k,])+
pk[k]*ukst[,k,i]%*%t(uk[k,])+pk[k]*uk[k,]%*%t(ukst[,k,i])}}
m3kst=array(0,c(p,h,n)); for(k in 1:h) for(i in 1:n){
m3kst[,k,i]=(1/2)*pk[k]^(-1/2)*pkst[i,k]*(tr(m2))^(1/2)*uk[k,]+
        (1/2)*pk[k]^(1/2)*(tr(m2))^(-1/2)*tr(m2st[,,i])*uk[k,]+
        pk[k]^(1/2)*(tr(m2))^(1/2)*ukst[,k,i]}
mst=array(0,c(p,p*h+p+h,n)); for(i in 1:n){
mst1=numeric(); for(k in 1:h) mst1=cbind(mst1,m1kst[,,k,i])
mst3=numeric(); for(k in 1:h) mst3=cbind(mst3,m3kst[,k,i])
mst[,,i]=cbind(mst1,m2st[,,i],mst3)}
if(r==0) qlef=diag(p);qrig=diag(p*h+p+h)
svdm=svd(m);u1=svdm$u[,1:r];v1=svdm$v[,1:r]
qlef=diag(p)-u1%*%t(u1);qrig=diag(p*h+p+h)-v1%*%t(v1)
```

```
qmstq=array(0,c(p,p*h+p+h,n)); for(i in 1:n) qmstq[,,i]=
                                    qlef%*%mst[,,i]%*%qrig
vqmstq=matrix(0,n,p*(p*h+p+h)); for(i in 1:n) vqmstq[i,]=c(qmstq[,,i])
s=var(vqmstq);omega=eigen(s)$values[1:((p-r)*(p*h+p+h-r))]
teststat=n*sum(eigen(m%*%t(m))$values[(r+1):p])
obs=numeric();nsim=20000
for(isim in 1:nsim) {
ki=rnorm(length(omega))^2;obs=c(obs,sum(omega*ki))}
pval=length((1:nsim)[obs>teststat])/nsim;return(pval)
}
```

9.11 Applications

In this section we apply the sequential order determination methods to two data sets: the Big Mac data and the pen digit data.

Example 9.5 We apply the sequential tests for SIR, SAVE, and DR to the Big Mac data. For sequential SIR, we take the number of slices to be 8; for sequential SAVE and DR, we take the number of slices to be 3 and 4, respectively. The next table gives p-values for a range of r. Because the limiting distribution for sequential SIR is chi-square with $(p-r)(h-1-r)$ degrees of freedom, the largest r allowed is $\min(p-1,h-2)=6$. That is why the table only contains the cases $r=0,\ldots,6$. Though this is not the entire range for sequential SAVE and DR, these 7 columns contain enough information to determine the order.

Table 9.1 *p-values for sequential tests applied to the Big Mac data*

r	0	1	2	3	4	5	6
SIR	0.064	0.612	0.865	0.931	0.957	0.978	0.943
SAVE	0.015	0.111	0.125	0.197	0.229	0.407	0.471
DR	0.000	0.596	0.154	0.384	0.872	0.904	0.767

Table 9.1 indicates that all three tests have a substantial jump in p-value at $r=1$, with DR having the sharpest jump. Thus, if we take $d=0.1$, then the dimension d is estimated to be 1 for all three methods. This is also consistent with the scatter plots such as those shown in Figures 3.4 and 5.4, which shows a monotone pattern of the response versus the first sufficient predictors, but nearly no pattern for the response versus the second predictors. □

Example 9.6 Next, we applied sequential SAVE and DR to the pen digit data, with digits 0, 6, 9 as before, and with the combined cases of training and testing data sets. Because Y is categorical with 3 classes, we take these classes as the slices. Hence h is inherently 3. That means the r for SIR is limited to $\min(p-1,h-2)=1$. That is why we did not apply sequential SIR to this data set. This also reveals a limitation of SIR: the dimension of the central subspace cannot exceed $h-1$. Table 9.2 shows the p-values for sequential SAVE and

Table 9.2 *p-values for sequential tests applied to the pen digit data*

r	$0 \sim 7$	8	$9 \sim 13$	14	15
SAVE	0.000	0.000	0.000	0.008	0.089
DR	0.000	0.066	≥ 0.25	0.624	0.724

DR for $r = 0, \ldots, 15$. For $r = 0, \ldots, 7$, the p-values do not show in the third digit; so the table is compressed for these cases. As r further increases, the p-value shows a significant jump at 9 (≥ 0.25) for sequential DR, but remains 0 up to the third digit until $r = 14$. Columns 9 through 13 are also compressed for this reason. As a result, if we use $\alpha = 0.1$, then the dimension d is estimated as 9 by sequential DR, and 16 by sequential SAVE.   □

Chapter 10

Other Methods for Order Determination

10.1 BIC Type Criteria for Order Determination   141
10.2 Bootstrapped Eigenvector Variation   147
10.3 Eigenvalue Magnitude and Eigenvector Variation   150
10.4 Ladle Estimator   152
10.5 Consistency of the Ladle Estimator   156
10.6 Application: Identification of Wine Cultivars   156

Continuing on the theme of the last chapter, in this chapter we introduce three other order determination methods. The first is a class of BIC-type criteria that maximizes the information while penalizing the number of parameters. Similar to the sequential test, this method relies on the magnitude of the eigenvalues. The second method, in contrast, relies on the bootstrapped eigenvector variation, which tends to be larger in the null space of the candidate matrix than in the range space of the matrix. The third method combines the advantages of the first two methods by exploiting both the magnitude of eigenvalues and the bootstrapped eigenvector variations.

10.1 BIC Type Criteria for Order Determination

In this section we introduce a class of order determination criteria that we call the BIC-type criteria. Though having different appearance, these criteria are all similar in nature, and they stem from a criterion proposed in Zhu et al. (2006). See also Wang and Yin (2008), Li et al. (2010b), and Guo et al. (2015b).

To motivate the general form of these criteria, we first consider two special forms, proposed by Zhu et al. (2006) and Li et al. (2011). As before, let $\hat{\Lambda} = \Lambda(F_n)$ be the sample-level candidate matrix of a certain dimension reduction method, such as SIR, SAVE, CR, and DR, and let $\Lambda = \Lambda(F_0)$ be the true candidate matrix. Suppose that $\Lambda(F_0)$ has rank r under the null hypothesis

$$H_0^{(r)} : \dim(\mathscr{S}_{Y|X}) = r.$$

As shown in Chapter 10, many SDR methods have the asymptotic property

$$\sqrt{n}(\hat{\Lambda} - \Lambda) \xrightarrow{\mathscr{D}} N(0,V).$$

for some covariance matrix V. By perturbation theory (Kato, 1980), this implies

$$\hat{\lambda}_i - \lambda_i = O_P(n^{-1/2}),$$

where $\hat{\lambda}_i$ is the ith eigenvalue of $\hat{\Lambda}$, and λ_i is the ith eigenvalue of Λ. The criterion proposed by Zhu et al. (2006) takes the following form

$$L_n(k) = \sum_{i=r+1}^{p} [\log(\hat{\lambda}_i + 1) - \hat{\lambda}_i] - C_n k(2p - k + 1), \quad k = 0, \ldots, p, \tag{10.1}$$

where C_n is a positive sequence that converges to 0 at a certain rate. The rank estimator $\hat{r}$ is defined to be the maximizer of the above criterion over $k \in \{0, \ldots, p\}$.

In the same spirit, Li et al. (2011) proposed the following criterion

$$G_n(k) = \sum_{i=0}^{k} \hat{\lambda}_i + c_1(n)c_2(k), \quad k = 0, \ldots, k, \tag{10.2}$$

where $\hat{\lambda}_0$ is defined to be 0, $c_1(n)$ is a positive sequence that converges to 0 at a certain rate, and $c_2(k)$ is a positive, increasing function of k. The rank estimator is again defined as the maximizer of this function.

In the following we refer to the criterion (10.1) as ZMP, and the criterion in (10.2) as LAL, which are the initials of the authors' names of the two papers. We now give the general definition BIC-type criterion that covers both cases.

Definition 10.1 Let $\rho_k(\lambda_1, \ldots, \lambda_p)$, $k = 0, \ldots, p$, be differentiable functions of $\lambda_1, \ldots, \lambda_p$ satisfying the following conditions

$$\rho_0(\lambda_1, \ldots, \lambda_p) < \cdots < \rho_r(\lambda_1, \ldots, \lambda_p)$$
$$= \rho_{r+1}(\lambda_1, \ldots, \lambda_p) = \cdots = \rho_p(\lambda_1, \ldots, \lambda_p). \tag{10.3}$$

When $r = 0$, this is to be understood as a string of p equalities. Let $c_1(n)$ be a sequence of positive (possibly random) numbers, and $c_2(k)$ be an increasing function of k with $c_2(0) = 0$. Then the following criterion

$$B_n(k) = \rho_k(\lambda_1, \ldots, \lambda_p) + c_1(n)c_2(k)$$

is called the BIC-type criterion for order determination for SDR. Correspondingly, the BIC-type order estimator for SDR is

$$\hat{r} = \operatorname{argmax}\{B_n(k) : k = 0, \ldots, p\}.$$

Note that in this definition we allow $c_1(n)$ to be random variables. This is to accommodate the situations where this constant is data dependent. It is obvious that

$$\rho_k(\lambda_1, \ldots, \lambda_p) = \begin{cases} 0 & \text{if } k = 0 \\ \sum_{i=1}^{k} \lambda_i & \text{if } k = 1, \ldots, p. \end{cases} \tag{10.4}$$

satisfies the conditions in Definition 10.1, as long as all the non-zero eigenvalues $\lambda_1, \ldots, \lambda_r$ are distinct. The next proposition shows that the first term in (10.1) also satisfies these conditions.

Proposition 10.1 *If $\lambda_1 > \cdots > \lambda_r = \lambda_{r+1} = \cdots \lambda_p$, then the functions*

$$\rho_k(\lambda_1,\ldots,\lambda_p) = \begin{cases} \sum_{i=k+1}^{p}[\log(1+\lambda_i) - \lambda_i], & \text{if } k = 0,\ldots,p-1 \\ 0, & \text{if } k = p \end{cases} \quad (10.5)$$

satisfy the condition in Definition 10.1.

PROOF. Let $f(t) = -\log(1+t) + t$. Then $f'(t) = t/(1+t)$, which is positive for all $t \geq 0$. This implies f is monotone increasing in t over $t \geq 0$. In particular, $f(t) > 0$ whenever $t > 0$. Now for any $k = 1,\ldots,r$,

$$\rho_k(\lambda_1,\ldots,\lambda_p) - \rho_{k-1}(\lambda_1,\ldots,\lambda_p) = -\log(1+\lambda_k) + \lambda_k = f(\lambda_k) > 0.$$

For any $k = r,\ldots,p$, we have

$$\rho_k(\lambda_1,\ldots,\lambda_p) = \sum_{i=k+1}^{p} \log(1) = 0.$$

Thus (10.3) is satisfied. □

The intuition behind the BIC-type criterion is as follows. The first term of $B_n(k)$, $\rho_k(\hat\lambda_1,\ldots,\hat\lambda_k)$, increases with k, and the second term $-c_1(n)c_2(k)$ decreases with k. So, for an appropriately chosen penalty coefficient $c_1(n)$, the function peaks at a certain k. Note that before k reaches r, the first term converges to an increasingly larger positive constant. So, as long as we choose $c_1(n)$ to converge to 0, the first term dominates the second term and so the function increases in k. After k reaches r, however, the increasing rate of the first term is expected to slow down substantially, because the $\hat\lambda_i$'s after this point is of the order $O_P(n^{-1/2})$. If we choose $c_1(n)$ to converge to 0 slowly so that the second term dominates the increment of ρ caused by these $\hat\lambda_i$'s, then $B_n(k)$ tends to peak at the rank r of Λ. We now prove the consistency of the BIC-type order estimator, which is essentially writing the above intuition in a mathematical form. In the following, we say that a random sequence U_n converges in probability to ∞ if, for any $K > 0$,

$$\limsup_{n\to\infty} P(U_n > K) = 1.$$

In this case, we write $U_n \overset{P}{\to} \infty$.

Theorem 10.1 *Suppose*

1. $\rho_k(\lambda_1,\ldots,\lambda_p)$ *satisfies the condition in Definition 10.1;*

2. *for any $k = 0,\ldots,p$,*

$$\rho_k(\hat\lambda_1,\ldots,\hat\lambda_p) - \rho_k(\lambda_1,\ldots,\lambda_p) \overset{P}{\to} 0;$$

3. *for any $k = r+1,\ldots,p$,*

$$\rho_k(\hat\lambda_1,\ldots,\hat\lambda_p) - \rho_{k-1}(\hat\lambda_1,\ldots,\hat\lambda_p) = O_P(a_n),$$

where a_n is a positive sequence that converges to 0 as $n \to \infty$;

4. $c_1(n) \overset{P}{\to} 0$, $a_n^{-1}c_1(n) \overset{P}{\to} \infty$, $P(c_1(n) > 1) = 0$.

Then $P(\hat r = r) \to 1$.

PROOF. By Definition 10.1,

$$\rho_k(\lambda_1,\ldots,\lambda_p) - \rho_r(\lambda_1,\ldots,\lambda_p) \begin{cases} < 0 & k < r \\ = 0 & k > r \end{cases}$$

If $k < r$, then by condition 2,

$$\begin{aligned} B_n(k) - B_n(r) &= \rho_k(\hat{\lambda}_1,\ldots,\hat{\lambda}_p) - \rho_r(\hat{\lambda}_1,\ldots,\hat{\lambda}_p) - c_1(n)[c_2(k) - c_2(r)] \\ &= \rho_k(\lambda_1,\ldots,\lambda_p) - \rho_r(\lambda_1,\ldots,\lambda_p) - c_1(n)[c_2(k) - c_2(r)] + o_P(1). \end{aligned}$$

Also, since $c_n(n) \xrightarrow{P} 0$, we have $c_1(n)[c_2(k) - c_2(r)] \xrightarrow{P} 0$. Hence, in this case,

$$P[B_n(k) < B_n(r)] \to 1.$$

If $k > r$, then by condition 3,

$$\rho_k(\hat{\lambda}_1,\ldots,\hat{\lambda}_p) - \rho_r(\hat{\lambda}_1,\ldots,\hat{\lambda}_p) = O_p(a_n).$$

Then

$$a_n^{-1}[B_n(k) - B_n(r)] = -a_n^{-1}c_1(n)[c_2(k) - c_2(r)] + O_P(1).$$

By condition 4, $c_2(k)$ increases in k. Hence $c_2(k) - c_2(r) > 0$, which, together with $a_n^{-1}c_1(n) \xrightarrow{P} \infty$, implies that, for any $K > 0$,

$$P\left(-n^{1/2}c_1(n)[c_2(k) - c_2(r)] < -K \right) \to 1.$$

Therefore $P(a_n^{-1}[B_n(k) - B_n(r)] < 0) \to 0$, which implies

$$P(B_n(k) < B_n(r) < 0) \to 1.$$

Finally,

$$\begin{aligned} P(\hat{r} = r) &= P\left(\mathrm{argmax}\{B_n(k) : k = 1,\ldots,p\} = r \right) \\ &= P\left(\bigcap_{k=1}^{r-1}\{B_n(k) < B_n(r)\} \right) + P\left(\bigcap_{k=r+1}^{p}\{B_n(k) < B_n(r)\} \right) \to 1, \end{aligned}$$

which completes the proof.   □

In the special case where $B_n(k)$ is the criterion proposed by Zhu et al. (2006), condition 1 holds because $\hat{\lambda}_i = \lambda_i + O_p(n^{-1/2})$ holds for any $i = 1,\ldots,p$. To check the second condition, note that

$$\rho_k(\hat{\lambda}_1,\ldots,\hat{\lambda}_p) - \rho_{k-1}(\hat{\lambda}_1,\ldots,\hat{\lambda}_p) = \log(\hat{\lambda}_k + 1) - \hat{\lambda}_k.$$

Since the right-hand side has its first derivative equal to 0 at $\lambda_i = 0$, and is twice differentiable, it is of the order $O_p(n^{-1})$. Thus Condition 2 is satisfied for $a_n = n^{-1}$. Regarding the choices of $c_1(n)$ and $c_2(k)$, Zhu et al. (2006) proposed the following sequences

$$c_0[0.5\log(n) + 0.1n^{1/3}]/(2n),$$

where c_0 is the reciprocal of the number of observations in each slice. Since Zhu et al. (2006)

is concerned with the situation where p and the number of slices h increase with n, and we are focused on the setting where both p and h are fixed, we need to make some adjustment to this $c_1(n)$. In particular, we can no longer use c_0, because when h is fixed c_0 is in fact of the order n^{-1}, which brings the term $c_1(n)$ down to $O(n^{-2})$, and the condition $nc_1(n) \to \infty$ is not satisfied. Thus, we replace c_0 by $\hat{\lambda}_1/3$. That is, we propose the following $c_1(n)$:

$$c_1(n) = (\hat{\lambda}_1/3)[\log(n) + n^{1/2}]/(2n).$$

Note that here, $c_1(n)$ is random, but it satisfies the requirement $c_1(n) \to 0$ and $nc_1(n) \xrightarrow{P} \infty$. The choice of $c_2(k)$ in Zhu et al. (2006) is

$$c_2(k) = k(2p - k + 1),$$

which we adopt. To see that this is an increasing function of k in the range $k = 0, \ldots, p$, consider the function $c_2(t)$ defined on $[0, p]$, and differentiate with respect to t:

$$dc_2(t)/dt = 2p - 2t + 1.$$

This derivative is positive for $t \in [0, p]$.

For the criterion proposed by Li et al. (2011), condition 1 holds for the same reason. In this case, condition 2 holds with $a_n = n^{-1/2}$, because, for $k > r$,

$$\rho_k(\hat{\lambda}_1, \ldots, \hat{\lambda}_p) - \rho_{k-1}(\hat{\lambda}_1, \ldots, \hat{\lambda}_p) = \hat{\lambda}_k = O_P(n^{-1/2}).$$

For the choices of $c_1(n)$ and $c_2(k)$, Li et al. (2011) proposed the following sequences

$$c_0 \hat{\lambda}_1 n^{-1/2} \log(n), \quad k,$$

where c_0 was taken to be $1/3$ or $1/5$. In the examples and in the data analysis in this chapter, we have used $c_0 = 2$. This choice satisfies $c_1(n) \to 0$ and $n^{1/2} c_1(n) \to \infty$.

We now summarize the procedures for BIC, coupled with a dimension reduction estimator, as the following algorithm.

Algorithm 10.1 BIC-type order determination procedures

1. Compute the candidate matrix using the standardized X; for example, if SIR is used, then the candidate matrix is

$$\hat{\Lambda} = \hat{M}\hat{M}^{\mathsf{T}}, \quad \hat{M} = \left(\hat{p}_1^{-1/2}\hat{\Sigma}^{-1/2}(\hat{\mu}_1 - \hat{\mu}), \ldots, \hat{p}_h^{-1/2}\hat{\Sigma}^{-1/2}(\hat{\mu}_h - \hat{\mu}) \right).$$

2. Let $\hat{\lambda}_1, \ldots, \hat{\lambda}_p$ be the eigenvalues of $\hat{\Lambda}$, and let $\rho_k(\hat{\lambda}_1, \ldots, \hat{\lambda}_p)$ be either the criterion (10.4) or the criterion (10.5). Then estimate the rank of $\hat{\Lambda}$ by maximizing this criterion over $k \in \{0, \ldots, p\}$.

In the following R-code, the inputs x, y, h, and ytype have the same meaning as those in the previous R-codes. The input criterion refers to the penalty criteria, which have two choices: "zmp", the ZMP criterion, and "lal", the LAL criterion. The input method refers to the name of the underlying sufficient dimension reduction methods, which have three options: "sir" for SIR, "save" for SAVE, and "dr" for DR. Readers can add to the R-codes other sufficient dimension reduction methods according to their needs.

R-code to implement sequential test for SIR

```
bic=function(x,y,h,ytype,criterion,method){
maximizer=function(x,y) return(x[order(y)[length(y)]])
r=2;n=dim(x)[1];p=dim(x)[2]
if(method=="sir") candmat=sir(x,y,h,r,ytype)$sirmat
if(method=="save") candmat=save(x,y,h,r,ytype)$savemat
if(method=="dr") candmat=dr(x,y,h,r,ytype)$drmat
out=eigen(candmat);lam=out$values
if(criterion=="lal"){
gn=numeric();for(k in 0:p){
if(k==0) gn=c(gn,0) else
gn=c(gn,sum(lam[1:k])-(2*lam[1])*n^(-1/2)*(log(n))^(1/2)*k)}}
if(criterion=="zmp"){
gn=numeric();for(k in 0:(p-1)){
c1=(lam[1]/3)*(0.5* log(n)+0.1* n^(1/3))/(2*n)
c2=k*(2*p-k+1)
gn=c(gn,sum(log(lam[(k+1):p]+1)-lam[(k+1):p])-c1*c2)}
gn=c(gn,-c1*p*(2*p-p+1))}
return(list(rhat=maximizer(0:p,gn),rcurve=gn))
}
```

Example 10.1 We applied the two BIC-type criteria, ZMP and LAL, in conjunction with three methods, SIR, SAVE, and DR, to model (9.18) in Example 9.1. We chose $h = 8, 3, 5$ for SIR, SAVE, and DR, respectively. We took the sample size n to be 500, 1000, 1500, and 2000. For each of the two criteria, three methods, and four sample sizes, we simulated the data 100 times, and report the percentage of correct estimation of the true dimension ($r = 2$) in Table 10.1.

Table 10.1 *Percentage (with symbol % omitted) of correct estimation by BIC*

		ZMP				LAL		
n	500	1000	1500	2000	500	1000	1500	2000
SIR	8	23	64	90	38	83	96	100
SAVE	57	51	76	87	22	77	97	99
DR	75	100	100	100	73	95	100	100

As predicted by the asymptotic theory, the percentage of correct estimation for each method and each criterion converge to 100 percent, with DR performing the best under different BIC-type criteria.   □

A disadvantage of the BIC-type method is that the proportional constant in $c_i(n)$ tends to vary from model to model, and it is hard to find universally effective proportional constant for different models. Indeed, the original $c_1(n)$ for LAL proposed in Li et al. (2011) performed rather poorly in the above example. On the plus side, the BIC-type criteria are very easy to use, because it does not rely on asymptotic expansions which, as we saw from Section 9.9, for example, can be quite complicated for a particular SDR estimator. These disadvantages are overcome by the sequential methods and the two subsequent methods to various degrees.

10.2 Bootstrapped Eigenvector Variation

Both the sequential test in Chapter 9 and BIC methods in Section 10.1 rely on the magnitude of the eigenvalues $\hat{\lambda}_i$. A rather different idea for order determination was proposed by Ye and Weiss (2003), which is based on the variation of eigenvectors rather than the magnitude of eigenvalues. The intuition is as follows. Suppose for the time being that the candidate matrix Λ has distinct nonzero eigenvalues. That is,

$$\lambda_1 > \lambda_2 > \cdots > \lambda_r > \lambda_{r+1} = \cdots = \lambda_p = 0.$$

The eigenvectors $v_{r+1}, \ldots, v_p$ corresponding to the last $p - r$ zero eigenvalues are arbitrary as long as they span the same subspace of $\mathbb{R}^p$. Thus, if $\hat{\Lambda}$ is an estimator of Λ that is, say, $\sqrt{n}$ consistent, and $\hat{v}_1, \ldots, \hat{v}_r, \hat{v}_{r+1}, \ldots, \hat{v}_p$ are eigenvectors of $\hat{\Lambda}$, then $\hat{v}_{r+1}, \ldots, \hat{v}_p$ should vary greatly from sample to sample, as they do not have fixed targets to converge to; whereas $\hat{v}_1, \ldots, \hat{v}_r$ should have substantially small variations as they converge to $v_1, \ldots, v_r$ at the $\sqrt{n}$-rate.

This observation motivated Ye-Weiss's method, which is based on the bootstrapped variation of $\hat{v}_{r+1}, \ldots, \hat{v}_p$. Specifically, suppose $(X_1, Y_1), \ldots, (X_n, Y_n)$ are an i.i.d. sample from (X, Y). Then we take repeated bootstrap samples

$$(X_1^*, Y_1^*), \ldots, (X_n^*, Y_n^*)$$

and use these samples to estimate the candidate matrix Λ. For each fixed $k = 1, 2, \ldots, p - 1$, if $k \le r$, the subspace spanned by $\{v_1^*, \ldots, v_k^*\}$, say $\mathscr{S}_k^*$, would converge to $\mathscr{S}_k = \text{span}\{v_1, \ldots, v_k\}$ at a $\sqrt{n}$-rate. So, it should have a small bootstrap variation. If $k > r$, however, the eigenvalues $\lambda_{r+1}, \ldots, \lambda_p$ are all 0, and the subspace spanned by $v_{r+1}^*, \ldots, v_k^*$, which contains v_k^* that has no fixed target to converge to, should have a large variation.

To proceed further we must have a measure of variation of a subspace — or more specifically, a measure of the difference between two subspaces of $\mathbb{R}^p$. There are many possibilities for such a measure. Ye and Weiss (2003) proposed the following one. Let B_1 be a basis matrix of $\mathscr{S}_1$ and B_2 be a basis matrix $\mathscr{S}_2$, and assume both subspaces are of dimension k. Assume the columns of B_1 and B_2 are orthonormal sets, and consider

$$d(\mathscr{S}_1, \mathscr{S}_2) = \{1 - |\det(B_1^\mathsf{T} B_2)|\}.$$

If $\mathscr{S}_1 = \mathscr{S}_2$, then $B_1^\mathsf{T} B_2 = I_k$ and $d(\mathscr{S}_1, \mathscr{S}_2) = 0$. If $\mathscr{S}_1 \perp \mathscr{S}_2$ then $B_1^\mathsf{T} B_2 = 0$, and consequently $d(\mathscr{S}_1, \mathscr{S}_2) = 1$. In general, $\det(B_1^\mathsf{T} B_2)$, which can be viewed as a generalization of the absolute value of cosine of the angle between two vectors, measures the discrepancy between the two subspaces.

We now generate m bootstrap samples, and compute $B_1^*(k), \ldots, B_h^*(k)$. Let $\hat{B}(k)$ be the estimate on the full sample. We define

$$C(k) = \frac{1}{m} \sum_{\ell=1}^{m} \{1 - |\det(\hat{B}(k)^\mathsf{T} B_\ell^*(k))|\},$$

which measures the average variability of $B_\ell^*(k)$ about $\hat{B}(k)$. We call this criterion the bootstrapped eigenvector variability (BEV). The criterion should be small at the beginning, and should have a jump at $k = r + 1$. It then seems reasonable to estimate r as k if $C(k)$ has a significant jump at $k + 1$. We refer to this method as the Ye-Weiss method for order determination.

When applying the Ye-Weiss method, one should pay attention to two important points. The first is that as k approaches p, the boostrapped variation of the eigenvectors tend to become smaller again. This is because, when $k = p$, $\hat{v}_1, \ldots, \hat{v}_p$ always span the whole space $\mathbb{R}^p$, leaving no room for variation. So we should stop at a reasonably large k, so that we are confident the range $\{1,\dots,k\}$ covers the true rank r. The second is that, depending on the behavior of the nonzero true eigenvalues $\lambda_1,\dots,\lambda_r$, the pattern of eigenvector variation can be different. In the above we only discussed the case where the nonzero eigenvalues are all distinct. However, if some or all of the nonzero eigenvalues are the same, then the variation pattern is more complicated. For example, if the true eigenvalues are

$$\lambda_1 = \cdots = \lambda_r > \lambda_{r+1} = \cdots = \lambda_p = 0,$$

then, before k reaches r, the space $\text{span}\{\hat{v}_1,\dots,\hat{v}_k\}$ also does not have a definite target to converge to, and therefore would tend to display larger variation. As a result, the eigenvector variation tends to be minimized at the true rank r. In the same vain, if the downward trend of $\lambda_1,\dots,\lambda_r$ contains several flat spots — for example, if the eigenvalues are

$$3,3,3,2,2,2,1,1,0,\dots,0,$$

then the eigenvector variation tends to display the following pattern,

$$\text{large, large, small, large, large, small, large, small, large, large, }\dots$$

As a result, it is not always clear whether one should use the minimizer or the sudden jumping point of the BEV curve to estimate the order. Nevertheless, the idea of using the pattern of BEV for order determination is a very good one, and it leads to the development of the ladle estimator described in the next section, which remedies this drawback. The Ye-Weiss method for order determination is implemented by the following algorithm.

Algorithm 10.2 Ye-Weiss method for order determination

---

1. Generate m bootstrap samples of size n, using the discrete uniform distribution on $X_1,\dots,X_n$. Denote these samples by

$$\{X_{ib}^* : i = 1,\dots,n, b = 1,\dots,m\}.$$

2. For each bootstrap sample $X_{1:n,b} = \{X_{ib} : i = 1,\dots,n\}$, compute the candidate matrix (in Z-scale) using one of the SDR methods introduced in the earlier chapters. Compute the eigenvectors $v_{1b}^*,\dots,v_{pb}^*$, and form the matrix $B_b^*(k) = (v_{1b}^*,\dots,v_{kb}^*)$. Compute the matrix B_k for the full sample.
3. Evaluate the criterion

$$C(k) = m^{-1} \sum_{b=1}^{m} [1 - \det(B_k^\mathsf{T} B_{kb}^*)]$$

and then plot $C(k)$ against k.

In the following R-code, the input variables x, y, h, ytype, and method have the same meanings and options as those in the R-code for the BIC-type criteria. The input variable kmat means the maximum k in $C(k)$, and nboot refers to the bootstrap sample size.

R-code for Ye-Weiss method

```
yw=function(x,y,h,kmax,nboot,ytype,method){
r=2;n=dim(x)[1];p=dim(x)[2]
candmat=function(x,y,h,r,ytype,method){
if(method=="sir") mat=sir(x,y,h,r,ytype)$sirmat
if(method=="save") mat=save(x,y,h,r,ytype)$savemat
if(method=="dr") mat=dr(x,y,h,r,ytype)$drmat
return(mat)}
out=candmat(x,y,h,r,ytype,method)
eval.full=eigen(out)$values;evec.full=eigen(out)$vectors
prepare=function(kmax,evec1,evec2){
out=numeric();for(k in 1:kmax){
if(k==1) out=c(out,1-abs(t(evec1[,1])%*%evec2[,1]))
if(k!=1) out=c(out,1-abs(det(t(evec1[,1:k])%*%evec2[,1:k])))}
return(out)}
yw.out=0;for(iboot in 1:nboot){
bootindex=round(runif(n,min=-0.5,max=n+0.5))
xs=x[bootindex,];ys=y[bootindex]
mat=candmat(xs,ys,h,r,ytype,method);evec=eigen(mat)$vectors
yw.out=yw.out+prepare(kmax,evec.full,evec)/nboot}
return(yw.out)
}
```

Example 10.2 We applied Ye-Weiss method together with SIR to the model in Example 9.1, with three sample sizes, $n = 200, 300, 400$. We chose 8 slices for SIR. For each sample size n we simulated the data 50 times, and for each simulated sample we computed the bootstrap eigenvector variation, resulting in 50 bootstrap values. The bootstrap sample size (m in Algorithm 10.2, step 3) is taken to be 200. These values are presented as boxplots shown in Figure 10.1.

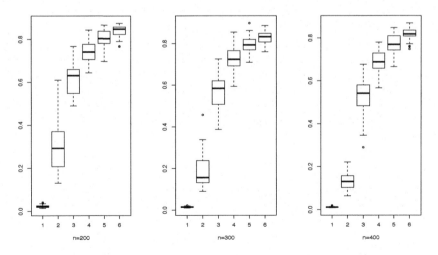

Figure 10.1 *Bootstrapped eigenvector variation based on SIR for $k = 1, \ldots, 6$, as applied to the model in Example 9.1.*

From the boxplots we see that, as the sample size n is increased from 200 to 400, the bootstrap eigenvector variation displays an increasingly pronounced jump at $k = 3$, indicating that the true rank is 2. This pattern is due to the fact that the true nonzero eigenvalues for the population-level candidate matrix Λ are all distinct. In fact, by Monte Carlo, the two nonzero eigenvalues are 0.618 and 0.111 (the Monte Carlo sample size is 50,000, sufficiently large to make all the eigenvalues beyond the second to be smaller than $1/1000$). □

10.3 Eigenvalue Magnitude and Eigenvector Variation

From the discussion of the last section, we see that the pattern of bootstrapped eigenvector variation depends on the pattern of flat regions in the scree plot, which can cause ambiguity in the region where the eigenvalues are positive. This motivated Luo and Li (2016) to propose a new method for order determination called the "ladle estimator", which exploits the patterns of both the eigenvalue magnitude and the bootstrapped eigenvector variation. That is, when the eigenvalues are close together, their eigenvectors tend to vary greatly; when the eigenvalues are far apart, their variability tends to be small. The combination of both helps to pinpoint the rank of a matrix more precisely than using the eigenvalue or the eigenvector variation alone. In this section we take a closer look at this intrinsic relation between eigenvalue magnitude and bootstrapped eigenvector variation, thus preparing the way for the introduction of the ladle estimator in the next section. To see this pattern more clearly, let us consider the two scenarios:

(i) $\lambda_k > \lambda_{k+1}$. In this case v_k and v_{k+1} resides in different subspaces. Then all the bootstrap estimates of v_k are close to v_k (up to a sign) because of the convergence of $\hat{\Lambda}$ to Λ. In particular, this applies to $k = r$, because $\lambda_r > 0$ and $\lambda_{r+1} = 0$;

(ii) $\lambda_k = \lambda_{k+1}$. In this case, all we know is that v_k and v_{k+1} reside in the same eigenspace, but these two vectors themselves are not identifiable. As a result, the bootstrap estimates of v_k and those of v_{k+1} do not have definite target to converge to, though all are close to the common eigenspace. This lack of definite targets causes large variation in the directions of these estimates.

Because $\lambda_{r+1} = \cdots = \lambda_r = 0$, scenario (ii) arises when $k = r+1, \ldots, p$. For $k = 1, \ldots, r$, however, either scenario may occur, as we demonstrated in the last section. This change in behavior in the bootstrapped variation of an eigenvector in accordance with its eigenvalue and the next eigenvalue can be characterized mathematically, as was done in Luo and Li (2016). Here we present the result without proof. We first introduce some regularity assumptions needed for this result.

Assumption 10.1 *There is a random matrix $H(X,Y)$ with mean 0 and finite second moment such that*

$$\hat{\Lambda} = \Lambda + E_n H(X,Y) + o_P(n^{-1/2}). \tag{10.6}$$

Assumption 10.2 *The bootstrap estimator M^* satisfies*

$$n^{1/2}[\text{vech}(\Lambda^*) - \text{vech}(\hat{\Lambda})] \overset{\mathscr{D}}{\to} N\left(0, \text{var}_F[\text{vech}\{H(X,Y)\}]\right) \quad a.s. \quad P \tag{10.7}$$

where $\text{vech}(\cdot)$ is the vectorization of the upper triangular part of a matrix and $\text{var}_F[\text{vech}\{H(X,Y)\}]$ is positive definite.

The convergence in the above assumption is called "convergence in distribution almost surely." In the bootstrap context there are two layers of randomness. Let $S =$

$\{(X_1, Y_1), (X_2, Y_2), \ldots\}$ be a sequence of independent copies of (X, Y). Let F be the distribution of (X, Y), and let F_n be the empirical distribution based on $(X_1, Y_1), \ldots, (X_n, Y_n)$. Conditioning on the sequence S, let $(X_{1,n}^*, Y_{1,n}^*), \ldots, (X_{n,n}^*, Y_{n,n}^*)$ be an independent and identically distributed bootstrap sample from F_n, and let F_n^* be the empirical distribution based on the bootstrap sample. Both F_n and F_n^* are random measures, but once S is given, F_n becomes a fixed measure while F_n^* remains random. Let T be a statistical functional. Then we say that $W_n = \sqrt{n}(T(F_n^*) - T(F_n))$ converges in distribution almost surely to a distribution G if $P(\mathfrak{A}) = 1$, where $\mathfrak{A}$ is the collection of all sequences s such that the distribution of W_n converges to G conditioning on $S = s$. By the development in Chapter 9, Assumption 10.1 implies that

$$n^{1/2}[\mathrm{vech}(\hat{\Lambda}) - \mathrm{vech}(\Lambda)] \xrightarrow{\mathscr{D}} N(0, \mathrm{var}_F[\mathrm{vech}\{H(X, Y)\}]). \tag{10.8}$$

Thus, comparing (10.7) and (10.8) we see that Assumption 10.2 simply says that the sequences of random elements

$$n^{1/2}[\mathrm{vech}(\hat{\Lambda}) - \mathrm{vech}(\Lambda)] \quad \text{and} \quad n^{1/2}[\mathrm{vech}(\Lambda^*) - \mathrm{vech}(\hat{\Lambda})]$$

have the same limit (albeit in different convergence mode). This is a kind of self-similarity because, if one think of F as the parent, F_n as the child, and F_n^* as the grandchild, then the relation between the child and the grandchild is similar to that between the parent and the child.

The validity of Assumption 10.1 is already discussed in Chapter 9. The validity of Assumption 10.2 is discussed, for example, in Bickel and Freedman (1981), Parr (1985), Liu et al. (1989) and Gill (1989). The next assumption is quite mild: it simply says that if a random sequence W_n is of the order $O_P(a_n)$, then the nonrandom sequence $E(W_n)$ is of the order $O(a_n)$. This kind of condition can often be proved by the Dominated Convergence Theorem together with the Skorohod representation (see, for example, Billingsley (1995)).

Assumption 10.3 *For any sequence of nonnegative random variables $\{Z_n : n = 1, 2, \ldots\}$ involved hereafter, if $Z_n = O_P(c_n)$ for some sequence $\{c_n : n \in \mathbb{N}\}$ with $c_n > 0$, then $E(c_n^{-1} Z_n)$ exists for each n and $E(c_n^{-1} Z_n) = O(1)$.*

To state our result we need the following notation. We say that a sequence of random variables Z_n is of the order $O_P^+(1)$ if, for any sequence $\varepsilon_n \to 0$, $\varepsilon_n > 0$, we have $P(Z_n > \varepsilon_n) \to 1$. If $c_n > 0$ is a sequence and $Z_n/c_n = O_P^+(1)$, then we write $Z_n = O_P^+(c_n)$. Roughly, this concept is the asymptotic analogue of a random variable Z taking positive values with probability 1. With these, we are ready to state the result that characterizes the mentioned eigenvalue-eigenvector pattern.

Theorem 10.2 *Let $c_n = [\log\{\log(n)\}]^{-2}$. If Assumptions 10.1, 10.2, and 10.3 hold, and $M \in \mathbb{R}^{p \times p}$ is a positive semi-definite matrix of rank $d \in \{0, \ldots, p-1\}$, then for any $k = 1, \ldots, p-1$, the following relation holds for almost every sequence $S = \{(X_n, Y_n) : n = 1, 2, \ldots\}$:*

$$f_n(k) = \begin{cases} O_P(n^{-1}), & \lambda_k > \lambda_{k+1}, \\ O_P^+(c_n), & \lambda_k = \lambda_{k+1}, \end{cases}$$

where the probability in O_P and O_P^+ is the conditional probability given S.

Although $\{c_n\}$ converges to zero, the convergence rate is very slow. For example, $\{\log(\log 10^4)\}^{-2} \approx 0.2$. Hence, in practice $O_P^+(c_n)$ can be nearly treated as $O_P^+(1)$. Thus the above theorem says that if $\lambda_k > \lambda_{k+1}$, then the bootstrapped eigenvector variability at k is of the order $O_P(n^{-1})$; if $\lambda_k = \lambda_{k+1}$, then this variability is of the order almost as large as $O_P(1)$. This asymptotic behavior is the basis of our new order determination method.

10.4   Ladle Estimator

As in Section 10.2, based on m bootstrap samples of the full sample, define the bootstrapped eigenvector variation function as

$$f_n^0(k) = \begin{cases} 0, & k = 0, \\ m^{-1}\sum_{b=1}^{m}\{1 - |\det(\hat{B}_k^T B_{k,b}^*)|\}, & k = 1,\dots,p-1. \end{cases} \tag{10.9}$$

Here, we have augmented the domain of $C(k)$ to include 0 so as to cover the possibility that the rank of the true candidate matrix is 0, as in the case where $Y \perp\!\!\!\perp X$. We renormalize f_n^0 to be

$$f_n : \{0,\dots,p-1\} \to \mathbb{R}, \quad f_n(k) = f_n^0(k) / \{1 + \sum_{i=0}^{p-1} f_n^0(i)\}, \tag{10.10}$$

where the constant 1 in the denominator is introduced for the following reason: when $d = p - 1$, and if all the nonzero eigenvalues of Λ are distinct, then all $f_n^0(i)$ are small, which makes the criterion unstable. Adding 1 to the denominator helps to stabilize the criterion.

Next, we define a criterion that reflects the variation pattern of the eigenvalues. Parallel to f_n, we renormalize the sample eigenvalues and define the function

$$\phi_n : \{0,\dots,p-1\} \to \mathbb{R}, \quad \phi_n(k) = \hat{\lambda}_{k+1}/(1 + \sum_{i=0}^{p-1}\hat{\lambda}_{i+1}), \tag{10.11}$$

where, similar to the criterion f_n, the constant 1 in the denominator is introduced to stabilize the performance of the criterion when $d = 0$. Note that we also shifted the eigenvalues so that ϕ_n takes a small value at $k = r$ instead of at $k = r+1$. We define the objective function of our estimator as

$$g_n : \{0,\dots,p-1\} \to \mathbb{R}, \quad g_n(k) = f_n(k) + \phi_n(k), \tag{10.12}$$

which utilizes information from both the eigenvectors and the eigenvalues.

Intuitively, when $k < r$, all the $\hat{\lambda}_{k+1}$ are significantly greater than 0 as they all converge to positive constants a the $\sqrt{n}$-rate. Meanwhile, for this range of k, although the bootstrapped eigenvector variation may not have a definite pattern, the combined function $g_n(k)$ should be significantly larger than 0 due to the eigenvalue contribution. When $k = r$, the eigenvalue $\hat{\lambda}_{k+1}$ is closed to 0, while at the same time, the bootstrapped variation at $k = r$ should be small because we know that $\lambda_r > \lambda_{r+1} = 0$, so that $\text{span}(\hat{v}_1,\dots,\hat{v}_r)$ converges in probability to $\text{span}(v_1,\dots,v_r)$. When $k > r$, because $\lambda_{r+1} = \cdots = \lambda_p$, all the bootstrapped eigenvector variations tend to be large, whereas $\hat{\lambda}_{r+1},\dots,\hat{\lambda}_p$ converge to 0. Hence, the combined function tends to be minimized at the true rank r.

From the above discussion we see that the combined function g_n achieves two goals at the same time. First, it eliminated the ambiguity of the bootstrapped eigenvector variability before k reaches r, because the eigenvalues are large and they bring up the combined function in this region. Second, it effectively employs the information contained in the portion of the bootstrapped eigenvector variation when k passes r, which is unambiguous and useful.

To further enhance the performance of this criterion we make the following modification. As mentioned in the last section, $f_n^0(p)$ is identically 0 because both $\hat{B}_p$ and $B_{p,i}^*$ in (10.9) span $\mathbb{R}^p$. This causes the function $f_n^0(k)$ to bend downwards when k gets close to p, resulting in a shape similar to the handle of a ladle shown in Figure 10.2. Moreover, as p increases, so does the denominator $1 + \sum_{i=0}^{p-1} f_n^0(i)$ of f_n, resulting in a smaller weight for the bootstrap component. Although these artifacts do not affect the asymptotic consistency of the ladle estimator, in finite samples it makes sense to define and minimize g_n for k up to a certain integer $k_{\max}(p) <$

$p-1$, rather than go all the way to $k = p - 1$. Correspondingly, the range of the sums in the denominators in (10.10) and (10.11) should be changed to $\{0, \ldots, k_{\max}(p)\}$, which leads to

$$g_n(k) = f_n(k) + \phi_n(k) \equiv \frac{f_n^0(k)}{1 + \sum_{i=0}^{k_{\max}(p)} f_n^0(i)} + \frac{\hat{\lambda}_{k+1}}{1 + \sum_{i=0}^{k_{\max}(p)} \hat{\lambda}_{i+1}}. \tag{10.13}$$

Of course, as long as one is confident that the true rank r is smaller than $k_{\max}(p)$, the smaller the $k_{\max}(p)$, the more accurate the estimator tends to be. So there is a balance between choosing a small $k_{\max}(p)$ and risking missing true value of r. As a rule of thumb, for p reasonably large, it seems reasonable to assume that $r < \lfloor p/\log(p) \rfloor$, where $\lfloor a \rfloor$ denotes the greatest integer smaller than or equal to a. For smaller p, it recommends using $k_{\max}(p) = p - 2$. (In Luo and Li (2016), it was recommended that $k_{\max}(p)$ be $\lfloor p/\log(p) \rfloor$ for $p > 10$ and $p - 1$ for $p \leq 10$.)

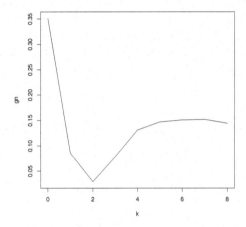

Figure 10.2 *Ladle plot for the model in Example 9.1.*

To provide more insights about this estimator, we computed $g_n(k)$ using SIR for the model in Example 9.1 with $p = 10$ and $n = 200$. The number of slices for SIR is taken to be $h = 8$, the bootstrap sample size is $n_{\text{BOOT}} = 200$. Figure 10.2 shows the plot of $g_n(k)$ for $k = 0, \ldots, 8$. We can see that the curve is roughly divided into three sections: in the first section, $\{0, 1, 2\}$, it decreases to the true rank $r = 2$; in the second section, $\{2, 3, 4, 5\}$, it increases to a certain level, and then in the third section, $\{5, 6, 7, 8\}$, it bends downwards again. The downward trend before $r = 2$ is due to the decrease of the eigenvalues; the upward trend after $r = 2$ is due to the increase of the bootstrapped eigenvector variations; the downward trend toward the right end is due to the fact that the bootstrapped eigenvector variation is 0 at p. Since this shape mimics a ladle with handle pointing to the right and it is quite common in many simulated and real data sets, we refer to the plot of g_n as the ladle plot, and the minimizer of the plot as the ladle estimator, as formally defined below.

Definition 10.2 *The ladle estimator of the rank r of a candidate matrix Λ is*

$$\hat{r} = \arg\min \{g_n(k) : k = 0, 1, \ldots, k_{\max}(p)\}, \tag{10.14}$$

where $k_{\max}(p)$ is $p - 2$ if $p \leq 10$ and $\lfloor p/\log(p) \rfloor$ if $p > 10$.

Another way to look at the ladle estimator is to view the second term in (10.13) as a data driven penalty term: instead of using an artificial penalty term, the ladle estimator uses the data-driven bootstrapped eigenvector variation as a natural counterbalance of model complexity. Moreover, the ladle estimator does not rely on any tuning parameters, and our simulation experiments indicate that its performance is quite stable for different models.

Algorithm 10.3 Ladle estimator

---

1. For a given SDR method and a sample $(X_1, Y_1), \ldots, (X_n, Y_n)$, calculate the candidate matrix $\hat{\Lambda}$ in the Z-scale. Perform spectral decomposition to obtain the eigenvalues $\hat{\lambda}_1, \ldots, \hat{\lambda}_n$ and eigenvectors $\hat{v}_1, \ldots, \hat{v}_p$. Form the matrices

$$\hat{B}_k = (\hat{v}_1, \ldots, \hat{v}_k), \quad k = 1, \ldots, k_{\max}(p),$$

where $k_{\max}(p)$ is as defined in Definition 10.2. Also, compute $f_n(k)$ for $k = 0, \ldots, k_{\max}(p)$ according to (10.13).

2. Generate n_{BOOT} bootstrap samples from the uniform distribution on $(X_1, Y_1), \ldots, (X_n, Y_n)$. For the bth sample, compute the candidate matrix Λ_b^*, and record its eigenvectors $v_{b1}^*, \ldots, v_{bp}^*$. Form the matrix

$$B_{k,b}^* = (v_{1,b}^*, \ldots, v_{k,b}^*), \quad k = 1, \ldots, k_{\max}(p).$$

Then form the matrix $D = \{\delta_{bk} : b = 1, \ldots, m, k = 1, \ldots, k_{\max}(p)\}$, where $\delta_{k,b} = 1 - |\det(\hat{B}_k^\mathsf{T} B_{k,b}^*)|$.

3. Take the column average of D to form $f_n^0(k)$ in (10.13), then form $f_n(k)$ in (10.13) for $k = 1, \ldots, k_{\max}(p)$. Set $f_n(0) = 0$.

4. Minimize $\phi_n(k) + f_n(k)$ to obtain the estimate of r.

For the choice of n_{BOOT} in step 2, we only need it to be large enough so that $f_n^0(k)$ are relatively stable. In most cases $n_{\text{BOOT}} = 200$ is sufficient. The R-code for implementing ladle estimator is given below. The input variables have the same meaning as those in the R-code for Ye-Weiss estimator. The output variables are the set $\{0, \ldots, k_{\max}\}$, the ladle function $g_n(k)$, and the ladle estimate $\hat{r}$.

R-code for ladle estimator

```
ladle=function(x,y,h,nboot,method,ytype){
r=2;n=dim(x)[1];p=dim(x)[2]
if(p<=10) kmax=p-2;if(p>10) kmax=round(p/log(p))
candmat=function(x,y,h,r,ytype,method){
if(method=="sir") mat=sir(x,y,h,r,ytype)$sirmat
if(method=="save") mat=save(x,y,h,r,ytype)$savemat
if(method=="dr") mat=dr(x,y,h,r,ytype)$drmat
return(mat)}
phin=function(kmax,eval){
den=1+sum(eval[1:(kmax+1)]);return(eval[1:(kmax+1)]/den)}
out=candmat(x,y,h,r,ytype,method)
eval.full=eigen(out)$values;evec.full=eigen(out)$vectors
pn=phin(kmax,eval.full)
```

```
prefn0=function(kmax,evec1,evec2){
out=numeric();for(k in 0:kmax){
if(k==0) out=c(out,0)
if(k==1) out=c(out,1-abs(t(evec1[,1])%*%evec2[,1]))
if(k!=0&k!=1) out=c(out,1-abs(det(t(evec1[,1:k])%*%evec2[,1:k])))}
return(out)}
fn0=0
for(iboot in 1:nboot){
bootindex=round(runif(n,min=-0.5,max=n+0.5))
xs=x[bootindex,];ys=y[bootindex]
mat=candmat(xs,ys,h,r,ytype,method)
eval=eigen(mat)$values;evec=eigen(mat)$vectors
fn0=fn0+prefn0(kmax,evec.full,evec)/nboot}
minimizer=function(a,b) return(a[order(b)][1])
fn=fn0/(1+sum(fn0));gn=pn+fn;rhat=minimizer(0:kmax,gn)
return(list(kset=(0:kmax),gn=gn,rhat=rhat))
}
```

Example 10.3 We now apply the ladle estimator with SIR, SAVE, and DR to two models, the model (9.18) in Example 9.1 and the model (5.8) in Example 5.1. We choose two models to show that ladle estimator performs competently under different models. Also, recall that model (5.8) contains a U-shaped trend that cannot be estimated by SIR. So this example also serves to demonstrate that an order determination procedure is inevitably affected by its accompanying SDR method. We used $n_{BOOT} = 200$ throughout this simulation. For SIR, we took the number of slices to be $h = 8$; for SAVE, $h = 3$; for DR, $h = 5$. For each of the 3 (methods) × 4 (sample sizes) × 2 (models) scenarios, we simulated the data 100 times and computed the percentages of the correct estimation of r, which is 2 for both models. These percentages are presented in Table 10.2. As a comparison, we also computed the corresponding percentages for BIC coupled with the ZMP criterion and the same dimension reduction methods for the same sample sizes, which are presented in the bottom half of the table.

Table 10.2 *Percentage (with symbol % omitted) of correct estimation by ladle estimator and BIC estimator coupled with the ZMP criterion*

Model		Model (9.18)				Model (5.8)			
n		200	300	400	500	200	300	400	500
	SIR	53	70	93	97	4	2	0	0
LADLE	SAVE	3	15	19	33	39	78	100	100
	DR	28	42	73	93	99	100	100	100
	SIR	5	5	4	7	0	0	0	0
BIC	SAVE	41	53	51	53	67	90	98	100
	DR	0	7	39	74	0	0	4	16

From the table we see that the ladle estimators converge to the true order faster than the BIC-type methods in most cases: the ladle estimators with SIR and DR perform significantly better than the corresponding BIC methods; the only cases where BIC performs significantly better is with SAVE and for Model (9.18). The second point to note is that the ladle estimator performs quite consistently for different models. In contrast, the BIC-type criteria display a significant difference in its performance under Model (9.18) and Model (5.8). The third point to note is that the performance of any order determination method is inevitably influenced by the performance of the underlying sufficient dimension reduction method. In particular, since Model (5.8) contains a U-shaped term, which cannot be estimated SIR, both order determination methods perform very poorly when applied with SIR.   □

The example clearly demonstrates the benefit of combining the information from both the magnitude of eigenvalues and the variation of eigenvectors for order determination.

10.5   Consistency of the Ladle Estimator

In this section we state a theorem concerning the consistency of the ladle estimator as defined in Definition 10.2. A proof can be found in Luo and Li (2016), and is omitted here. As mentioned earlier, consistency in the bootstrap context also needs to reflect the two layers of randomness. More specifically, since ϕ_n is a function of the sample $(X_1, Y_1), \ldots, (X_n, Y_n)$, it is nonrandom once the sequence $S = \{(X_k, Y_k) : k = 1, 2, \ldots\}$ is given. Bootstrap introduces another layer of randomness on top of this. Correspondingly, the minimizer $\hat{r}$ of g_n also has two layers of randomness. Reflecting this composite probability structure, the consistency in our context means that in the event that the conditional probability of $\hat{r} = r$ given S converges to 1 happens almost surely in the probability space of S.

Theorem 10.3 *Under Assumptions 10.1, 10.2, and 10.3, the ladle estimator (10.14) satisfies:*

$$P\{\lim_{n \to \infty} P(\hat{r} = r \mid S) = 1\} = 1. \tag{10.15}$$

10.6   Application: Identification of Wine Cultivars

We now compare the sequential test, the BIC criterion, and the ladle estimator combined with Directional Regression to the wine cultivar data in Forina et al. (1988). The data set can be found at UCI Machine Learning Repository at the website

        https://archive.ics.uci.edu/ml/datasets/wine

The data set consists of 178 wine samples from three cultivars. For each wine sample, the name of the cultivar and 13 covariates are recorded. The 13 covariates are

1. alcohol
2. malic acid
3. ash
4. alcalinity of ash
5. magnesium
6. total phenols
7. flavanoids
8. nonflavanoid phenols
9. oroanthocyanins
10. color intensity
11. hue
12. OD280/OD315 of diluted wines
13. proline.

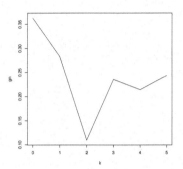

Figure 10.3 *Ladle plot for the wine data.*

Of the 178 samples, 59 belong to the first cultivar, 71 the second, and 48 the third. The dimension reduction problem here is to extract the most important sufficient predictors from the original 13 predictors that help to identify the cultivars. Our goal is to use the order determination methods to determine the number of important sufficient predictors.

For the ladle estimator, we choose the bootstrap sample size to be $n_{\mathrm{BOOT}} = 200$. The ladle plot for this data set is given in Figure 10.3, which determines the order to be 2. This is consistent sequential test, whose p-values are given in Table 10.3, which, if we use significance level $\alpha = 0.05$, also estimates the dimension d as 2.

Table 10.3 *p-values for DR-sequential test for the wine data*

r	0	1	2	3	4	5
p-value	0.000	0.001	0.000	0.542	1.000	1.000

The BIC combined with Directional Regression and the ZMP criterion estimates the dimension as 5, whereas that with the LAL criterion estimates the dimension as 2.

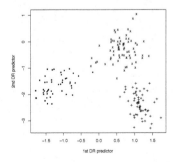 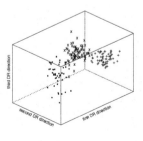

Figure 10.4 *Comparing the 2-d and 3-d plots to see the effect of the third sufficient predictor in the wind data.*

Our analysis suggests strongly that the dimension of this problem is 2. To reconfirm this, in Figure 10.4 we show the scatter plot of the first two DR predictors and the perspective plot of the first three DR predictors. We see that the 2-d plot provides clear enough separation of the three cultivars, whereas the 3-d plot does not seem to provide further discrimination of the three groups, either in terms of their locations or in terms of their scales.

Chapter 11

Forward Regressions for Dimension Reduction

11.1 Outer Product of Gradients   160
11.2 Fisher Consistency of Gradient Estimate   163
11.3 Minimum Average Variance Estimate   167
11.4 Refined MAVE and refined OPG   170
11.5 From Central Mean Subspace to Central Subspace   173
11.6 dOPG and Its Refinement   173
11.7 dMAVE and Its Refinement   178
11.8 Ensemble Estimators   180
11.9 Simulation Studies and Applications   184
11.10 Summary   188

The Sufficient Dimension Reduction methods described in Chapter 3 through Chapter 6 are what can be called inverse regression methods, as they depend on conditional moments given the response Y. This name is justified because conditional expectations are in fact regression in L_2-space in the following sense: if $f(X)$ is a function of X then $E[f(X)|Y]$ minimizes $E\{[f(X) - g(Y)]^2|Y\}$ over all square-integrable $g(Y)$. The advantage of these approaches is that they borrow the elliptical symmetry of X to avoid conditioning on a high-dimensional variable. It is easier to estimate conditional expectation given a scalar than to estimate the conditional expectation given a high-dimensional vector. In this chapter we introduce a group of methods that may be called the "forward regression", because they all rely on the conditional expectations given the predictor X. The advantage of these forward methods is that they do not require strong predictor assumptions such as elliptical symmetry; the disadvantage is that they do involve high-dimensional kernel — at least for initial estimates, and that they require the response Y to be continuous random variables. This group of estimators consist of two basic methods, the Outer Product of Gradients, the Minimum Average Variance Estimator, and their extensions into two directions. The first extension is to use an updated and dimensionally reduced kernel after an initial estimate; the second extension is to replace estimation of the gradient of conditional mean by that of conditional p.d.f., conditional c.d.f., or an ensemble of functions of the response variable. The development of this chapter traces a sequence of papers on forward regressions, including Xia et al. (2002) , Hristache et al. (2001), Xia (2007), Wang and Xia (2008), and Yin and Li (2011).

11.1   Outer Product of Gradients

The basic ingredient of the conditional expectation in forward regression, which was inspired by the Average Derivative Estimation (Hardle and Stoker (1989)) and Local Linear Regression (Fan and Gijbels (1993), Fan (1993), and Ruppert and Wand (1994)). The idea is to perform linear regression in a local region of the parameter space. Let

$$\kappa : \mathbb{R}^p \times \mathbb{R}^p \to \mathbb{R}$$

be a kernel function. We often assume it to be a positive function. For example, it can take the form

$$\kappa(x_1, x_2) = K\left(\frac{x_1 - x_2}{h}\right)$$

where $K(\cdot)$ is a probability density function, such as that of $N(0,1)$. The symbol h indicates a positive number, representing the bandwidth. It is easier to think of K as a disc of radius h centered at 0 — that is, $K(u) = I(\|u/h\| \leq 1)$ — so that one only collects data within that disc when performing the regression. At the population level, local linear regression around a point x_0 is performed by minimizing the objective function

$$L(a,b) = E[(Y - a - b^\mathsf{T}(X - x_0))^2 K((X - x_0)/h)]$$

among $a \in \mathbb{R}$, $b \in \mathbb{R}^p$.

We can use Theorem 1.1 to write down the solution explicitly.

Proposition 11.1 *If the elements of the following matrix and vector*

$$\begin{pmatrix} 1 \\ X - x_0 \end{pmatrix} \begin{pmatrix} 1 \\ X - x_0 \end{pmatrix}^\mathsf{T} K\left(\frac{X - x_0}{h}\right), \quad \begin{pmatrix} 1 \\ X - x_0 \end{pmatrix} Y K\left(\frac{X - x_0}{h}\right)$$

are integrable, then $L(a,b)$ has the unique minimizer

$$\begin{pmatrix} a_h^*(x_0) \\ b_h^*(x_0) \end{pmatrix} = \left\{ E\left[\begin{pmatrix} 1 \\ X - x_0 \end{pmatrix} \begin{pmatrix} 1 \\ X - x_0 \end{pmatrix}^\mathsf{T} K\left(\frac{X - x_0}{h}\right) \right] \right\}^{-1}$$

$$E\left[\begin{pmatrix} 1 \\ X - x_0 \end{pmatrix} Y K\left(\frac{X - x_0}{h}\right) \right].$$

PROOF. Let

$$U = Y K^{1/2}\left(\frac{X - x_0}{h}\right), \quad V = \begin{pmatrix} 1 \\ X - x_0 \end{pmatrix} K^{1/2}\left(\frac{X - x_0}{h}\right).$$

Then apply Theorem 1.1 to obtain the desired result.   □

In local linear regression, the parameter of interest is a^*, because it can be shown that $a_h^*(x_0)$ is close to $E(Y|x_0)$ when $h \to 0$. For Sufficient Dimension Reduction, however, the parameter of interest is $b_h^*(x_0)$ because, as we will show in the next section, as $h \to 0$,

$$b^* = \left.\frac{\partial E(Y|X = x)}{\partial x}\right|_{x=x_0} + O(h^2). \tag{11.1}$$

Moreover, as the following lemma shows, the derivatives on the right-hand side always belong to the central mean subspace.

Theorem 11.1 *If is the function $x \mapsto E(Y|X = x)$ is differentiable, then*

$$\partial E(Y|X = x)/\partial x$$

belongs to $\mathscr{S}_{E(Y|X)}$. Furthermore, if the support of $U = \beta^\mathsf{T} X$ is convex, then

$$\text{span}\left\{ E\left[\left(\frac{\partial E(Y|X)}{\partial X} \right) \left(\frac{\partial E(Y|X)}{\partial X^\mathsf{T}} \right) \right] \right\} = \mathscr{S}_{E(Y|X)}, \tag{11.2}$$

where $\partial E(Y|X)/\partial X$ stands for the random vector $g(X)$, g being the function $x \mapsto \partial E(Y|X = x)/\partial x$.

PROOF. Let β denote a basis matrix of $\mathscr{S}_{E(Y|X)}$. Then $E(Y|X) = E(Y|\beta^\mathsf{T} X)$. Hence

$$\frac{\partial E(Y|x)}{\partial x} = \frac{\partial E(Y|\beta^\mathsf{T} x)}{\partial x} = \frac{(\partial \beta^\mathsf{T} x)^\mathsf{T}}{\partial x} \frac{\partial E(Y|\beta^\mathsf{T} x)}{\partial \beta^\mathsf{T} x} = \beta \frac{\partial E(Y|\beta^\mathsf{T} x)}{\partial \beta^\mathsf{T} x},$$

where the right-hand side is a vector in $\text{span}(\beta) = \mathscr{S}_{E(Y|X)}$.

To prove exhaustiveness, suppose (11.2) is not true. Then there is an $\alpha \in \mathscr{S}_{E(Y|X)}$ such that

$$\alpha^\mathsf{T} g(x) = 0$$

for all x, which implies $\alpha^\mathsf{T} \beta h(u) = 0$ for all $u \in \text{supp}(U)$, where $h = E(Y|U)$. Because $\alpha \in \mathscr{S}_{E(Y|X)}$, we have $\alpha^\mathsf{T} \beta = \gamma \neq 0$. Let $u_1, u_2 \in \text{supp}(U)$ and suppose $u_2 - u_1$ is parallel to γ. Then

$$\frac{dh[(1-\varepsilon)u_0 + \varepsilon u_1]}{\partial \varepsilon} = \frac{\partial h(u_0)}{\partial u^\mathsf{T}}(u_1 - u_0) = 0$$

for all ε, which implies $h(u_1) = h(u_2)$. In other words, h does not change along the direction γ, contradicting the fact that $\mathscr{S}_{E(Y|X)}$ is the smallest dimension reduction subspace for conditional mean. $\qquad \square$

By Theorem 11.1 and approximation (11.1), it is reasonable to use the sample estimate of

$$E[b_h^*(X)(b_h^*(X))^\mathsf{T}]$$

as the candidate matrix for estimating the central mean space $\mathscr{S}_{E(Y|X)}$. This is, indeed, the first method in Xia et al. (2002), which is called the Outer Product of Gradients (OPG).

At the sample level, suppose $(X_1, Y_1), \ldots, (X_n, Y_n)$ are an i.i.d. sample of (X, Y). The OPG is constructed by mimicking the above population-level candidate matrix, by replacing the population-level moments with sample moments. The procedure is described in detail by the following algorithm. As in Section 9.4 we use X_a^i to denote the ith component of X_a, which is the observation on the ath subject from the sample.

Algorithm 11.1 Outer Product Gradient

---

1. Standardize $X_1, \ldots, X_n$ by

$$Z_a^i = (X_a^i - \bar{X}^i)/\hat{\sigma}_i.$$

$\hat{\sigma}_i = \sqrt{\text{var}_n(X^i)}$. This is to guarantee that all the predictors are on the same scale, so that it makes sense to use a spherically-shaped kernel K.

2. Choose a spherically symmetric kernel K. For example, it can be the standard Normal density $K(u) = \exp(-\|u\|^2/2)$. Since the kernel appears in both the numerator and the denominator, the proportionality constant can be ignored. Select a bandwidth h using the cross-validation method described in a later section.

3. For each $i = 1, \ldots, n$ compute

$$\hat{A}_h(Z_i) = \sum_{j=1}^{n} \left[\begin{pmatrix} 1 \\ Z_j - Z_i \end{pmatrix} \begin{pmatrix} 1 \\ Z_j - Z_i \end{pmatrix}^{\mathsf{T}} K\left(\frac{Z_j - Z_i}{h} \right) \right]$$

$$\hat{B}_h(Z_i) = \sum_{j=1}^{n} \left[\begin{pmatrix} 1 \\ Z_j - Z_i \end{pmatrix} Y_j K\left(\frac{Z_j - Z_i}{h} \right) \right].$$

Form the vector $\hat{b}_h(Z_i)$, which consists of the 2nd, $\ldots$, $(p+1)$th entries the $(p+1)$-dimensional vector $[\hat{A}_h(Z_i)]^{-1}\hat{B}_h(Z_i)$.

4. Perform spectral decomposition on $\sum_{i=1}^{n} \hat{b}_h(Z_i)\hat{b}_h(Z_i)^{\mathsf{T}}$ to obtain the first d eigenvectors $v_1, \ldots, v_d$. Let $\hat{D}$ be the diagonal matrix $\text{diag}(\hat{\sigma}_1, \ldots, \hat{\sigma}_p)$. A basis of the central mean subspace $\mathscr{S}_{E(Y|X)}$ is estimated by $\{\hat{D}^{-1/2}v_1, \ldots, \hat{D}^{-1/2}v_d\}$.

For larger sample sizes, and if the Gaussian kernel is used, it is more efficient to compute the matrix $\mathbb{K} = \{K[(Z_j - Z_i)/h]\}_{i,j=1}^{n}$ in a matrix form, rather than entrywise, as follows. Note that

$$(Z_i - Z_j)^{\mathsf{T}}(Z_i - Z_j) = Z_i^{\mathsf{T}}Z_i - Z_i^{\mathsf{T}}Z_j - Z_j^{\mathsf{T}}Z_i + Z_j^{\mathsf{T}}Z_j$$
$$= Z_i^{\mathsf{T}}Z_i - 2Z_i^{\mathsf{T}}Z_j + Z_j^{\mathsf{T}}Z_j. \tag{11.3}$$

So if we let $\mathbb{Z} = (Z_1, \ldots, Z_n)^{\mathsf{T}}$, then the matrix $\mathbb{U} = \mathbb{Z}\mathbb{Z}^{\mathsf{T}}$ contains all the information needed to construct the kernel matrix $\mathbb{K}$. Specifically, let R be the vector consists of the diagonal elements of $\mathbb{U}$, and let $\mathbb{M}$ be $n \times n$ matrix whose n rows are all R, then, by (11.3), $\mathbb{K}$ can be written as

$$\mathbb{M} + \mathbb{M}^{\mathsf{T}} - 2\mathbb{U}. \tag{11.4}$$

The following are the R-codes to compute the OPG estimator. For the function `opg`, the input variables are: `x`, the $n \times p$ predictor matrix, `y`, the n-dimensional vector of the responses, and `d`, the dimension of the central subspace. The R-codes also contain three subfunctions: `wls` is a function to compute the regression coefficients for weighted least squares: in our context the kernel matrix is the weight; `kern` is a function to compute the Gaussian kernel matrix; `standvec` is a function to standardize a sample of vectors marginally, as described in step 1 in the algorithm for OPG. For the bandwidth h, we use the one proposed in Xia (2007), which is

$$h = 2.34 n^{-1/(\max(p,3)+6)}. \tag{11.5}$$

The bandwidth can also be determined by cross-validation. However, if we use the refined OPG described in next section, then this bandwidth becomes less important.

R-code for OPG

```
opg=function(x,y,d){
p=dim(x)[2];n=dim(x)[1]
c0=2.34;p0=max(p,3);rn=n^(-1/(2*(p0+6)));h=c0*n^(-(1/(p0+6)))
sig=diag(var(x));x=apply(x,2,standvec)
kmat=kern(x,h);bmat=numeric()
for(i in 1:dim(x)[1]){
wi=kmat[,i];xi=cbind(1,t(t(x)-x[i,]))
bmat=cbind(bmat,wls(xi,y,wi)$b)}
beta=eigen(bmat%*%t(bmat))$vectors[,1:d]
return(diag(sig^(-1/2))%*%beta)
}

wls=function(x,y,w){
n=dim(x)[1];p=dim(x)[2]-1
out=c(solve(t(x*w)%*%x/n)%*%apply(x*y*w,2,mean))
return(list(a=out[1],b=out[2:(p+1)]))}

kern=function(x,h){
x=as.matrix(x);n=dim(x)[1]
k2=x%*%t(x);k1=t(matrix(diag(k2),n,n));k3=t(k1);k=k1-2*k2+k3
return(exp(-(1/(2*h^2))*(k1-2*k2+k3)))}

standvec=function(x) return((x-mean(x))/sd(x))
```

11.2   Fisher Consistency of Gradient Estimate

While in this book we will not develop a fully-fledged asymptotic theory for the OPG-type methods, we would like to prove the relation (11.1), which gives us intuitions on why this type of methods work. The type of asymptotic statement in relation (11.1) plays the role of Fisher consistency defined in Chapter 2. Note, however, the notion of Fisher consistency in the sense of Chapter 2 no longer applies to the current context in the strict sense, because our statistical functional involves a tuning parameter. The asymptotic development here roughly corresponds to the bias part of the convergence rate of OPG as developed in Xia et al. (2002). We first expand the notation of Fisher consistency to the situations where an extra tuning parameter is involved.

As in Chapter 2, let $\mathfrak{F}$ be the class of all distributions of (X, Y), and let F_n be the empirical distribution based on an i.i.d. sample of (X, Y). Let $\mathbb{H}$ be a subset of a Euclidean space, say $\mathbb{R}^k$, which indicates the family of tuning parameters. Let $h_0 \in \mathbb{H}$ represent the limit to which the tuning parameter converges. In our context, h is the bandwidth and it converges to 0. A *tuned statistical functional T* is a mapping from $\mathfrak{F} \times \mathbb{H}$ to a set $\mathbb{S}$, which, for example, can be a collection of vectors or matrices. Fisher consistency of such functionals is defined as follows.

Definition 11.1 *A tuned statistical functional* $T(F_n, h)$ *is Fisher consistent for estimating a parameter* $\theta \in \mathbb{S}$ *if*

$$\lim_{h \to h_0} T(F_0, h) = \theta,$$

where F_0 *is the true distribution.*

Our goal is to show that $b_h^*(x)$ is a Fisher consistent of the gradient $\partial E(Y|x_0)/\partial x$. In fact, we will do more than this: we will also provide the rate at which the error goes to 0. We now re-express $b_h^*(x)$ in terms of the transformed random variable $U = (X - x_0)/h$. Note that, if X has p.d.f. $f(x)$, then U has p.d.f.

$$f(x_0 + hu)h^p.$$

In this notation $(a_h^*(x), b_h^{*\mathsf{T}}(x))^{\mathsf{T}}$ can be re-expressed as

$$\binom{a_h^*(x)}{b_h^*(x)} = \left\{ E\left[\binom{1}{hU} \binom{1}{hU}^{\mathsf{T}} K(U) \right] \right\}^{-1} E\left[\binom{1}{hU} YK(U) \right]$$

$$= \left[E\begin{pmatrix} K(U) & hU^{\mathsf{T}}K(U) \\ hUK(U) & h^2 UU^{\mathsf{T}}K(U) \end{pmatrix} \right]^{-1} E\begin{pmatrix} YK(U) \\ hUYK(U) \end{pmatrix}.$$

Using the formula for block matrix inversion we can easily write down the explicit expression for $b_h^*(x)$ as follows:

$$b_h^*(x) = h^{-1} \left\{ E[UU^{\mathsf{T}}K(U)] - \frac{E[UK(U)]E[U^{\mathsf{T}}K(U)]}{E[K(U)]} \right\}^{-1}$$

$$\left\{ E[UYK(U)] - \frac{E[UK(U)]E[YK(U)]}{E[K(U)]} \right\}.$$

Next, we develop approximations to the moments involved in the above expression. To do so, we first make some assumptions.

Assumption 11.1 *The kernel function is a spherically-contoured p.d.f. with finite fourth moments.*

By "spherically-contoured" we mean that the p.d.f. is a function of the norm of its variable.

Assumption 11.2 *The p.d.f. of X is supported on* $\mathbb{R}^p$ *and has continuous second derivatives such that*

$$\sup_{x \in \mathbb{R}^p} \lambda_{\max}[\ddot{f}(x)] < \infty, \quad \inf_{x \in \mathbb{R}^p} \lambda_{\min}[\ddot{f}(x)] > -\infty,$$

where $\lambda_{\min}(\cdot)$ *and* $\lambda_{\max}(\cdot)$ *are, as before, the maximum and minimum eigenvalues of a matrix.*

This and the next assumptions are made to control the remainder of the Taylor expansions.

Assumption 11.3 *For each* $y \in \Omega_Y$, *the joint p.d.f.* $f(y,x)$ *of* (Y,X) *has continuous second derivative (hessian matrix), and*

$$-\infty < \inf_{x \in \mathbb{R}^p} \{\lambda_{\min}[\partial^2 f(y,x)/\partial x \partial x^{\mathsf{T}}]\} \leq \sup_{x \in \mathbb{R}^p}\{\lambda_{\max}[\partial^2 f(y,x)/\partial x \partial x^{\mathsf{T}}]\} < \infty.$$

Lemma 11.1 *Under Assumptions 11.1 and 11.2, we have*

$$E[K(U)] = h^p f(x_0) \int K(u)du + O(h^{p+2}).$$

PROOF. By Taylor's mean value theorem,

$$
\begin{aligned}
EK(U) &= \int K(u)h^p f(x_0 + hu)du \\
&= \int K(u)h^p[f(x_0) + hf(x_0)^\mathsf{T}u + (1/2)h^2 u^\mathsf{T}\ddot{f}(x^\dagger(u))u]du \\
&= h^p \int K(u)f(x_0)du + h^{p+1}\int K(u)\dot{f}(x_0)^\mathsf{T}udu + \tfrac{1}{2}h^{p+2}\int K(u)u^\mathsf{T}\ddot{f}(x^\dagger)udu,
\end{aligned}
\tag{11.6}
$$

where $x^\dagger(u)$, which can depend on u, is a point that satisfies $\|x^\dagger(u) - x_0\| \leq h\|u\|$. Because, by Assumption 11.1,

$$\int uK(u)du = 0,$$

the second term on the right-hand side of (11.6) is 0. By Assumption 11.2,

$$\left| \int K(u)h^p[(1/2)h^2 u^\mathsf{T}\ddot{f}(x^\dagger(u))u]du \right| \leq (1/2)\lambda_0 h^{p+2}\left| \int K(u)u^\mathsf{T}udu \right|,$$

where

$$\lambda_0 = \max\left\{ \sup_{x \in \mathbb{R}^p}|\lambda_{\max}(\ddot{f}(x))|, \sup_{x \in \mathbb{R}^p}|\lambda_{\min}(\ddot{f}(x))| \right\}.$$

Hence the third term on the right-hand side is of the order $O(h^{p+2})$, as to be proved. $\qquad\square$

Lemma 11.2 *Under Assumptions 11.1 and 11.2,*

$$E[UK(U)] = h^{p+1}\left(\int uu^\mathsf{T}K(u)du \right)\dot{f}(x_0) + O(h^{p+2}).$$

PROOF. By Taylor's mean value theorem

$$
\begin{aligned}
&E[UK(U)] \\
&= h^p \int uK(u)f(x_0)du + h^{p+1}\int uK(u)[\dot{f}(x_0)^\mathsf{T}u]du + h^{p+2}\int uK(u)u^\mathsf{T}\ddot{f}(x^\dagger)udu.
\end{aligned}
$$

Because, by Assumption 11.1, $K(u)$ is a spherically symmetric p.d.f., the first term on the right-hand side of the above equality is 0. Furthermore, using the argument similar to that in the proof of Lemma 11.1 and the assumption that all the third moments of U are finite, we can show that the fourth term is of the order $O(h^{p+2})$. $\qquad\square$

The derivations of the two expansions in the next lemma are rather similar to those in Lemmas 11.1 and 11.2, and are omitted.

Lemma 11.3 *Under Assumptions 11.1 and 11.2,*

$$E[YK(U)] = h^p f(x_0)\left(\int K(u)du \right)E(Y|x_0) + O(h^{p+2}),$$

$$E[UU^\mathsf{T}K(U)] = h^p f(x_0)\left(\int uu^\mathsf{T}K(u)du \right) + O(h^{p+2}).$$

The expansion of $E[UYK(U)]$ is slightly more complicated. So we give the proof below.

Lemma 11.4 *Under Assumptions 11.1 and 11.3, we have the following expansion*

$$E[UYK(U)] = h^{p+1} f(x_0) \left(\int uu^\mathsf{T} K(u) du \right) \partial E(Y|x_0)/\partial x$$
$$+ h^{p+1} E(Y|x_0) \left(\int uu^\mathsf{T} K(u) du \right) \dot{f}(x_0) + O(h^{p+2}).$$

PROOF. By definition,

$$E[UYK(U)] = h^p \int uy K(u) f(y, x_0) du dy + h^{p+1} \int uy K(u) \frac{\partial f(y, x_0)}{\partial x^\mathsf{T}} u du dy$$
$$+ h^{p+2} \int uy K(u) h^2 u^\mathsf{T} \frac{\partial^2 f(y, x^\dagger)}{\partial x \partial x^\mathsf{T}} u du dy. \tag{11.7}$$

Because, by Assumption 11.1, K is a spherical density, the first term on the right-hand side is

$$h^p \left(\int uK(u) du \right) \left(\int y f(y, x_0) dy \right) = 0.$$

The second term on the right-hand side of (11.7) is

$$h^{p+1} \int uy K(u) [\partial f(y|x_0)/\partial x^\mathsf{T} f(x_0) + f(y|x_0) \dot{f}(x_0)^\mathsf{T}] u du dy$$
$$= h^{p+1} \int uy K(u) [\partial f(y|x_0)/\partial x^\mathsf{T} f(x_0)] u du dy + h^{p+1} \int uy K(u) [f(y|x_0) \dot{f}(x_0)^\mathsf{T}] u du dy$$
$$= h^{p+1} \left(\int uu^\mathsf{T} K(u) du \right) \frac{\partial E(Y|x_0)}{\partial x} + h^{p+1} \left(\int uu^\mathsf{T} K(u) du \right) \dot{f}(x_0) E(Y|x_0).$$

By the similar argument used in the proof of Lemma 11.1, we can easily show that under Assumption 11.3 the third term on the right-hand side of (11.7) is of the order $O(h^{p+2})$. This proves the desired result.   □

Next, we use the results of Lemma 11.1 through Lemma 11.4 to prove the Fisher consistency of $b^*(h)$.

Theorem 11.2 *Under Assumptions 11.1 and 11.3, we have, as $h \to 0$,*

$$b^*(h) = \frac{\partial E(Y|x_0)}{\partial x} + O(h^2).$$

PROOF. Let

$$A(h) = E[UU^\mathsf{T} K(U)] - \frac{E[UK(U)]E[U^\mathsf{T} K(U)]}{E[K(U)]}$$
$$B(h) = \left\{ E[UYK(U)] - \frac{E[UK(U)]E[YK(U)]}{E[K(U)]} \right\},$$

where we write these quantities as functions of h because the distribution of U depends on h. In these notations,

$$b^*(h) = h^{-1} A^{-1}(h) B(h). \tag{11.8}$$

By Lemmas 11.1, 11.2, and 11.3, $A(h)$ can be expanded as

$$A(h) = h^p f(x_0) \left(\int uu^\mathsf{T} K(u) du \right)$$

$$- \frac{h^{p+1} \left(\int uu^\mathsf{T} K(u) du \right) \dot{f}(x_0) \dot{f}(x_0)^\mathsf{T} h^{p+1} \left(\int uu^\mathsf{T} K(u) du \right)}{h^p f(x_0) \int K(u) du} + O(h^{p+2}).$$

Because the second term on the right-hand side is of the order $O(h^{p+2})$, we have

$$A(h) = h^p f(x_0) \left(\int uu^\mathsf{T} K(u) du \right) + O(h^{p+2}). \tag{11.9}$$

By Lemmas 11.1, 11.2, 11.3, and 11.4,

$$B(h) = h^{p+1} f(x_0) \left(\int uu^\mathsf{T} K(u) du \right) \frac{\partial E(Y|x_0)}{\partial x}$$

$$+ h^{p+1} E(Y|x_0) \left(\int uu^\mathsf{T} K(u) du \right) \dot{f}(x_0)$$

$$- \frac{h^{p+1} \left(\int uu^\mathsf{T} K(u) du \right) \dot{f}(x_0) h^p f(x_0) \left(\int K(u) du \right) E(Y|x_0)}{h^p f(x_0) \int K(u) du} + O(h^{p+2})$$

Note that, after canceling out the common factors in the denominator and numerator of the third term, it becomes identical to the second term. Hence we have

$$B(h) = h^{p+1} f(x_0) \left(\int uu^\mathsf{T} K(u) du \right) \frac{\partial E(Y|x_0)}{\partial x} + O(h^{p+2}) \tag{11.10}$$

Now substitute (11.9) and (11.10) into (11.8) to obtain

$$b^*(h) = h^{-1} \left[h^p f(x_0) \left(\int uu^\mathsf{T} K(u) du \right) + O(h^{p+2}) \right]^{-1}$$

$$\left[h^{p+1} f(x_0) \left(\int uu^\mathsf{T} K(u) du \right) \frac{\partial E(Y|x_0)}{\partial x} + O(h^{p+2}) \right]$$

$$= \frac{\partial E(Y|x_0)}{\partial x} + O(h^2),$$

where the second equality holds because, by the assumption that K is spherically contoured, $\int uu^\mathsf{T} K(u) du$ is proportional to the identity matrix I_p. □

11.3 Minimum Average Variance Estimate

Minimum Average Variance Estimation is, in a sense, a more efficient way of performing the same task done by the OPG. Because $b_h^*(x)$ are approximately in $\mathscr{S}_{E(Y|X)}$ for all $x \in \Omega_X$, we have the approximate relation

$$b_h^*(x) = \beta c_h^*(x),$$

where β is any basis matrix for $\mathscr{S}_{E(Y|X)}$ and $c_h^*(x)$ is a vector in $\mathbb{R}^d$ that depends on x. Consequently, $a_h * (x), b_h^*(x)$, and β should minimize

$$\frac{E\{[Y - a - c^\mathsf{T} B(X - x)]^2 K[(X - x)/h]\}}{EK[(X - x)/h]}$$

among all $a \in \mathbb{R}$, $c \in \mathbb{R}^d$, and $B \in \mathbb{R}^{p \times d}$. Moreover, this is true for all $x \in \Omega_X$. Equivalently, $a_h^*(x)$, $b_h^*(x)$, and β are the minimizer of the objective function

$$E\left\{\frac{[Y - a(X') - c^\top(X')B^\top(X - X')]^2 K[(X - X')/h]}{E\{K[(X - X')/h]\}}\right\}, \tag{11.11}$$

where X' is an independent copy of X, among all $a : \Omega_X \to \mathbb{R}$, $b : \Omega_X \to \mathbb{R}^d$, and $B \in \mathbb{R}^{p \times d}$. The MAVE proposed in Xia et al. (2002) is based on minimization of the sample-level counterpart of the above function. Since (11.11) is an approximation of

$$E[Y - E(Y|B^\top X)]^2 = E\left\{E\left[E\left(Y - E(Y|B^\top X)\right)^2 | B^\top X\right]\right\} = E[\mathrm{var}(Y|B^\top X)],$$

the procedure approximately minimizes the averaged conditional variance $\mathrm{var}(Y|B^\top X)$ over B, and hence the name *Minimum Average Variance Estimation*.

At the sample level, we simply replace all expectations in (11.11) by sample averages. For more compact notation, let

$$W_h(x_1, x_2) = \frac{K[(x_2 - x_1)/h]}{E_n\{K[(X - x_1)/h]\}}.$$

Our objective function is

$$L(a_1, \ldots, a_n, c_1, \ldots, c_n, B) = \sum_{i=1}^n \sum_{j=1}^n [Y_j - a_i - c_i^\top B^\top(X_j - X_i)]^2 W_h(X_i, X_j) \tag{11.12}$$

among all $a_1, \ldots, a_n \in \mathbb{R}$, $c_1, \ldots, c_n \in \mathbb{R}^p$, and $B \in \mathbb{R}^{p \times d}$. Let $(\hat{a}_1, \ldots, \hat{a}_n, \hat{c}_1, \cdots, \hat{c}_n, \hat{B})$ be the minimizer of this objective function. The subspace $\mathrm{span}(\hat{B})$ is used as estimator of the central mean subspace $\mathscr{S}_{E(Y|X)}$.

This optimization can be performed iteratively between $\{a_i\}$, $\{c_i\}$ and B. That is, first fix a B and minimize the objective function over $\{a_i\}$ and $\{c_i\}$, and then fix $\{a_i\}$, $\{c_i\}$ and minimize the objective function over B. It turns out that each of the two steps can be formulated as a weighted least squares problem, for which the minimizer can be explicitly expressed.

Specifically, let us first fix a $B \in \mathbb{R}^{p \times d}$ and minimize $L(a_1, \ldots, a_n, c_1, \ldots, c_n, B)$ over $\{a_i\}$ and $\{c_i\}$, and let $(\hat{a}_1(B), \ldots, \hat{a}_n(B), \hat{c}_1(B), \ldots, \hat{c}_n(B))$ denote such a minimizer. Since the objective function is the sum of n terms whose ith term only involves (a_i, c_i), we can simply minimize the ith term of the objective function to obtain $\hat{a}_i(B), \hat{c}_i(B)$. That is,

$$L(a_1, \ldots, a_n; c_1, \ldots, c_n, B) = \sum_{i=1}^n L_i(a_i, b_i, B),$$

where

$$L_i(a_i, b_i, B) = \sum_{j=1}^n [Y_j - a_i - c_i^\top B^\top(X_j - X_i)]^2 W_h(X_i, X_j).$$

The vector $(\hat{a}_i, \hat{c}_i^\top)^\top$ is simply the minimizer of $L_i(a_i, b_i, B)$ over $a_i \in \mathbb{R}$, $c_i \in \mathbb{R}^d$. Rewrite $L_i(a_i, b_i, B)$ as the weighted least squares problem

$$\sum_{j=1}^n \left[Y_j - (a_i, c_i^\top)\begin{pmatrix} 1 \\ B^\top(X_j - X_i) \end{pmatrix}\right]^2 W_h(X_i, X_j).$$

By Theorem 1.1, the explicit minimizer of this function is

$$\begin{pmatrix} \hat{a}_i(B) \\ \hat{c}_i(B) \end{pmatrix} = \left[\sum_{j=1}^{n} \begin{pmatrix} 1 \\ B^{\mathsf{T}}(X_j - X_i) \end{pmatrix} \begin{pmatrix} 1 \\ B^{\mathsf{T}}(X_j - X_i) \end{pmatrix}^{\mathsf{T}} W_h(X_i, X_j) \right]^{-1}$$
$$\left[\sum_{j=1}^{n} \begin{pmatrix} 1 \\ B^{\mathsf{T}}(X_j - X_i) \end{pmatrix} Y_j W_h(X_i, X_j) \right]. \tag{11.13}$$

Next, we fix $\{a_i\}$, $\{c_i\}$ and minimize the objective function over B. Rewrite the objective function in (11.12) as

$$L(a_1, \ldots, a_n, c_1, \ldots, c_n, B) = \sum_{i=1}^{n} \sum_{j=1}^{n} [Y_j - a_i - ((X_j - X_i) \otimes c_i)^{\mathsf{T}} \gamma]^2 W_h(X_i, X_j),$$

where γ is the vector $\text{vec}(B^{\mathsf{T}})$. Take derivative with respect to γ to obtain the equation

$$\sum_{i=1}^{n} \sum_{j=1}^{n} [(X_j - X_i) \otimes c_i)(Y_j - a_i - ((X_j - X_i) \otimes c_i)^{\mathsf{T}} \gamma) W_h(X_i, X_j)] = 0,$$

which has the explicit solution

$$\hat{\gamma} = \left\{ \sum_{i=1}^{n} \sum_{j=1}^{n} [W_h(X_i, X_j)((X_j - X_i) \otimes c_i)((X_j - X_i) \otimes c_i)^{\mathsf{T}}] \right\}^{-1}$$
$$\left\{ \sum_{i=1}^{n} \sum_{j=1}^{n} [W_h(X_i, X_j)((X_j - X_i) \otimes c_i)(Y_j - a_i)] \right\}. \tag{11.14}$$

We then iterative between these two steps until $\{a_i\}$, $\{b_i\}$, and B converge.

To initiate this iteration we need a starting value of B, for which we recommend the OPG estimate, which usually works very well. Also, it is preferred to perform MAVE on the marginally standardized variables X^i, as was done for the OPG. This is for the same reason that we need the scale of each component of X to be roughly the same to justify the use of a single bandwidth for all the components of X. Alternatively, it is entirely reasonable to use separate bandwidths, say $h_1, \ldots, h_p$, for different components of X.

We should mention at this point that the parameter β in MAVE, as it appears in the objective function $L(a_1, \ldots, a_n, b_1, \ldots, b_n, \beta)$, is not identifiable, because one can multiply any constant to it, and divide b_i by that constant, without changing the value of the objective function. However, this does not concern us because we start with a reasonably good approximation by OPG, and the iterative usually converges quite quickly to a solution near the OPG estimate — typically within 5 or 6 iterations.

In the following, we use $\text{mat}(v, p, d)$ to denote the inverse operation of vec. That is, if v is a pd-dimensional vector, then $\text{mat}(v, p, d)$ is a $p \times d$ matrix such that $\text{vec}[\text{mat}(v, p, d)] = v$. The algorithm for MAVE is summarized as follows.

Algorithm 11.2 Minimum Average Variance Estimation

1. Marginally standardize $X_1, \ldots, X_n$ to $Z_1, \ldots, Z_n$ as in the algorithm for OPG. Choose a small number ε, say 10^{-4}.

2. In the Z-scale, use the OPG to compute the initial $\hat{B}$, which is the $p \times d$ matrix $(v_1, \ldots, v_d)$ in step 4 of the OPG algorithm.

3. Use (11.13) to compute $(\hat{a}_i(\hat{B}), \hat{c}_i^{\mathsf{T}}(\hat{B}))^{\mathsf{T}}$.

4. Use (11.14) to compute $\hat{\gamma}$. Set $\hat{B} = \mathrm{mat}(\hat{\gamma}, p, d)$. Let $\hat{B}_1$ be the $\hat{B}$ in step 3, and $\hat{B}_2$ be the $\hat{B}$ in step 4. If $\|P_{\hat{B}_2} - P_{\hat{B}_1}\| > \varepsilon$ then return to step 3. Otherwise go to step 5.

5. Use $\mathrm{span}(\hat{D}^{-1/2}\hat{B})$ as the estimate of $\mathscr{S}_{E(Y|X)}$, where $\hat{D}$ is as defined in the OPG algorithm.

As an alternative to setting a convergence criterion ε, we can also set the number of iterations, as we do in the R-codes. Usually it is reasonably safe to set the number of iterations to 10, as the algorithm usually converges by then. The inputs for the following R-code are similar to those for OPG. The extra input `nit` is the number of iterations between steps 3 and 4 in the above algorithm. The R-code calls the functions `standvec`, `opg`, `kern`, and `wls` listed in under the R-codes for OPG. The Gaussian kernel is used.

R-code for MAVE

```
mave=function(x,y,h,d,nit){
sig=diag(var(x));n=dim(x)[1];p=dim(x)[2]
x=apply(x,2,standvec);beta=opg(x,y,h,d);kermat=kern(x,h)
for(iit in 1:nit){
b=numeric();a=numeric(); for(i in 1:n){
wi=kermat[,i]/(apply(kermat,2,mean)[i])
ui=cbind(1,t(t(x)-x[i,])%*%beta)
out=wls(ui,y,wi);a=c(a,out$a);b=cbind(b,out$b)}
out=0;out1=0; for(i in 1:n){
xi=kronecker(t(t(x)-x[i,]),t(b[,i]))
yi=y-a[i];wi=kermat[,i]/apply(kermat,2,mean)[i]
out=out+apply(xi*yi*wi,2,mean)
out1=out1+t(xi*wi)%*%xi/n}
beta=t(matrix(solve(out1)%*%out,d,p))}
return(diag(sig^(-1/2))%*%beta)}
```

11.4   Refined MAVE and refined OPG

As discussed in the last section, MAVE minimizes an objective function that approximates $E[(Y - E(Y|B^{\mathsf{T}}X))^2]$. More precisely, it minimizes the sample version of

$$E\{E[(Y - E(Y|B^{\mathsf{T}}X))^2|X]\}.$$

An alternative way to write the above expectation is

$$E\{E[(Y - E(Y|B^{\mathsf{T}}X))^2|B^{\mathsf{T}}X]\}.$$

The refined MAVE, or rMAVE, minimizes the sample version of this form of the loss function. The advantage of this form is that the d-dimensional $B^{\mathsf{T}}X$ — instead of the p-dimensional X — appears as the conditioned random variable. The sample estimate of the former employs a d-dimensional kernel, which sharpens the focus of the p-dimensional kernel employed by MAVE. At the sample level, rMAVE is defined through the minimization of the objective function

$$L(a_1,\ldots,a_n,c_1,\ldots,c_n,B) = E_n\{[Y - a_i - c_i^{\mathsf{T}}B^{\mathsf{T}}(X - X_i)]^2 W_h(B^{\mathsf{T}}X_i, B^{\mathsf{T}}X_j)\}, \qquad (11.15)$$

where

$$W_h(u_1, u_2) = K[(u_2 - u_1)/h]/E_n\{K[(U - u_1)/h]\}.$$

One way to minimize this objective function is through iteration among two steps. In the first step, we fix B and minimize the objective function over $\{a_i\}$ and $\{c_i\}$. In the second step, we fix $\{a_i\}$, $\{b_i\}$, and the B in $W_h(B^{\mathsf{T}}X_i, B^{\mathsf{T}}X_j)$ to minimize the objective function (11.15), treating the B in $c_i^{\mathsf{T}}B(X - X_i)$ alone as the variable. We then go back to step 1 with (11.15) updated. Again, minimizations for step 1 and step 2 can be formulated a weighted least squares, as we describe in detail below.

The solution $(\hat{a}_i(B), \hat{c}_i(B))$ for rMAVE is essentially the same as (11.13), except that $W_h(X_i, X_j)$ is replaced by $W_h(B^{\mathsf{T}}X_i, B^{\mathsf{T}}X_j)$. That is,

$$\begin{pmatrix} \hat{a}_i(B) \\ \hat{c}_i(B) \end{pmatrix} = \left[\sum_{j=1}^{n} \begin{pmatrix} 1 \\ B^{\mathsf{T}}(X_j - X_i) \end{pmatrix} \begin{pmatrix} 1 \\ B^{\mathsf{T}}(X_j - X_i) \end{pmatrix}^{\mathsf{T}} W_h(B^{\mathsf{T}}X_i, B^{\mathsf{T}}X_j) \right]^{-1}$$
$$\left[\sum_{j=1}^{n} \begin{pmatrix} 1 \\ B^{\mathsf{T}}(X_j - X_i) \end{pmatrix} Y_j W_h(B^{\mathsf{T}}X_i, B^{\mathsf{T}}X_j) \right].$$

The solution for step 2 is essentially the same as (11.14) except that $W_h(X_i, X_j)$ is replaced by $W_h(B^{\mathsf{T}}X_i, B^{\mathsf{T}}X_i)$ where B is the estimate from the last iteration. That is,

$$\mathrm{vec}(B_{\mathrm{new}}^{\mathsf{T}}) = \left\{ \sum_{i=1}^{n}\sum_{j=1}^{n} [W_h(B^{\mathsf{T}}X_i, B^{\mathsf{T}}X_j)((X_j - X_i) \otimes c_i)((X_j - X_i) \otimes c_i)^{\mathsf{T}}] \right\}^{-1}$$
$$\left\{ \sum_{i=1}^{n}\sum_{j=1}^{n} [W_h(B^{\mathsf{T}}X_i, B^{\mathsf{T}}X_j)((X_j - X_i) \otimes c_i)(Y_j - a_i)] \right\}. \qquad (11.16)$$

As proposed by Xia (2007) in a related method, we update the bandwidth after each iteration, with initial bandwidth h_0 set to (11.5). After each iteration, we update the bandwidth by

$$h_t = \max(r_n h_{t-1}, c_0 n^{-1/(d+4)}),$$

where $r_n = n^{-1/[2(\max(p,3)+6)]}$.

An R-code for implementing rMAVE is provided below, where the initial value is also taken to be the OPG estimate. The R-code calls the functions `kern`, `standvec`, `opg`, and `wls` listed in the R-codes for OPG.

R-code for rMAVE

```
rmave=function(x,y,d,nit){
sig=diag(var(x));n=dim(x)[1];p=dim(x)[2]
```

```
x=apply(x,2,standvec)
c0=2.34;p0=max(p,3);h=c0*n^(-(1/(p0+6)));rn=n^(-1/(2*(p0+6)))
beta=opg(x,y,d)
for(iit in 1:nit){
kermat=kern(x%*%beta,h);mkermat=apply(kermat,2,mean)
b=numeric();a=numeric()
for(i in 1:n){
wi=kermat[,i]/mkermat[i];ui=cbind(1,t(t(x)-x[i,])%*%beta)
out=wls(ui,y,wi);a=c(a,out$a);b=cbind(b,out$b)}
out=0;out1=0
for(i in 1:n) {
xi=kronecker(t(t(x)-x[i,]),t(b[,i]));yi=y-a[i]
wi=kermat[,i]/mkermat[i]
out=out+apply(xi*yi*wi,2,mean)
out1=out1+t(xi*wi)%*%xi/n}
beta=t(matrix(solve(out1)%*%out,d,p))
h=max(rn*h,c0*n^((-1/(d+4))))
}
return(diag(sig^(-1/2))%*%beta)}
```

One can apply the similar refinement to OPG. Interestingly, as far as we know this natural extension was not developed in Xia et al. (2002) or any subsequent papers, such as Xia (2007), though this extension substantially improves the performance of OPG with little extra effort. In the rest of this section we develop this extension.

Like in rMAVE, we replace the kernel $K_h(X_j - X_i)$ by $K_h[\hat{\beta}^{\mathsf{T}}(X_j - X_i)]$. This reduces the dimension of the kernel from p to d, and effectively mitigates the "curse of dimensionality". Specifically, after the first round of OPG as described above, we form the matrix $\hat{B} = (v_1, \ldots, v_d)$, where $v_1, \ldots, v_d$ are as defined in the OPG algorithm. We then replace the kernel $K_h(X_j - X_i)$ by $K_h[\hat{B}^{\mathsf{T}}(X_j - X_i)]$ and complete the next round of iteration, which leads to an updated $\hat{B}$. We then iterate this process, each time with an updated bandwidth, until convergence. We call this procedure refined OPG, or rOPG. The update of bandwidth is done as in rMAVE. Below is the R-code to implement rOPG. The inputs are the same as the R-code for rMAVE. It also calls the same subfunctions as does rMAVE.

R-code for rOPG

```
ropg=function(x,y,d,nit){
sig=diag(var(x));x=apply(x,2,standvec);p=dim(x)[2];n=dim(x)[1]
c0=2.34;p0=max(p,3);rn=n^(-1/(2*(p0+6)));h=c0*n^(-(1/(p0+6)))
beta=diag(p)
for(iit in 1:nit){
kmat=kern(x%*%beta,h);bmat=numeric()
for(i in 1:dim(x)[1]){
wi=kmat[,i];xi=cbind(1,t(t(x)-x[i,]))
bmat=cbind(bmat,wls(xi,y,wi)$b)}
beta=eigen(bmat%*%t(bmat))$vectors[,1:d]
h=max(rn*h,c0*n^((-1/(d+4))))
}
```

```
beta.final=diag(sig^(-1/2))%*%beta
return(beta.final)
}
```

11.5   From Central Mean Subspace to Central Subspace

The MAVE-type estimators we developed so far (OPG, rOPG, MAVE, and rMAVE) all target the central mean subspace $\mathscr{S}_{E(Y|X)}$. This is because they estimate the gradient of $E(Y|x)$ for $x \in \Omega_x$, which are vectors in $\mathscr{S}_{E(Y|X)}$. In the next three sections we discuss how these methods can be augmented to estimate the central subspace $\mathscr{S}_{Y|X}$. We describe three augmentations. The first is an augmentation of OPG and rOPG; the second is an augmentation of MAVE and rMAVE; the third is a general class that accommodates the first two, and includes many other estimators. In the first two augmentations, the random variable Y is replaced by a kernel function of Y. These were proposed by Xia (2007). The third augmentation replaces Y by a class of functions of Y that is in a sense sufficiently rich to characterize the conditional distribution of Y given X. This was proposed by Yin and Li (2011), and it covers all the MAVE-type estimators as special cases. A related method introduced by Wang and Xia (2008), called Sliced Regression method, will also be discussed, under the framework of the third augmentation.

Since the developments in next three sections are in their broad outlines parallel to the those in the last three sections, we will focus on the ideas, algorithms, and their comparative advantages without dwelling on rigorous proofs of their theoretical properties.

11.6   dOPG and Its Refinement

The idea here is to replace the gradient of the conditional mean by the gradient of the conditional density. But since the conditional density function is not one function but a class of functions, we need to perform the OPG procedures many times.

Let η be a basis matrix of $\mathscr{S}_{Y|X}$. Under the premise that the conditional density $f_{Y|X}(y|x)$ of Y given X exists, we have

$$Y \perp\!\!\!\perp X | \eta^\mathsf{T} X \iff f_{Y|X}(y|x) = f_{Y|\eta^\mathsf{T}X}(y|\eta^\mathsf{T}x).$$

Furthermore, the gradient of the conditional density lies in span(η), because

$$\frac{\partial f_{Y|X}(y|x)}{\partial x} = \eta \frac{\partial f_{Y|\eta^\mathsf{T}X}(y|\eta^\mathsf{T}x)}{\partial (\eta^\mathsf{T}x)}.$$

Thus, to estimate the central subspace, we target the gradient of the conditional density $\partial f_{Y|X}(y|x)/\partial x$. Using the same argument for the proof of Theorem 11.1, we can show that

$$\text{span}\left\{ E\left[\frac{\partial f_{Y|X}(Y|X)}{\partial X} \frac{\partial f_{Y|X}(Y|X)}{\partial X^\mathsf{T}} \right] \right\} = \mathscr{S}_{Y|X}, \tag{11.17}$$

where, again, $\partial f_{Y|X}(Y|X)/\partial X$ stands for $g(Y,X)$ where $g(Y,x) = f_{Y|X}(Y|X=x)/\partial x$. Note that the expectation above is with respect to the joint distribution of (X,Y).

As proposed in Fan et al. (1996), the estimation of the gradient of the conditional density of $Y|X$ is achieved by combining local linear regression with the kernel estimator of conditional density — a method called the "double kernel estimate". Like the estimation of the gradient of $E(Y|x)$, we use the local linear regression to estimate this gradient of the conditional density using a kernel function of Y. Specifically, let $H(v)$ be a probability density function on $\mathbb{R}$. Let $H_b(v) = H(v/b)/b$, where $b > 0$ is the bandwidth for the kernel H. Observe that, for any $x \in \Omega_X$, as $b \to 0$,

$$
\begin{aligned}
E[H_b(Y-y)|x] &= b^{-1} \int H[(y'-y)/b] f_{Y|X}(y'|x) dy' \\
&= \int H(v) f_{Y|X}(y+bv|x) dv \\
&\approx f_{Y|X}(y|x) \int H(v) dv \\
&= f_{Y|X}(y|x).
\end{aligned}
\tag{11.18}
$$

At the same time, as $x \to x_0$,

$$
f_{Y|X}(y|x) \approx f_{Y|X}(y|x_0) + [\partial f_{Y|X}(y|x_0)/\partial x^\top](x-x_0).
\tag{11.19}
$$

Combining (11.18) and (11.19), we have the approximation

$$
E[H_b(Y-y)|x] \approx f_{Y|X}(y|x_0) + [\partial f_{Y|X}(y|x_0)/\partial x^\top](x-x_0).
\tag{11.20}
$$

To further develop the intuition, we prove the following lemma.

Lemma 11.5 *Suppose Y is a random variable, and $X \in \mathbb{R}^p$ is a random vector with $\mathrm{var}(X) > 0$. If $E(Y|X) = \alpha + \beta^\top(X-EX)$ for some $\alpha \in \mathbb{R}$, $\beta \in \mathbb{R}^p$, then $(\alpha, \beta^\top)^\top$ is the unique minimizer of*

$$
E\{[Y-a-b^\top(X-EX)]^2\}
\tag{11.21}
$$

over all $a \in \mathbb{R}$, $b \in \mathbb{R}^p$.

PROOF. For convenience, reset X to $X - EX$, so that $E(X) = 0$. By the definition of conditional expectation, we have, for any $h \in L_2(P_X)$,

$$
E[(Y - \alpha - \beta^\top X)h(X)] = 0.
$$

In particular, for any $a \in \mathbb{R}$, $b \in \mathbb{R}^p$,

$$
E[(Y - \alpha - \beta^\top X)(a + b^\top X)] = 0.
$$

Hence

$$
\begin{aligned}
E[(Y-a-b^\top X)^2] &= E[(Y-\alpha-\beta^\top X)^2] + E\{[(\alpha+\beta^\top X)-(a-b^\top X)]^2\} \\
&\geq E[(Y-\alpha-\beta^\top X)^2],
\end{aligned}
$$

which means $(\alpha, \beta^\top)^\top$ is a minimizer of (11.21). To see that it is the unique minimizer, note that, whenever $(a,b) \neq (\alpha, \beta)$,

$$
E\{[(\alpha+\beta^\top X)-(a-b^\top X)]^2\} = (a-\alpha)^2 + (b-\beta)^\top \mathrm{var}(X)(b-\beta) > 0,
$$

because $\mathrm{var}(X)$ is positive definite. $\qquad\square$

From Lemma 11.5 and the approximate relation (11.20) we deduce that, when x is near x_0, $(f_{Y|x_0}(y|x_0), \partial f_{Y|X}(y|x_0)/\partial x)$ is approximated by the unique minimizer to the expectation of

$[H_b(Y-y)-a-b^{\mathsf{T}}(X-x_0)]^2$. If we quantify the statement "when x is near x_0" by a kernel for the variable x, then $(f_{Y|x_0}(y|x_0), \partial f_{Y|X}(y|x_0)/\partial x)$ is approximated by the minimizer of

$$E\{[H_b(Y-y)-a-b^{\mathsf{T}}(X-x_0)]^2 W_h(x_0,X)\}, \tag{11.22}$$

where $W_h(x_0,x) = K_h(x-x_0)/EK_h(X-x_0)$. This is the population-level loss function of the double kernel estimator for conditional density. Note that the minimizer $(a^*(x_0,y), b^*(x_0,y))$ of this objective function is specific to (x_0,y).

Mimicking (11.22), we construct the sample-level objective function for dOPG as

$$\sum_{k=1}^{n}[H_b(Y_k-Y_j)-a-b^{\mathsf{T}}(X_k-X_i)]^2 W_h(X_i,X_k), \quad i,j=1,\ldots,n, \tag{11.23}$$

where $W_h(X_i,X_k)$ is the sample-level counterpart of the W_h in (11.22); that is, $K_h(X_k - X_i)/E_n[K_h(X-X_i)]$. Minimization of (11.23) is a weighted least squares problem with explicit solution

$$\begin{pmatrix} \hat{a}_{ij} \\ \hat{b}_{ij} \end{pmatrix} = \left[\sum_{k=1}^{n} W_h(W_i,X_k) \begin{pmatrix} 1 \\ X_k - X_i \end{pmatrix} \begin{pmatrix} 1 \\ X_k - X_i \end{pmatrix}^{\mathsf{T}} \right]^{-1}$$
$$\times \left[\sum_{k=1}^{n} W_h(X_i,X_k) \begin{pmatrix} 1 \\ X_k - X_i \end{pmatrix} H_b(Y_j - Y_i) \right]. \tag{11.24}$$

We then use $\hat{b}_{ij}$ to form the candidate matrix

$$\hat{C} = n^{-2} \sum_{i=1}^{n} \sum_{j=1}^{n} \hat{b}_{ij} \hat{b}_{ij}^{\mathsf{T}}. \tag{11.25}$$

The first d eigenvectors of this matrix, $v_1, \ldots, v_d$, are then used to estimate the central subspace. It is recommended that Y_i be standardized and X_i be standardized marginally so as to set the same baseline for determining the bandwidth for different data sets. Then the output of the vectors v_i should be replaced by $\hat{D}^{-1/2} v_i$, with $\hat{D}$ being defined in Section 11.1.

Note that to compute dOPG as described above amounts to performing weighted least squares n^2 times, each with responses $\{H_b(Y_k,Y_j) : k = 1,\ldots,n\}$ and predictors $\{X_k - X_i : k = 1,\ldots,n\}$. It is rather inefficient to perform these in a big loop in R, which is slow in processing loops. A much faster way is to perform these weighted least squares in batches as multivariate weighted least squares problems. Let

$$H_{..}(b) = \{H_b(Y_n,Y_j)\}_{i,j=1}^{n}, \quad A_i = \begin{pmatrix} a_{i1} & \cdots & a_{in} \\ b_{i1} & \cdots & b_{in} \end{pmatrix},$$
$$W_i(h) = \mathrm{diag}(W_h(X_i,X_1),\ldots,W_h(X_i,X_n)), \tag{11.26}$$
$$\Delta_{\cdot i} = (X_1 - X_i, \ldots, X_n - X_i)^{\mathsf{T}}.$$

Using these notations the n^2 objective functions in (11.23) can be written as the following n matrix-valued objective function

$$[(H_{..}(b) - (1_n, \Delta_{\cdot i})A_i]^{\mathsf{T}} W_i(h)[(H_{..}(b) - (1_n, \Delta_{\cdot i})A_i]. \tag{11.27}$$

This matrix-valued objective function has an explicit minimizer in terms of Louwner's ordering, as shown by the following lemma.

Lemma 11.6 *Suppose $V \in \mathbb{R}^{n \times p}$, $U \in \mathbb{R}^{n \times q}$, $W \in \mathbb{R}^{n \times n}$ are matrices, W is positive semidefinite, and $U^\mathsf{T} W U$ is positive definite. Let $A^* = (U^\mathsf{T} W U)^{-1} U^\mathsf{T} W V$. Then, for any $A \in \mathbb{R}^{p \times q}$,*

$$(V - UA^*)^\mathsf{T} W (V - UA^*) \leq (V - UA)^\mathsf{T} W (V - UA) \tag{11.28}$$

where the matrix inequality $\geq$ indicates Louwner's ordering (i.e. $A_1 \geq A_2$ iff $A_1 - A_2$ is positive semidefinite).

PROOF. First, note that

$$\begin{aligned}
&(V - UA^*)^\mathsf{T} W U \\
&= V^\mathsf{T} (I - W U (U^\mathsf{T} W U)^{-1} U^\mathsf{T}) W U = V^\mathsf{T} W U - V^\mathsf{T} W U = 0.
\end{aligned} \tag{11.29}$$

The right-hand side of the inequality (11.28) can be decomposed as four terms:

$$\begin{aligned}
(V - UA)^\mathsf{T} W (V - UA) &= (V - UA^* + UA^* - UA)^\mathsf{T} W (V - UA^* + UA^* - UA) \\
&= (V - UA^*)^\mathsf{T} W (V - UA^*) + (UA^* - UA)^\mathsf{T} W (UA^* - UA) \\
&\quad + (V - UA^*)^\mathsf{T} W (UA^* - UA) + (UA^* - UA)^\mathsf{T} W (V - UA^*).
\end{aligned}$$

By (11.29) the last two terms on the right-hand side are 0, leaving us with

$$\begin{aligned}
(V - UA)^\mathsf{T} W (V - UA) &= (V - UA^* + UA^* - UA)^\mathsf{T} W (V - UA^* + UA^* - UA) \\
&= (V - UA^*)^\mathsf{T} W (V - UA^*) + (UA^* - UA)^\mathsf{T} W (UA^* - UA),
\end{aligned}$$

which implies the desired inequality because the second matrix on the right is positive semidefinite. □

By this lemma, the minimizer of (11.27) over $A_i \in \mathbb{R}^{(p+1) \times n}$ is

$$\hat{A}_i = [(1_n, \Delta_{\cdot i})^\mathsf{T} W_i(h) (1_n, \Delta_{\cdot i})]^{-1} (1_n, \Delta_{\cdot i})^\mathsf{T} W_i(h) H_{\cdot \cdot}(b).$$

We can then read off $\hat{b}_{i1}, \ldots, \hat{b}_{in}$ from the bottom p rows of the matrix $\hat{A}_i$. This scheme of simultaneous weighted least squares usually results in more than a 90% reduction of computing time if R is used.

A refinement of dOPG was also proposed in Xia (2007) by rescaling $X_k - X_i$ inside the kernel K_h. Upon completing the dOPG, we obtain in the process the matrix $\hat{C}$. We can use this to reweight the $X_k - X_i$ in the kernel. Motivated by the development in Hristache et al. (2001), Xia (2007) proposed to replace $K_h(X_j - X_i)$ by $K_h[\hat{C}^{1/2}(X_j - X_i)]$. Note that $\hat{C}$ converges to a matrix whose columns span the same subspace as $\mathrm{span}(\eta)$. Hence its column space is increasingly focused on the directions in the central subspace. Using $\hat{C}^{1/2}(X_k - X_i)$ down weights the part of the predictor that is in the orthogonal complement of the central subspace, which, in effect, reduces the dimension of the kernel K_h.

Rather than pursuing this line, we develop a refinement that is parallel to rOPG in (11.4), which in our experience performs better that the refinement in Xia (2007) described above. After the first round of dOPG, we form the matrix $\hat{B} = (v_1, \ldots, v_d)$, and then replace the kernel $K_h(X_j - X_i)$ by $K_h[\hat{B}^\mathsf{T}(X_j - X_i)]$. Complete the next round of dOPG using this reduced kernel, which leads to an updated $\hat{B}$. We then iterative this process until convergence. We call this procedure refined dOPG, or rdOPG. We now summarize the rdOPG procedure as the following algorithm. We will omit the algorithm for the dOPG as it is just the first iteration of the refined dOPG. In each iteration we need a sequence of bandwidths for K_h and H_b. This will be detailed in the algorithm. In the following, we use the notation

$$W_i(h, B) = \mathrm{diag}(W_h(B^\mathsf{T} X_i, B^\mathsf{T} X_1), \ldots, W_h(B^\mathsf{T} X_i, B^\mathsf{T} X_n)), \tag{11.30}$$

where B is a matrix with p rows.

Algorithm 11.3 rdOPG

---

1. Marginally standardize $X_1, \ldots, X_n$ and, for convenience, still use X_i to denote the marginally standardized variable. Reset Y_i to its standardized version $(Y_i - \bar{Y}_i)/\hat{\sigma}_Y$, where $\hat{\sigma}_Y = \sqrt{\mathrm{var}_n(Y)}$. Select the initial bandwidths h_1 and b_1 according to

$$c_0 = 2.34, \quad p_0 = \max(p, 3), \quad h_0 = c_0 n^{-1/(p_0+6)}, \quad b_0 = c_0 n^{-1/(p_0+5)}.$$

   Choose a small $\varepsilon > 0$ (such as 0.0001) as the convergence criterion. Set $\hat{B}_0 = I_p$ to start the following loop. Alternatively, we can set the maximum number of iterations, say 10.

2. At iteration $t = 1, 2, \ldots$, compute $\hat{A}_i^{(t)}$ according to

$$\hat{A}_i^{(t)} = [(1_n, \Delta_{\cdot i})^\mathsf{T} W_i(h_{t-1}, \hat{B}^{(t-1)})(1_n, \Delta_{\cdot i})]^{-1}(1_n, \Delta_{\cdot i})^\mathsf{T} W_i(h_{t-1}, \hat{B}^{i-1}) H_{\cdot\cdot}(b_{t-1}).$$

   Then read off the vectors $\hat{b}_{i1}^{(t)}, \ldots, \hat{b}_{in}^{(t)}$ from the $2, \ldots, p+1$ rows of the $(p+1) \times n$ matrix $\hat{A}_i^{(t)}$.

3. Compute $\hat{C}_{(t)} = n^{-2} \sum_{i=1}^n \sum_{j=1}^n \hat{b}_{ij}^{(t)} (\hat{b}_{ij}^{(t)})^\mathsf{T}$. Extract the first d eigenvectors, say $v_1, \ldots, v_d$, and form the matrix $\hat{B}_{(t)} = (v_1, \ldots, v_d)$.

4. If $\|\hat{B}_{(t)} - \hat{B}_{(t-1)}\| > \varepsilon$ (the norm being the Frobenius matrix norm), reset

$$h_t = \max(r_n h_{t-1}, c_0 n^{-1/(d+4)}), \quad b_t = \max(r_n b_{t-1}, c_0 n^{-1/(d+3)}, c_0 n^{-1/5}),$$

   where $r_n = n^{-1/(2(p_0+6))}$, and return to step 2. Otherwise use $\hat{D}^{-1/2} \hat{B}_{(t)}$ as the estimates of a basis of $\mathcal{S}_{Y|X}$, $\hat{D}$ being the matrix described in the OPG section.

   The bandwidths used in the algorithm were proposed by Xia (2007). They decrease with the number of iterations, but stop decreasing to settle down at a fixed level after a certain number of iterations. The following is an R-code to implement the above algorithm. It calls the functions swls, kern, standvec, of which the latter two were listed in the R-codes for OPG; swls, which stands for *simultaneous weighted least squares*, is provided below.

R-code for rdOPG

```
rdopg=function(x,y,d,nit){
sig=diag(var(x));x=apply(x,2,standvec);y=(y-mean(y))/sd(y)
n=dim(x)[1];p=dim(x)[2]
c0=2.34;p0=max(p,3);rn=n^(-1/(2*(p0+6)))
h=c0*n^(-(1/(p0+6)));b=c0*n^(-(1/(p0+5)))
```

```
hmat=kern(y,b);bbt=diag(p)
for(iit in 1:nit){
kmat=kern(x%*%bbt,h);bmat=numeric()
for(i in 1:n) bmat=cbind(bmat,swls(hmat,kmat,x,i)$b)
bbt=bmat%*%t(bmat)/(n^2);bbt=eigen(bbt)$vectors[,1:d]
h=max(rn*h,c0*n^((-1/(d+4))))
b=max(rn*b,c0*n^(-1/((d+3))),c0*n^(-1/5))
}
beta.final=diag(sig^(-1/2))%*%bbt
return(beta.final)
}


swls=function(hmat,kmat,x,i){
wi=diag(kmat[,i]);xdi=t(t(x)-x[i,]);xdi1=cbind(1,xdi)
abmat=solve(t(xdi1)%*%wi%*%xdi1)%*%t(xdi1)%*%wi%*%hmat
return(list(a=abmat[1,],b=abmat[-1,],ab=abmat))}
```

11.7 dMAVE and Its Refinement

Interestingly, the term dMAVE was used in Xia (2007) to describe what is parallel to rMAVE with a reduced kernel, but here we adhere to the rigorous analogy to refer to the parallel of MAVE as dMAVE. As in dOPG, we replace Y by the kernel function of Y. Specifically, dMAVE minimizes the objective function

$$\sum_{i=1}^{n}\sum_{j=1}^{n}\sum_{k=1}^{n}[H_b(Y_k-Y_j)-a_{ij}-c_{ij}^{\mathsf{T}}B^{\mathsf{T}}(X_k-X_i)]^2 W_h(X_i,X_k).$$

As before, we minimize this by first fixing B and minimizing over $\{a_{ij}\}, \{c_{ij}\}$, and then fixing $\{a_{ij}\}, \{c_{ij}\}$ and minimizing over B. As in MAVE, the minimizations in each of the two steps can be written in the form of the weighted least squares, whose solutions have explicit expressions, which are detailed in the algorithm below.

A similar refinement can be made for dMAVE; that is, we minimize the objective function with reduced kernel

$$\sum_{i=1}^{n}\sum_{j=1}^{n}\sum_{k=1}^{n}[H_b(Y_k-Y_j)-a_{ij}-c_{ij}^{\mathsf{T}}B^{\mathsf{T}}(X_k-X_i)]^2 W_h(B^{\mathsf{T}}X_i,B^{\mathsf{T}}X_k).$$

As in rMAVE, we first fix the B in the kernel and perform the 2 steps similar to dMAVE to obtain the updated $\{a_{ij}\}, \{c_{ij}\}$, and B. Then substitute the new B into the kernel and repeat the above steps until convergence. We call this procedure the refined dMAVE, or rdMAVE.

As in rdOPG, it is more efficient to perform the weighted least squares simultaneously. Let $H_{\cdot\cdot}$ and $W_i(h,B)$ be as defined in (11.26) and (11.30), and let

$$\Delta_{\cdot i}(B) = (B^{\mathsf{T}}(X_1-X_i),\ldots,B^{\mathsf{T}}(X_n-X_i))^{\mathsf{T}}, \quad G_i = \begin{pmatrix} a_{i1} & \cdots & a_{in} \\ c_{i1} & \cdots & c_{in} \end{pmatrix}.$$

By Lemma 11.6, the formula for simultaneous weighted least squares for G_i when B is fixed is

$$\hat{G}_i(h,B) = [(1_n,\Delta_{\cdot i}(B))^{\mathsf{T}}W_i(h,B)(1_n,\Delta_{\cdot i})]^{-1}(1_n,\Delta_{\cdot i}(B))^{\mathsf{T}}W_i(h,B)H_{\cdot\cdot}(b). \tag{11.31}$$

When $\hat{G}_i$ is fixed, the minimizer for B can be derived in the similar fashion as we derived (11.16). Specifically, let

$$\Delta_{ki} = X_k - X_i, \quad \nabla_{ki} = Y_k - Y_i.$$

Then the minimizer over B for the next iteration is given by the relation

$$\text{vec}(B_{\text{new}}^{\mathsf{T}}(h,b,B)) = \left\{ \sum_{i=1}^{n}\sum_{j=1}^{n}\sum_{k=1}^{n} [W_h(B^{\mathsf{T}}X_i, B^{\mathsf{T}}X_k)(\Delta_{ki}\otimes c_{ij})(\Delta_{ki}\otimes c_{ij})^{\mathsf{T}}] \right\}^{-1}$$
$$\left\{ \sum_{i=1}^{n}\sum_{j=1}^{n}\sum_{k=1}^{n} [W_h(B^{\mathsf{T}}X_i, B^{\mathsf{T}}X_k)(\Delta_{ki}\otimes c_{ij})[H_b(\nabla_{kj}) - a_{ij}]] \right\}. \tag{11.32}$$

The triple sum can be quite time consuming for R. For example, if $n = 1000$, the loop would involve a billion operations. To release the looping burden for R, we can write the inner loop (the $k = 1, \ldots, n$ loop) as matrix product. Let

$$M_i(h,B) = \text{diag}(W_h(B^{\mathsf{T}}X_i, B^{\mathsf{T}}X_1), \ldots, W_h(B^{\mathsf{T}}X_i, B^{\mathsf{T}}X_n)),$$
$$L_{ij}(h,b,B) \tag{11.33}$$
$$= (W_h(B^{\mathsf{T}}X_i, B^{\mathsf{T}}X_1)[H_b(\nabla_{1j} - a_{ij})], \ldots, W_h(B^{\mathsf{T}}X_i, B^{\mathsf{T}}X_n)[H_b(\nabla_{nj} - a_{ij})])^{\mathsf{T}}.$$

Then it is easy to check that

$$\sum_{k=1}^{n} W_h(B^{\mathsf{T}}X_i, B^{\mathsf{T}}X_k)(\Delta_{ki}\otimes c_{ij})(\Delta_{ki}\otimes c_{ij})^{\mathsf{T}} = (\Delta_{\cdot i}^{\mathsf{T}} M_i(h,B)\Delta_{\cdot i}) \otimes (c_{ij}c_{ij}^{\mathsf{T}})$$

$$\sum_{k=1}^{n} [W_h(B^{\mathsf{T}}X_i, B^{\mathsf{T}}X_k)(\Delta_{ki}\otimes c_{ij})[H_b(\nabla_{kj}) - a_{ij}] = (\Delta_{\cdot i}^{\mathsf{T}}\otimes c_{ij}) L_{ij}(h,b,B).$$

Hence (11.32) becomes

$$\text{vec}(B_{\text{new}}^{\mathsf{T}}(h,b,B)) = \left\{ \sum_{i=1}^{n}\sum_{j=1}^{n} [(\Delta_{\cdot i}^{\mathsf{T}} M_i(h,B)\Delta_{\cdot i}) \otimes (c_{ij}c_{ij}^{\mathsf{T}})] \right\}^{-1}$$
$$\left\{ \sum_{i=1}^{n}\sum_{j=1}^{n} [(\Delta_{\cdot i}^{\mathsf{T}}\otimes c_{ij}) L_{ij}(h,b,B)] \right\}. \tag{11.34}$$

Below we only present the algorithm for rdMAVE, omitting that of dMAVE as it is just one iteration of dMAVE with kernel $W_h(B^{\mathsf{T}}X_i, B^{\mathsf{T}}X_k)$ replaced by $W_h(X_i, X_k)$.

Algorithm 11.4 rdMAVE

1. Marginally standardize X_i, standardize Y_i, and choose the initial bandwidth h_1, b_1 as in step 1 of the rdOPG algorithm. Use dOPG (i.e. the first rdOPG iteration) to compute the initial $\hat{B}_0$. Then perform the loop below.

2. At iteration $t = 1, 2, \ldots$, compute $\hat{G}_i^{(t)} = G_i(h_{t-1}, \hat{B}^{(t-1)})$ according to (11.31), and read off $\{\hat{c}_{i1}^{(t)}, \ldots, \hat{c}_{in}^{(t)}\}$ from the bottom d rows of the $(d+1)\times n$ matrix $\hat{G}_i^{(t)}$.

3. Compute the updated $\hat{B}_{(t)} = B_{\text{NEW}}(h_{t-1}, b_{t-1}, \hat{B}_{(t-1)})$ according to (11.34).

4. Reset $\hat{B}_{(t)}$ to $\hat{B}_{(t)}(\hat{B}_{(t)}^{\mathsf{T}}\hat{B}_{(t)})^{-1/2}$ so that its columns form an orthonormal set. If $\|\hat{B}_{(t)} - \hat{B}_{(t-1)}\| > \varepsilon$, then compute h_t, b_t as in step 4 of the rdOPG algorithm and return to step 2; otherwise, use $\hat{D}^{-1/2}\hat{B}_{(t)}$ as the estimate of a basis in $\mathscr{S}_{Y|X}$.

The following R-codes for rdMAVE calls the functions `rdopg` (first iteration), `kern`, `standvec`, and `swls` listed in the R-codes for OPG and rdOPG.

R-code for rdMAVE

```
rdmave=function(x,y,d,nit){
sig=diag(var(x));x=apply(x,2,standvec);y=(y-mean(y))/sd(y)
n=dim(x)[1];p=dim(x)[2]
c0=2.34;p0=max(p,3);rn=n^(-1/(2*(p0+6)))
h=c0*n^(-(1/(p0+6)));b=c0*n^(-(1/(p0+5)))
hmat=kern(y,b);bbt=rdopg(x,y,d,1)
for(iit in 1:nit){
kmat=kern(x%*%bbt,h);mkmat=apply(kmat,2,mean);kmat=t(kmat*(1/mkmat))
acarray=array(0,c(1+d,n,n))
for(i in 1:n) acarray[,,i]=swls(hmat,kmat,x%*%bbt,i)$ab
vecb1=0;vecb2=0; for(i in 1:n) for(j in 1:n) {
tmp=t(t(x)-x[i,]);tmp1=kmat[,i];tmp3=acarray[2:(1+d),j,i]
vecb1=vecb1+kronecker(t(tmp*tmp1)%*%tmp,tmp3%*%t(tmp3))
tmp4=kmat[,i]*(hmat[,j]-acarray[1,j,i])
vecb2=vecb2+kronecker(t(tmp),tmp3)%*%tmp4}
bbt=t(matrix(solve(vecb1)%*%vecb2,2,p))
h=max(rn*h,c0*n^((-1/(d+4))))
b=max(rn*b,c0*n^(-1/((d+3))),c0*n^(-1/5))
}
beta.final=diag(sig^(-1/2))%*%bbt
return(beta.final)
}
```

11.8   Ensemble Estimators

Following the spirit of dMAVE, Wang and Xia (2008) proposed another forward regression method called Sliced Regression, which, instead of estimating the gradient of the conditional density of Y given X, targets the gradient of the conditional cumulative distribution of Y given X. This is achieved by replacing the kernel $H_b(Y - y)$ in dMAVE by the indicator function $I(Y \le y)$. One immediate advantage of this idea is that one does not need to choose the bandwidth b; another is that the indicator function is bounded, making the method robust against the outliers in Y.

The ideas of dMAVE and Sliced Regression were further developed by Yin and Li (2011) into a general class of *ensemble estimators*. This method is based on the fact that, for a sufficiently rich class of functions of y, say $\mathfrak{F}$, the collection of conditional expectations $\{E[f(Y)|X] : f \in \mathfrak{F}\}$ completely characterizes the conditional distribution of $Y|X$. This means the collection of the central mean subspaces $\{\mathscr{S}_{E[f(Y)|X]} : f \in \mathfrak{F}\}$ span the central subspace $\mathscr{S}_{Y|X}$. Consequently, we can perform a collection of dimension reductions for the conditional mean to recover the central subspace.

Apart from dMAVE and Sliced Regression, two other sources for the idea of the ensemble estimator are the Fourier transformation approach to SDR (zhu; Zeng, 2008; Zhou and Zhu, 2010) and the projective resampling method for SDR with multivariate responses (Li et al., 2008). The former puts together the central mean subspaces of e^{itY}, as estimated by Fourier transform of the gradient of $E(e^{uY}|X)$, to recover the central subspace. The latter recovers the central subspace by projecting a multivariate Y repeatedly on to random directions and estimating the central mean subspace for each randomly projected response. In both cases the central subspace is estimated by performing dimension reduction for the conditional mean of a family of functions of Y.

The ensemble estimator is applicable to *any* existing dimension reduction method that targets the conditional mean — such as those discussed in Chapter 8 — not restricted to the MAVE-type estimators. Nevertheless, here we shall focus on the MAVE-type methods to illustrate ensemble estimation. When applied with the MAVE-type methods, the ensemble estimators include dMAVE and Sliced Regression as special cases: in the former case, the class of functions is $\{H_b(Y-y) : y \in \Omega_Y\}$; in the latter case, the class of functions is $\{I(Y \leq y) : y \in \Omega_Y\}$. In the following, $\mathbb{F}$ is a field that can be the set of real numbers $\mathbb{R}$, or the set of complex numbers $\mathbb{C}$.

Definition 11.2 *Let $\mathfrak{F}$ be a family of $\mathbb{F}$-valued functions defined on Ω_Y. We say that $\mathfrak{F}$ characterizes the central subspace if*

$$\mathscr{S}(\mathfrak{F}) \equiv \text{span}\{\mathscr{S}_{E[f(Y)|X]} : f \in \mathfrak{F}\} = \mathscr{S}_{Y|X}$$

It can be shown that, for example, if $\mathfrak{F}$ is dense in $L_2(P_Y)$, where P_Y is the distribution of Y, then $\mathfrak{F}$ characterizes the central subspace; see Yin and Li (2011). There are many such families. For example, the family $\{y \mapsto e^{uy} : t \in \mathbb{R}\}$, where $\iota = \sqrt{-1}$ because the characteristic function $E(e^{uY}|X)$ uniquely determines the conditional distribution of Y given X. The Gauss kernel family $\{y \mapsto H_b(y,u) : u \in \Omega_Y\}$ is also dense in $L_2(P_Y)$; see Fukumizu et al. (2009). If the moment generating function of Y exists, then the class of polynomials $\{y^m : m = 0, 1, \ldots\}$ is dense in $L_2(P_Y)$. Finally, the class of indicator functions $\{y \mapsto I(y \leq u) : u \in \Omega_Y\}$ is dense in $L_2(P_X)$ and hence also characterizes $\mathscr{S}_{Y|X}$, justifying the Sliced Regression.

Furthermore, even if $\mathfrak{F}$ is not characteristic, it still makes sense to carry out the ensemble estimate because $Y \perp\!\!\!\perp X|\beta^{\mathsf{T}}X$ implies $E[f(Y)|X] = E[f(Y)|\beta^{\mathsf{T}}X]$, which implies $\mathscr{S}(\mathfrak{F})$ is always a subset of $\mathscr{S}_{Y|X}$, whether or not $\mathscr{F}$ characterizes $\mathscr{S}_{Y|X}$. Also note that, if $\mathfrak{F}_1 \subseteq \mathfrak{F}_2$, then $\mathscr{S}(\mathfrak{F}_1) \subseteq \mathscr{S}(\mathfrak{F}_2)$. Thus, we can estimate a larger portion of $\mathscr{S}_{Y|X}$ if $\mathfrak{F}$ is a richer class. Two classes were investigated in Yin and Li (2011): the class $\{y \mapsto e^{uy} : t \in \mathbb{R}\}$, which is written as $\mathfrak{F}_C$, and the class of Box-Cox transformations:

$$f(y;t) = \begin{cases} (y^t - 1)/t & t \neq 0 \\ \log(y) & t = 0 \end{cases}$$

where t ranges over an interval in $\mathbb{R}$, say $[-2,2]$. We denote the Box-Cox family by $\mathfrak{F}_B$. Yin and Li (2011) recommended the family $\{f(\cdot;t) : t = -2, -1.5, -1, -0.5, 0, 0.5, 1, 1.5, 2\}$. In addition, the Haar wavelets (see, for example, Donoho and Johnstone (1994)) were also mentioned as a possible $\mathfrak{F}$. Henceforth we refer to the class $\mathfrak{F}$ as an *ensemble*.

The ensemble estimation not only gives us a wider family of functions y to choose from, but also allows us to substantially cut down the computing time. For larger sample size n, say $n \geq 1000$, the triple sums in (11.32) for rdMAVE can be quite time consuming even if we use the fast algorithm based on (11.34). Based on our experience, ensemble estimators can achieve accuracy comparable to methods such as rdMAVE using far fewer functions than n (i.e. $m \ll n$). Another advantage of the ensemble approach is we often do not need to choose a bandwidth for the kernel of y. This, of course, depends on the family $\mathfrak{F}$ used, as $\{y' \mapsto H_b(y'-y) : y \in \Omega_Y\}$ itself can be regarded as an ensemble.

The ensemble method can be combined with OPG, MAVE, rOPG, and rMAVE. We denote these combinations as eOPG, eMAVE, reOPG, and reMAVE. Here, we only present the procedures reOPG and reMAVE, as their simplified versions eOPG and eMAVE can be easily derived from them.

For ensemble estimator with $\mathfrak{F}_C$, assuming that the Y_i's are standardized to have mean 0 and standard deviation 1, we generate $t_1, \ldots, t_m$ from $N(0, 1)$ distribution and form the family

$$\mathfrak{F}_C = \{e^{\iota t_k y} : k = 1, \ldots, m\}$$

Let $\mathbb{C}$ be the collection of complex numbers. For each $i = 1, \ldots, n$, $j = 1, \ldots, m$, the objective function for reOPG is

$$L(a, b) = \sum_{k=1}^{n} |e^{\iota t_j Y_k} - a - b^{\mathsf{T}}(X_k - X_i)|^2 K_h(B^{\mathsf{T}}(X_k - X_i))$$

where $a \in \mathbb{C}$, $b \in \mathbb{C}^p$, and $|\cdot|$ is the absolute value of a complex number; that is $|r_1 + \iota r_2| = (r_1^2 + r_2^2)^{1/2}$. As in rdOPG, the B in the kernel K_h above is I_p in the first iteration and is a $p \times d$ dimensional matrix after the first iteration. For a complex number or vector u, let $u(1)$ be the real part of u, and $u(2)$ be the imaginary part of u. The above objective function can be written as the sum of two terms according to the real and imaginary parts:

$$L(a, b) = \sum_{\ell=1}^{2} \sum_{k=1}^{n} [f_\ell(t_j Y_k) - a(\ell) - b^{\mathsf{T}}(\ell)(X_k - X_i)]^2 K_h(B^{\mathsf{T}}(X_k - X_i)), \qquad (11.35)$$

where $f_1(u) = \cos(u)$, $f_2(u) = \sin(u)$. As in the algorithm for rdOPG, to speed up computation, we write (11.35) in a matrix form. Let

$$A_i = \begin{pmatrix} a_{i1}(1) & \cdots & a_{im}(1) & a_{i1}(2) & \cdots & a_{im}(2) \\ b_{i1}(1) & \cdots & b_{im}(1) & b_{i1}(2) & \cdots & b_{im}(2) \end{pmatrix},$$

$$F.. = \begin{pmatrix} f_1(t_1 y_1) & \cdots & f_1(t_m y_1) & f_2(t_1 y_1) & \cdots & f_2(t_m y_1) \\ \vdots & & & & & \vdots \\ f_1(t_1 y_n) & \cdots & f_1(t_m y_n) & f_2(t_1 y_n) & \cdots & f_2(t_m y_n) \end{pmatrix}.$$

The objective function (11.35) can be rewritten as

$$L(a, b) = [F.. - (1_n, \Delta._i)A_i]^{\mathsf{T}} W_i(h, B)[F.. - (1_n, \Delta._i)A_i],$$

where $W_i(h, B)$ is as defined in (11.30). The explicit minimizer is

$$\hat{A}_i = [(1_n, \Delta._i)^{\mathsf{T}} W_i(h, B)(1_n, \Delta._i)]^{-1}(1_n, \Delta._i)^{\mathsf{T}} W_i(h, B)F..$$

Then read off the vectors $\hat{b}_{i1}, \ldots, \hat{b}_{im}$ from the $2, \ldots, p+1$ rows of $\hat{A}_i$ to form the updated candidate matrix $\hat{C} = \sum_{i=1}^{n} \sum_{j=1}^{m} \hat{b}_{ij} \hat{b}_{ij}$. Finally, update the matrix $\hat{B}$ by extracting the first d eigenvectors of $\hat{C}$ and repeat the process until convergence.

For ensemble estimate with $\mathfrak{F}_B$, the algorithm is the same except that A_i and $F..$ are redefined as

$$A_i = \begin{pmatrix} a_{i1} & \cdots & a_{im} \\ b_{i1} & \cdots & b_{im} \end{pmatrix}, \quad F.. = \begin{pmatrix} f(y_1, t_1) & \cdots & f(y_1, t_m) \\ \vdots & & \vdots \\ f(y_n, t_1) & \cdots & f(y_n, t_m), \end{pmatrix},$$

and $W_i(h, B)$ is as defined in (11.30). The explicit minimizer is

$$\hat{A}_i = [(1_n, \Delta._i)^{\mathsf{T}} W_i(h, B)(1_n, \Delta._i)]^{-1}(1_n, \Delta._i)^{\mathsf{T}} W_i(h, B)F..$$

The following is the R-code for reOPG, which calls the functions `kern`, `standvec`, and `swls` listed in the R-codes for OPG and rdOPG. The inputs are the same as those for rdOPG, with the additional input `ensemble`, which represents the ensemble family $\mathfrak{F}$. It has two values: `fc` indicates $\mathfrak{F}_C$; `fb` indicates $\mathfrak{F}_B$.

R-code for reOPG

```
reopg=function(x,y,d,m,nit,ensemble){
sig=diag(var(x));x=apply(x,2,standvec);y=(y-mean(y))/sd(y)
n=dim(x)[1];p=dim(x)[2]
c0=2.34;p0=max(p,3);rn=n^(-1/(2*(p0+6)))
h=c0*n^(-(1/(p0+6)));b=c0*n^(-(1/(p0+5)))
if(ensemble=="fc")
{tt=rnorm(m);hmat=cbind(cos(y%*%t(tt)),sin(y%*%t(tt)))}
if(ensemble=="fb")
{tt=seq(from=-2,to=2,by=0.5);tt=tt[-5]
y1=y-min(y)+0.5;hmat=cbind((y1%*%t(tt)-1)%*%diag(1/tt),log(y1))}
bbt=diag(p)
for(iit in 1:nit){
kmat=kern(x%*%bbt,h);bmat=numeric()
for(i in 1:n) bmat=cbind(bmat,swls(hmat,kmat,x,i)$b)
bbt=bmat%*%t(bmat)/(n^2);bbt=eigen(bbt)$vectors[,1:d]
h=max(rn*h,c0*n^((-1/(d+4))))
b=max(rn*b,c0*n^(-1/((d+3))),c0*n^(-1/5))
}
beta.final=diag(sig^(-1/2))%*%bbt
return(beta.final)
}
```

To develop the algorithm for reMAVE, let us use the generic notation

$$\mathfrak{F} = \{f(\cdot;\theta) : \theta \in \Theta\}$$

to denote the ensemble, where Θ is a finite set $\{\theta_1,\ldots,\theta_m\}$. Thus, for $\mathfrak{F}_C$, $\Theta = \{1,2\} \times \{t_1,\ldots,t_m\}$, and $f(y;(1,t_i)) = \cos(t_i y)$, and $f(y;(2,t_i)) = \sin(t_i y)$. For $\mathfrak{F}_B$, Θ is simply $\{-2,-1.5,\ldots,2\}$ as described before. Let $F_{..}$ be the $n \times m$ matrix $\{f(y_k;\theta_j) : k = 1,\ldots,n, j = 1,\ldots,m\}$. Let $M_i(h,B)$ be the same as defined in (11.33), and let

$$L_{ij}(h,b,B) = (W_h(B^\mathsf{T}X_i,B^\mathsf{T}X_1)f(y_1;\theta_j),\ldots,W_h(B^\mathsf{T}X_i,B^\mathsf{T}X_n))f(y_n;\theta_j))^\mathsf{T}.$$

Then the algorithm for reMAVE consists of iterations between the two equations

$$\hat{G}_i(h,B) = [(1_n,\Delta_{\cdot i}(B))^\mathsf{T}W_i(h,B)(1_n,\Delta_{\cdot i})]^{-1}(1_n,\Delta_{\cdot i}(B))^\mathsf{T}W_i(h,B)F_{..}$$

$$\text{vec}[B_{\text{new}}^\mathsf{T}(h,B)) = \left\{ \sum_{i=1}^{n}\sum_{j=1}^{m} [(\Delta_{\cdot i}^\mathsf{T}M_i(h,B)\Delta_{\cdot i}) \otimes (c_{ij}c_{ij}^\mathsf{T})] \right\}^{-1}$$

$$\left\{ \sum_{i=1}^{n}\sum_{j=1}^{m} (\Delta_{\cdot i}^\mathsf{T} \otimes c_{ij})L_{ij}(h,B) \right\}.$$

The R-code for reMAVE is given below, where the inputs are the same as reOPG.

R-code for reMAVE

```
remave=function(x,y,d,m,nit,ensemble){
sig=diag(var(x));x=apply(x,2,standvec);y=(y-mean(y))/sd(y)
n=dim(x)[1];p=dim(x)[2]
c0=2.34;p0=max(p,3);rn=n^(-1/(2*(p0+6)))
h=c0*n^(-(1/(p0+6)));b=c0*n^(-(1/(p0+5)))
if(ensemble=="fc")
{tt=rnorm(m);hmat=cbind(cos(y%*%t(tt)),sin(y%*%t(tt)))}
if(ensemble=="fb")
{tt=seq(from=-2,to=2,by=0.5);tt=tt[-5]
y1=y-min(y)+0.5;hmat=cbind((y1%*%t(tt)-1)%*%diag(1/tt),log(y1))}
bbt=rdopg(x,y,d,1)
for(iit in 1:nit){
kmat=kern(x%*%bbt,h);mkmat=apply(kmat,2,mean);kmat=t(kmat*(1/mkmat))
if(ensemble=="fc") {acarray=array(0,c(1+d,2*m,n));jlim=2*m}
if(ensemble=="fb") {acarray=array(0,c(1+d,9,n));jlim=9}
for(i in 1:n) acarray[,,i]=swls(hmat,kmat,x%*%bbt,i)$ab
vecb1=0;vecb2=0; for(i in 1:n) for(j in 1:jlim) {
tmp=t(t(x)-x[i,]);tmp1=kmat[,i];tmp3=acarray[2:(1+d),j,i]
vecb1=vecb1+kronecker(t(tmp*tmp1)%*%tmp,tmp3%*%t(tmp3))
tmp4=kmat[,i]*(hmat[,j]-acarray[1,j,i])
vecb2=vecb2+kronecker(t(tmp),tmp3)%*%tmp4}
bbt=t(matrix(solve(vecb1)%*%vecb2,2,p))
h=max(rn*h,c0*n^((-1/(d+4))))
b=max(rn*b,c0*n^(-1/((d+3))),c0*n^(-1/5))
}
beta.final=diag(sig^(-1/2))%*%bbt
return(beta.final)
}
```

11.9 Simulation Studies and Applications

In this section we compare performances of the methods described in Sections 11.1 through 11.8: OPG, MAVE, rOPG, rMAVE, rdOPG, rdMAVE, reOPG, and reMAVE, under two models. The ensemble for reOPG and reMAVE is taken to be $\mathfrak{F}_c$, with ten t independently generated from $N(0,1)$.

Example 11.1 The first model is

$$Y = 2\sin(1.4X_1) + (X_2 + 1)^2 + 0.4\varepsilon, \tag{11.36}$$

where $\varepsilon \sim N(0,1)$, $X \sim N(0, I_p)$, $p = 10$, and $\varepsilon \perp\!\!\!\perp X$. The sample size is $n = 200$. In this model the central subspace and the central mean subspace are the same subspace, spanned by the vectors e_1, e_2, so that all eight methods estimate the same subspace. The points of this comparison are two: the first is to see the effect of refinement; that is, how much better rOPG and rMAVE are than OPG and MAVE; the second is to see what is the cost of estimating the central subspace when it is in fact the central mean subspace. Note that rdOPG, rdMAVE, reOPG, and reMAVE are designed to estimate the central subspace $\mathscr{S}_{Y|X}$, whereas OPG, rOPG, MAVE, and rMAVE are designed to estimate the central mean subspace $\mathscr{S}_{E(Y|X)}$. Thus we expect that when these two subspaces are the same, the rdOPG, rdMAVE, reOPG, and reMAVE may be less efficient than the OPG, rOPG, MAVE, and rMAVE. By comparing the eight methods under the above model, we can get a rough idea of the efficiency loss incurred by employing a more complex method to estimate the same object.

To evaluate the performance of each estimator, we used the distance measure in Li et al. (2005). That is, if $\mathscr{S}_1$ and $\mathscr{S}_2$ are two subspaces of $\mathbb{R}^p$ of the same dimension, then their distance is defined as

$$\|P_{\mathscr{S}_1} - P_{\mathscr{S}_2}\|,$$

where $P_{\mathscr{S}}$ is the projection on to $\mathscr{S}$, and $\|\cdot\|$ is the Frobenius matrix norm.

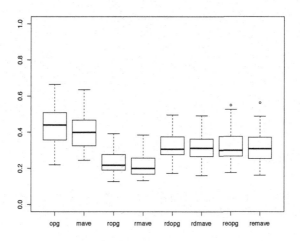

Figure 11.1 *Comparison of OPG, MAVE, rOPG, rMAVE, rdOPG, rdMAVE, reOPG, reMAVE for the model in Example 11.1.*

For each of the eight methods, we generate 50 samples from (11.36) and compute the above distances between the true β and the estimated $\hat{\beta}$'s. A small distance indicates a more accurate estimate. We show in Figure 11.1 the boxplots of these distances for the eight estimators. We note that rOPG and rMAVE are substantially better than OPG and MAVE. Thus the refinement does lead to significant improvement in accuracy. We also note that rOPG and rMAVE perform significantly better than rdOPG, rdMAVE, reOPG and reMAVE, which indicates that targeting the central mean subspace does lead to improved efficiency. □

Example 11.2 Next, we compare the eight methods under the model

$$Y = 2(\beta_1^{\mathsf{T}}X)^2 + 2\exp(\beta_2^{\mathsf{T}}X)\varepsilon,$$

where X is a 10-dimensional random vector whose components are i.i.d. $\sim U(-\sqrt{3}, \sqrt{3})$ variables, $\varepsilon \sim N(0,1)$, $X \perp\!\!\!\perp \varepsilon$, and

$$\beta_1^{\mathsf{T}} = (1,2,0,\ldots,0,2)/3, \quad \beta_1^{\mathsf{T}} = (0,0,3,4,0,\ldots,0)/5.$$

In this model, the central mean subspace is different from the central subspace: the former is spanned by β_1; the latter by β_1 and β_2. The points of this comparison are: first, to see whether rdOPG, rdMAVE, reOPG, and reMAVE successfully recover the central subspace as they are designed to do, and second, to show that these forward regression methods do not rely on the elliptical distribution assumption.

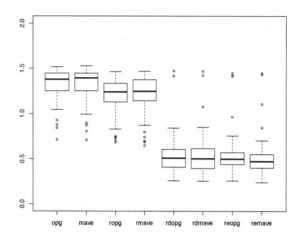

Figure 11.2 *Comparison of OPG, MAVE, rOPG, rMAVE, rdOPG, rdMAVE, reOPG, reMAVE for the model in Example 11.2.*

Like in the previous example, we show in Figure 11.2 the 50 distances of the estimates from the true central subspace for each of the eight methods. We can see a sharp drop in the distances by the density and ensemble based methods. OPG and MAVE and their refinements can only estimate β_1, but missed β_2 entirely. On the other hand the density and ensemble based methods estimate the subspace spanned by both β_1 and β_2. The high performance of these methods show that they are not affected by the shape of the distribution of X.   □

The advantages of the ensemble estimates are clearly shown in the above two examples. In both cases, the performances of the two ensemble estimators are comparable with — or even slightly better than — those of rdOPG and rdMAVE, even though they only take a fraction of the computing times of rdOPG and rdMAVE. In fact, because we chose $m = 10$ for the ensemble $\mathfrak{F}_C$, there are only 20 functions in the ensemble, as compared with 200 functions for rdOPG and rdMAVE. As a result, the computing times for reOPG and reMAVE are roughly 1/10 of computing times for drOPG and rdMAVE.

Example 11.3 In this example we apply the forward regression methods to a data set concerning the ages and various measurements of abalones. The response variable is the ages (as measured by the number of rings) of the abalones; the seven predictors are

1. Length
2. Diameter
3. Height

4. Whole weight
5. Shucked weight
6. Viscera weight
7. Shell weight

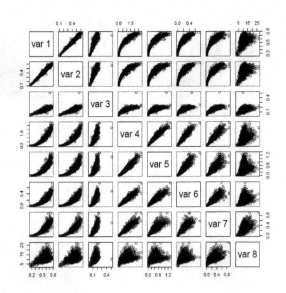

Figure 11.3 *Scatter plot matrix for the abalone data.*

The data consists of observations from 4177 abalones, of which 1528 are male, 1307 are female, and 1342 are infant. For simplicity we only used data from the male abalones. The scatter plot matrix of the eight variable (with the response as the eighth variable) is shown in Figure 11.3. We observe that some scatter plots are severely skewed or non-elliptical, such as X_2 versus X_7; there is heteroscedasticity among the predictors, such as X_5 versus X_7; there is heteroscedasticity in the response, such as Y versus X_1, X_2, X_4, X_5, X_6 and X_7.

We used two forward regression methods to fit this data set: rOPG and reOPG. For reOPG, we used the ensemble $\mathfrak{F}_C$ with ten t's generated independently from $N(0, 1)$. Figure 11.4 shows the response versus the first two rOPG predictors (upper panels) and versus the first two reOPG predictors (lower panels). The sufficient predictor plots do not show a significant difference between rOPG and reOPG. This indicates that the central mean subspace and central subspace are the same for this data set. On the other hand, we do see strong heteroscedasticity in Y. This seems to indicate that the heteroscedasticity is a function of the mean. A possible explanation is the following. Note that the predictors in this data set are all related to the weights and sizes of the abalones. It is conceivable that, after a certain age, abalones stops growing in size or weight — or such growth slows down considerably. This tendency would cause the age to have larger variation in the abalones who have grown to larger weights and sizes. Consequently, both the conditional mean and conditional variance of age vary in the same direction.

Another point worth mentioning here is that, even for this modest sample size, it is already infeasible to perform a fully fledged dMAVE or rdMAVE, because the large ensemble size makes computing painstakingly slow when n is in the thousand or tens of thousands. On the other hand, the twenty-member ensemble $\mathfrak{F}_c$ renders the computing time to a few minutes.

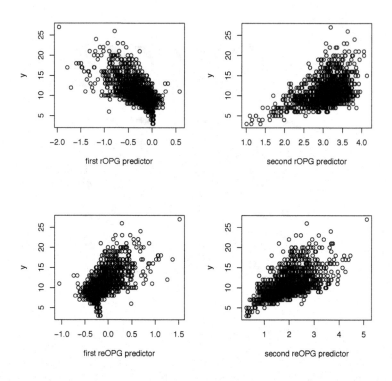

Figure 11.4 *Age versus the first two predictors from rOPG and reOPG for the abalone data.*

□

11.10   Summary

In this chapter we systematically develop theories and algorithms for the forward regression methods for Sufficient Dimension Reduction. The ideas are centered around gradient estimation by local linear regression. The basic methods are OPG and MAVE. Extensions are made in two directions: by refinement through updated low-dimensional kernels, and by ensembles of functions to recover the central subspace.

This multitude of forward regression methods can be categorized by whether refinement is used and what type of ensemble is used. For example, OPG and MAVE can be regarded as special cases of the ensemble estimates with the ensemble consisting of the single function $f(y) = y$. For convenience, let sOPG and sMAVE denote the the methods developed in Xia (2007), where the prefix s stands for "slice"; let rsOPG and rsMAVE denote the corresponding refinement. Let cOPG and cMAVE denote the ensemble estimates using the ensemble $\mathfrak{F}_c$, where prefix c stands for "characteristic function"; let rcOPG and rcMAVE denote the corresponding refinement. Similarly define bOPG, bMAVE, rbOPG, and rbMAVE, where the prefix b stands for "Box-Cox transformations". With these abbreviations, the following table summarizes all the methods in this chapter according to the presence of refinement and types of ensembles.

Table 11.1 *Types of forward regression estimators*

ensemble type	no refinement		refinement	
	OPG	MAVE	OPG	MAVE
$\{y\}$	OPG	MAVE	rOPG	rMAVE
$\{H_b(y'-y) : y \in \Omega_Y\}$	dOPG	dMAVE	rdOPG	rdMAVE
$\{I(y \leq b) : b \in \mathbb{R}\}$	sOPG	sMAVE	rsOPG	rsMAVE
$\{e^{\iota t y} : t \in \mathbb{R}\}$	cOPG	cMAVE	rcOPG	rcMAVE
Box-Cox	bOPG	bMAVE	rbOPG	rbMAVE

Chapter 12

Nonlinear Sufficient Dimension Reduction

12.1 Reproducing Kernel Hilbert Space   192
12.2 Covariance Operators in RKHS   193
12.3 Coordinate Mapping   199
12.4 Coordinate of Covariance Operators   200
12.5 Kernel Principal Component Analysis   202
12.6 Sufficient and Central σ-Field for Nonlinear SDR   204
12.7 Complete Sub σ-Field for Nonlinear SDR   206
12.8 Converting σ-Fields to Function Classes for Estimation   208

In this chapter we outline the general theory for nonlinear sufficient dimension reduction as formulated in Li et al. (2011) and Lee et al. (2013). The sufficient dimension reduction problems studied so far rely on a set of linear functions of X. That is, $Y \perp\!\!\!\perp X | \beta^{\mathsf{T}} X$. Nonlinear sufficient dimension reduction replaces the linear sufficient predictor $\beta^{\mathsf{T}} X$ by a nonlinear predictor $f(X)$. That is, we consider the following problem

$$Y \perp\!\!\!\perp X | f(X), \tag{12.1}$$

where f is a member of a general Hilbert space. This greatly expands the scope and applicability of the linear SDR, and also makes connections with several previously existing dimension reduction methods, such as the Kernel Sliced Inverse Regression by Wu (2008) and Kernel Canonical Correlation Analysis developed by Akaho (2001), Bach and Jordan (2002), and Fukumizu et al. (2007). The general problem (12.1) was also articulated in Cook (2007). Interestingly, this framework has much in common with the linear SDR, with concepts such as sufficiency, unbiasedness, exhaustiveness, inverse regression, all neatly carried over to the current setting. In addition, its generality allows us to bring more insights from classical statistical inference into this field, such as the relation between completeness and minimal sufficiency, the relation between inverse regression and completeness, which do not have counterparts in linear SDR.

Although the theoretical formulation of nonlinear Sufficient Dimension Reduction does not rely on the form of the Hilbert space that accommodates f in (12.1), in the special case when this Hilbert space is a reproducing kernel Hilbert space, the analogy between linear and nonlinear SDR is the most pronounced, because $f(x)$ can be written as $\langle f, \kappa(\cdot, x) \rangle$ — where $\kappa(\cdot, x)$ is the reproducing kernel — is a direct generalization of the Euclidean inner procuct $\beta^{\mathsf{T}} x$. For this reason, and also because RKHS will be heavily used in the next chapter, we start our exposition with reproducing kernel Hilbert spaces.

12.1 Reproducing Kernel Hilbert Space

In this section we give an introduction to the theory surrounding the Reproducing Kernel Hilbert Space, which is important for both this and the next chapters. For more information about this topic, see, for example, Berlinet and Thomas-Agnon (2004) and Hsing and Eubank (2015).

Let S be a set in a metric space. A positive definite kernel is any function $\kappa : S \times S \to \mathbb{R}$ such that, for any finite subset $s_1, \ldots, s_m$ of S, the $m \times m$ matrix $\{\kappa(s_i, s_j) : i, j = 1, \ldots, m\}$ is symmetric and positive definite. There are many such kernel functions; for example, the Gaussian Radial Basis function

$$\kappa(s_1, s_2) = e^{-\gamma \|s_1 - s_2\|^2}, \quad s_1, s_2 \in S, \tag{12.2}$$

is positive definite. Now consider the collection functions

$$\mathcal{H}_0 = \mathrm{span}\{\kappa(\cdot, s) : s \in S\};$$

that is, the collection of all functions of the form $a_1 \kappa(\cdot, s_1) + \cdots + a_m \kappa(\cdot, s_m)$ where $a_1, \ldots, a_m \in \mathbb{R}$ and $s_1, \ldots, s_m \in S$. It is easy to see that this is a linear manifold. Consider the bilinear form

$$u : \mathcal{H}_0 \times \mathcal{H}_0 \to \mathbb{R}, \quad \left(\sum_{i=1}^{m} a_i \kappa(\cdot, s_i), \sum_{j=1}^{n} b_j \kappa(\cdot, t_j) \right) = \sum_{i=1}^{m} \sum_{j=1}^{n} a_i b_j \kappa(s_i, t_j).$$

It can be easily shown that, under the assumption that κ is positive definite, the above bilinear form is an inner product. The linear manifold, $\mathcal{H}_0$, together with the bilinear form u, forms an inner product space, or pre-Hilbert space. Its completion, which can be viewed as the collection of functions of the form

$$\mathcal{H} = \left\{ \sum_{i=1}^{\infty} a_i \kappa(\cdot, s_i) : \sum_{i=1}^{\infty} a_i^2 \kappa(s_i, s_i) < \infty \right\},$$

is then a Hilbert space with inner product

$$\left\langle \sum_{i=1}^{\infty} a_i \kappa(\cdot, s_i), \sum_{j=1}^{\infty} b_j \kappa(\cdot, t_j) \right\rangle_{\mathcal{H}} = \sum_{i=1}^{\infty} \sum_{j=1}^{\infty} a_i b_j \langle \kappa(s_i, t_j).$$

The space $\mathcal{H}$ is called the Reproducing Kernel Hilbert Space (RKHS) generated by the kernel κ. Note that, for any $f(x) = \sum_{i=1}^{\infty} a_i \kappa(\cdot, s_i)$, we have

$$\langle f, \kappa(\cdot, s) \rangle_{\mathcal{H}} = \sum_{i=1}^{\infty} a_i \kappa(s, s_i) = f(s). \tag{12.3}$$

This property is called the "reproducing property", because the inner product between f with the kernel $\kappa(\cdot, s)$ "reproduces" the evaluation of f at s. This property implies that, for any $s \in S$, the *evaluation functional*

$$T_s : \mathcal{H} \to \mathbb{R}, \quad f \mapsto f(s)$$

is a bounded linear functional with $\|T_s\|$ no greater than $\kappa(s,s)$, because, by the Cauchy-Schwarz inequality,

$$|T_s(f)| = |f(s)| = |\langle f, \kappa(\cdot, s)\rangle_{\mathscr{H}}| \le \kappa(s,s)\|f\|_{\mathscr{H}}.$$

The property that the evaluation functional is a bounded linear functional turns out to be the defining property of an RKHS, in a sense that *any* Hilbert space with this property is generated by some reproducing kernel in the sense of (12.3). Indeed, if $\mathscr{H}$ is a Hilbert space of functions defined on a set S and if, for any $s \in S$, the evaluation functional T_s is bounded, then, by the Riesz representation theorem, there is a unique member f_s of $\mathscr{H}$ such that

$$T_s(f) = \langle f, f_s\rangle_{\mathscr{H}}$$

for all $f \in \mathscr{H}$. Let $\kappa : S \times S \to \mathbb{R}$ be the function $(s,t) = f_s(t)$. Then it can be shown that κ is a positive definite kernel with the reproducing property (12.3).

12.2 Covariance Operators in RKHS

In this section we generalize covariance matrix to covariance operator. The former characterizes linear features of X; the latter characterizes the nonlinear features of X. Let $(\Omega, \mathscr{F}, P)$ be a probability space, let Ω_X be a subset of $\mathbb{R}^p$, and let $X : \Omega \to \Omega_X$ be a random vector of dimension p. Let $\mathscr{H}_X$ be a reproducing kernel Hilbert space of functions of X generated by a positive kernel $\kappa : \Omega_X \times \Omega_X \to \mathbb{R}$. Consider the mapping

$$X : \quad \mathscr{H}_X \to \mathbb{R}, \quad f \mapsto E[f(X)]. \tag{12.4}$$

Since the right-hand side can be rewritten as $E\langle f, \kappa(\cdot, X)\rangle_{\mathscr{H}_X}$, by the Cauchy-Schwarz inequality we have,

$$|E[f(X)]| \le \|f\|_{\mathscr{H}_X} [E(\kappa(X,X))]^{1/2}.$$

Thus, if we assume $E[\kappa(X,X)] < \infty$, then the mapping (12.4) is a bounded linear functional. By Riesz representation theorem, there is a unique member $f_0 \in \mathscr{H}_X$ such that

$$\langle f, f_0\rangle = E\langle f, \kappa(\cdot, X)\rangle_{\mathscr{H}_X}.$$

We call this member f_0 the mean element of X, and denote it by $E[\kappa(\cdot, X)]$. Thus, by construction, we have the identity

$$\langle f, E[\kappa(\cdot, X)]\rangle_{\mathscr{H}_X} = E\langle f, \kappa(\cdot, X)\rangle_{\mathscr{H}_X}.$$

Having defined the expectation of a random function in $\mathscr{H}_X$, let us define the expectation of a random operator. Let $\mathscr{B}(\mathscr{H}_X, \mathscr{H}_X)$ be the class of all bounded linear operators from $\mathscr{H}_X$ to $\mathscr{H}_X$. A random operator is a mapping from Ω to $\mathscr{B}(\mathscr{H}_X, \mathscr{H}_X)$ that is measurable with respect to $\mathscr{F}$ and the Borel σ-field generated by the open sets in $\mathscr{B}(\mathscr{H}_X, \mathscr{H}_X)$. Thus, we can write a random operator as

$$A : \quad \Omega \to \mathscr{B}(\mathscr{H}_X, \mathscr{H}_X), \quad \omega \mapsto A_\omega.$$

For a random operator A, and $f, g \in \mathscr{H}_X$, $\langle f, Ag\rangle_{\mathscr{H}_X}$ is a well defined random variable, and the mapping

$$\mathscr{H}_X \times \mathscr{H}_X \to \mathbb{R}, \quad (f,g) \mapsto E(\langle f, Ag\rangle_{\mathscr{H}_X}) \tag{12.5}$$

is a well defined bilinear form. Moreover, if the random variable $\|A\|$ is integrable, where $\|\cdot\|$ is the norm in $\mathscr{B}(\mathscr{H}_X, \mathscr{H}_X)$, then, by the Cauchy-Schwarz inequality,

$$|\langle f, Ag \rangle_{\mathscr{H}_X}| \le (E\|A\|)\|f\|_{\mathscr{H}_X}\|g\|_{\mathscr{H}_X}.$$

In other words the bilinear form (12.5) is bounded. By Theorem 2.2 in Conway (1990), there is a bounded linear operator $B = \mathscr{B}(\mathscr{H}_X, \mathscr{H}_X)$ such that $\langle f, Bg \rangle_{\mathscr{H}_X} = E(\langle f, Ag \rangle_{\mathscr{H}_X})$. The operator B is called the expectation of A, and is written as $E(A)$. In other words, $E(A)$ is the unique element in $\mathscr{H}_X$ that satisfies the relation

$$\langle f, E(A)g \rangle_{\mathscr{H}_X} = E(\langle f, Ag \rangle_{\mathscr{H}_X}). \tag{12.6}$$

For members f and g of $\mathscr{H}_X$, the tensor product $f \otimes g$ is the operator on $\mathscr{H}_X$ such that $(f \otimes g)h = f\langle g, h \rangle_{\mathscr{H}_X}$ for all $h \in \mathscr{H}_X$. Because

$$\langle f', (f \otimes g)g' \rangle_{\mathscr{H}_X} = \langle f', f\langle g, g' \rangle_{\mathscr{H}_X} \rangle_{\mathscr{H}_X} = \langle f', f \rangle_{\mathscr{H}_X} \langle g, g' \rangle_{\mathscr{H}_X},$$
$$\langle (g \otimes f)f', g' \rangle_{\mathscr{H}_X} = \langle \langle f', f \rangle_{\mathscr{H}_X} g, g' \rangle_{\mathscr{H}_X} = \langle f', f \rangle_{\mathscr{H}_X} \langle g', g \rangle_{\mathscr{H}_X},$$

the operator $g \otimes f$ is the adjoint operator of $f \otimes g$. That is,

$$(f \otimes g)^* = g \otimes f. \tag{12.7}$$

Now consider the following specific random operator

$$\Omega \to \mathscr{B}(\mathscr{H}_X, \mathscr{H}_X), \quad \omega \mapsto \kappa(\cdot, X(\omega)) \otimes \kappa(\cdot, X(\omega)).$$

We denote this random operator as $\kappa(\cdot, X) \otimes \kappa(\cdot, X)$. The next lemma gives sufficient conditions for the expectations of $\kappa(\cdot, X)$ and $\kappa(\cdot, X) \otimes \kappa(\cdot, X)$ to be defined.

Lemma 12.1 *1. If $E[\kappa(X, X)]^{1/2} < \infty$, then $E[\kappa(\cdot, X)]$ is a well defined member of $\mathscr{H}_X$;*

2. If $E[\kappa(X, X)] < \infty$, then $E[\kappa(\cdot, X) \otimes \kappa(\cdot, X)]$ is a well defined member of $\mathscr{B}(\mathscr{H}_X, \mathscr{H}_X)$;

PROOF. *1.* Note that

$$E\|\kappa(\cdot, X)\|_{\mathscr{H}_X} = E\langle \kappa(\cdot, X), \kappa(\cdot, X) \rangle_{\mathscr{H}_X}^{1/2} = E[\kappa(X, X)^{1/2}] < \infty.$$

Hence $E[\kappa(\cdot, X)]$ is a well defined member of $\mathscr{H}_X$.
2. For any $f \in \mathscr{H}_X$ and $g \in \mathscr{H}_X$, we have

$$\begin{aligned}
|E\langle f, \kappa(\cdot, X) \otimes \kappa(\cdot, X)g \rangle_{\mathscr{H}_X}| &\le E|\langle f, \kappa(\cdot, X) \otimes \kappa(\cdot, X)g \rangle_{\mathscr{H}_X}| \\
&= E\left(|\langle f, \kappa(\cdot, X) \rangle_{\mathscr{H}_X}| \, |\langle g, \kappa(\cdot, X) \rangle_{\mathscr{H}_X}|\right) \\
&\le \|f\|_{\mathscr{H}_X}\|g\|_{\mathscr{H}_X} E\kappa(X, X) < \infty.
\end{aligned}$$

Hence the mapping

$$(f, g) \mapsto E\langle f, \kappa(\cdot, X) \otimes \kappa(\cdot, X)g \rangle_{\mathscr{H}_X}$$

is a bounded bilinear form. $\qquad\square$

By Jensen's inequality,

$$E[\kappa(X, X)^{1/2}] \le [E\kappa(X, X)]^{1/2} < \infty.$$

So in the rest of this chapter we make the following assumption, which is sufficient for the existence of both expectations in the above lemma.

Assumption 12.1 $E[\kappa(X,X)] < \infty$.

We now summarize the above discussions in a formal definition, which also includes the concept of the variance operator.

Definition 12.1 *Under Assumption 12.1, the function $E[\kappa(\cdot,X)]$ is called the mean element of X in $\mathscr{H}_X$; the bounded linear operator $E[\kappa(\cdot,X) \otimes \kappa(\cdot,X)]$ is called the second-moment operator of X in $\mathscr{H}_X$; the bounded linear operator*

$$E[\kappa(\cdot,X) \otimes \kappa(\cdot,X)] - [E\kappa(\cdot,X)] \otimes [E\kappa(\cdot,X)] \tag{12.8}$$

is called the variance operator of X in $\mathscr{H}_X$.

We adopt the following notations for these objects:

$$\mu_X = E[\kappa(\cdot,X)],$$
$$M_{XX} = E[\kappa(\cdot,X) \otimes \kappa(\cdot,X)],$$
$$\Sigma_{XX} = M_{XX} - \mu_X \otimes \mu_X.$$

Recall that, when X is a random vector, the variance matrix var(X) satisfies

$$\mathrm{cov}(\alpha^{\mathsf{T}}X, \beta^{\mathsf{T}}X) = \alpha^{\mathsf{T}}\mathrm{var}(X)\beta$$

for all $\alpha, \beta \in \mathbb{R}^p$. A parallel property holds for the covariance operator Σ_{XX}, as shown in the next proposition.

Proposition 12.1 *If Assumption 12.1 is satisfied, then for all $f, g \in \mathscr{H}_X$,*

$$E[f(X)g(Y)] = \langle f, M_{XX}g \rangle_{\mathscr{H}_X}, \quad \mathrm{cov}[f(X), g(X)] = \langle f, \Sigma_{XX}g \rangle_{\mathscr{H}_X}.$$

PROOF. By construction, for $f, g \in \mathscr{H}_X$,

$$\begin{aligned}
\langle f, M_{XX}g \rangle_{\mathscr{H}} &= \langle f, E[\kappa(\cdot,X) \otimes \kappa(\cdot,X)]g \rangle_{\mathscr{H}} \\
&= E[\langle f, \kappa(\cdot,X) \otimes \kappa(\cdot,X)g \rangle_{\mathscr{H}}] \\
&= E[g(X)h(X)].
\end{aligned}$$

Similarly,

$$\begin{aligned}
\langle f, \Sigma_{XX}g \rangle_{\mathscr{H}} &= \langle f, M_{XX}g \rangle_{\mathscr{H}} - \langle f, (\mu_X \otimes \mu_X)g \rangle_{\mathscr{H}} \\
&= E[f(X)g(X)] - E[f(X)]E[g(X)] \\
&= \mathrm{cov}[f(X), g(X)].
\end{aligned}$$

This completes the proof.   □

We can further generalize the covariance operator to involve two random elements X, Y defined on $(\Omega, \mathscr{F}, P)$. In this setting, let κ_X and κ_Y be the positive definite kernels on $\Omega_X \times \Omega_X$ and $\Omega_Y \times \Omega_Y$. Let $\mathscr{H}_X$ and $\mathscr{H}_Y$ be Hilbert spaces generated by κ_X and κ_Y. Then we define

$$\mu_X = E[\kappa_X(\cdot,X)], \quad \mu_Y = E[\kappa_Y(\cdot,Y)],$$
$$M_{XY} = E[\kappa(\cdot,X) \otimes \kappa(\cdot,Y)], \quad \Sigma_{XY} = M_{XY} - \mu_X \otimes \mu_Y.$$

By (12.7), the linear operators M_{YX} and Σ_{YX} are simply the adjoint operators of M_{XY} and Σ_{XY}; that is,

$$M_{YX} = M_{XY}^*, \quad \Sigma_{YX} = \Sigma_{XY}^*.$$

Proposition 12.1 can be easily generalized to this setting, as given by the next proposition, whose proof will be omitted.

Proposition 12.2 *If Assumption 12.1 is satisfied for κ_X and κ_Y, then for all $f \in \mathcal{H}_X$, $g \in \mathcal{H}_Y$,*

$$\langle f, M_{XY}g \rangle_{\mathcal{H}_X} = E[f(X)g(Y)], \quad \langle f, \Sigma_{XY}g \rangle_{\mathcal{H}_X} = \mathrm{cov}[f(X), g(Y)]$$

Another important consequence of Assumption 12.1 is that the RKHS generated by κ_X the kernel is contained in $L_2(P_X)$, the class of all functions that are square integrable with respect to P_X.

Proposition 12.3 *If Assumption 12.1 is satisfied for κ_X and κ_Y, then $\mathcal{H}_X \subseteq L_2(P_X)$ and $\mathcal{H}_Y \subseteq L_2(P_Y)$.*

PROOF. This is because, by the Cauchy-Schwarz inequality, for any $f \in \mathcal{H}_X$,

$$Ef^2(X) = E\langle f, \kappa(\cdot, X) \rangle_{\mathcal{H}_X}^2 \leq E\left(\|f\|_{\mathcal{H}_X}^2 \|\kappa(\cdot, X)\|_{\mathcal{H}_X}^2 \right) \leq \|f\|^2 E\kappa(X, X) < \infty$$

as desired.   □

Recall that two Gaussian random vectors X and Y are independent if and only if $\mathrm{cov}(\alpha^\mathsf{T}X, \beta^\mathsf{T}Y) = 0$ are 0 for all α and β. In this sense the covariance matrix $\mathrm{cov}(X, Y)$ characterizes independence under the Gaussian assumption. In comparison, the covariance operator Σ_{XX} can characterize independence without the Gaussian assumption, provided that the families $\mathcal{H}_X$ and $\mathcal{H}_Y$ are sufficiently rich. We use the notion of "dense modulo constants" to characterize the richness of a functional class.

Definition 12.2 *We say that a subset $\mathscr{A}$ of $L_2(P)$ is dense modulo constants if, for each $f \in \mathscr{A}$, there is a sequence $\{f_n\} \subseteq L_2(P)$ such that*

$$\mathrm{var}[f_n(X) - f(X)] \to 0$$

as $n \to \infty$.

In the machine learning literature, the notion of "dense in $L_2(P)$" is called "universality". Such kernels are quite common. For example, the reproducing kernel Hilbert space generated by the Gaussian radial basis function is dense in $L_2(P)$; see, for example, Sriperumbudur et al. (2010) and Carmeli et al. (2010). Dense modulo constants in $L_2(P)$ is not very different from dense in $L_2(P)$: the former simply disregards constants. It is easy to see that, if $E[f(X)g(Y)] = E[f(X)]E[g(Y)]$ for all $f \in L_2(P_X)$ and $g \in L_2(P_Y)$, then $X \perp\!\!\!\perp Y$. This is because $L_2(P_X)$ (or $L_2(P_Y)$) contains all the indicator functions $I_A(X)$ (or $I_B(Y)$) for $A \in \mathscr{F}_X$ or $(B \in \mathscr{F}_Y)$. The next proposition shows that Σ_{XY} characterizes independence between X and Y if $\mathcal{H}_X$ and $\mathcal{H}_Y$ are universal.

Proposition 12.4 *If $\mathcal{H}_X$ is dense in $L_2(P_X)$ modulo constants and $\mathcal{H}_Y$ is dense in $L_2(P_Y)$ modulo constants, then*

$$X \perp\!\!\!\perp Y \Leftrightarrow \Sigma_{XY} = 0.$$

PROOF. First, suppose $X \perp\!\!\!\perp Y$. Because $\mathscr{H}_X$ and $\mathscr{H}_Y$ are subsets of $L_2(P_X)$ and $L_2(P_Y)$, respectively, we have

$$\langle f, \Sigma_{XY} g \rangle_{\mathscr{H}_X} = \mathrm{cov}[f(X), g(Y)] = 0$$

for all $f \in \mathscr{H}_X$, and $g \in \mathscr{H}_Y$. Taking $f = \Sigma_{XY} g$, we then have $\Sigma_{XY} g = 0$ for all $g \in \mathscr{H}_Y$, which means $\Sigma_{XY} = 0$.

Next, suppose $\Sigma_{XY} = 0$. To prove $X \perp\!\!\!\perp Y$ it suffices to show that, for all $f \in L_2(P_X)$ and $g \in L_2(P_Y)$, we have $\mathrm{cov}[f(X), g(Y)] = 0$. By the denseness assumption, there exist $\{f_n\} \subseteq \mathscr{H}_X$ and $\{g_n\} \subseteq \mathscr{H}_Y$ such that

$$\mathrm{var}[f(X) - f_n(X)] \to 0, \quad \mathrm{var}[g(Y) - g_n(Y)] \to 0.$$

Hence,

$$\begin{aligned}
\mathrm{cov}[f(X), g(Y)] &= \mathrm{cov}[f(X) - f_n(X) + f_n(X), g(Y) - g_n(Y) + g_n(Y)] \\
&= \mathrm{cov}[f - f_n(X), g - g_n(Y)] + \mathrm{cov}[f - f_n(X), g_n(Y)] \\
&\quad \mathrm{cov}[f_n(X), g(Y) - g_n(Y)] + \mathrm{cov}[f_n(X), g_n(Y)].
\end{aligned}$$

Abbreviate the above equality by $A = A_1(n) + \cdots + A_4(n)$. Then $A_4(n) = 0$ because $f_n \in \mathscr{H}_X$, $g_n \in \mathscr{H}_Y$, and $\langle f_n, \Sigma_{XY} g_n \rangle_{\mathscr{H}_X} = 0$. Thus we have $|A| \le |A_1(n)| + \cdots + |A_3(n)|$. In the meantime, we have, for example

$$\begin{aligned}
|A_2(n)| &\le \{\mathrm{var}[f(X) - f_n(X)]\}^{1/2} \{\mathrm{var}[g_n(Y)]\}^{1/2} \\
&\le \{\mathrm{var}[f(X) - f_n(X)]\}^{1/2} \{\mathrm{var}[g_n(Y) - g(Y)]\}^{1/2} \\
&\quad + \{\mathrm{var}[f(X) - f_n(X)]\}^{1/2} \{\mathrm{var}[g(Y)]\}^{1/2},
\end{aligned}$$

where the second inequality follows from the triangular inequality $[\mathrm{var}(U + V)]^{1/2} \le [\mathrm{var}(U)]^{1/2} + [\mathrm{var}(V)]^{1/2}$. By the assumption that $\mathscr{H}_X$ and $\mathscr{H}_Y$ are respectively dense in $L_2(P_X)$ and $L_2(P_Y)$ modulo constants, the right-hand side goes to 0 as $n \to \infty$. Consequently, $|A_2(n)| \to 0$. By the similar argument we can show that $|A_1(n)| \to 0$ and $|A_3(n)| \to 0$. Hence

$$|A| = \lim_{n \to \infty} |A| \le \lim_{n \to \infty} |A_1(n)| + \lim_{n \to \infty} |A_2(n)| + \lim_{n \to \infty} |A_3(n)| = 0,$$

as desired. □

The next proposition describes the relations between the range of the operator Σ_{XY} and that of the operators Σ_{XX}.

Proposition 12.5 *When Σ_{XX}, Σ_{YY}, and Σ_{XY} are defined, we have*

$$\mathrm{ran}(\Sigma_{XY}) \subseteq \overline{\mathrm{ran}}(\Sigma_{XX}), \quad \mathrm{ran}(\Sigma_{YX}) \subseteq \overline{\mathrm{ran}}(\Sigma_{YY}).$$

PROOF. Because the two relations imply each other if we switch the roles of X and Y, we only need to prove the first relation. For this to hold it suffices to show that

$$\overline{\mathrm{ran}}(\Sigma_{YY})^\perp \subseteq \mathrm{ran}(\Sigma_{YX})^\perp.$$

This is equivalent to $\ker(\Sigma_{YY}) \subseteq \ker(\Sigma_{XY})$ because, for any bounded linear operator A we have $\mathrm{ran}(A)^\perp = \overline{\mathrm{ran}}(A)^\perp = \ker(A^*)$ — see, for example, Conway (1990), Theorem 2.19. If $g \in$

$\ker(\Sigma_{YY})$, then $\langle g, \Sigma_{YY}g\rangle_{\mathscr{H}_Y} = \mathrm{var}[g(Y)] = 0$. So $g(Y)$ is almost surely constant. This implies, for any $f \in \mathscr{H}_X$, $\mathrm{cov}[g(Y), f(X)] = \langle \Sigma_{XY}g, f\rangle_{\mathscr{H}_X} = 0$. Taking $f = \Sigma_{XY}g$, we have

$$\langle \Sigma_{XY}g, \Sigma_{XY}g\rangle_{\mathscr{H}_X} = 0,$$

which implies $g \in \ker(\Sigma_{XY})$.   □

An inequality somewhat resembles the Cauchy-Schwarz inequality also applies to Σ_{YX}, as shown in the next proposition.

Proposition 12.6 *When Σ_{XX}, Σ_{YY}, and Σ_{XY} are defined, we have*

$$\|\Sigma_{XY}\| \le \|\Sigma_{XX}\|^{1/2} \|\Sigma_{YY}\|^{1/2},$$

where the norms are the operator norm.

PROOF. Let $f \in \mathscr{H}_X$, $g \in \mathscr{H}_Y$. By the Cauchy-Schwarz inequality,

$$\begin{aligned}
\langle f, \Sigma_{XY}g\rangle^2_{\mathscr{H}_X} &= \mathrm{cov}^2[f(X), g(Y)] \\
&\le \mathrm{var}[f(X)]\mathrm{var}[g(Y)] \\
&\le \langle f, \Sigma_{XX}f\rangle_{\mathscr{H}_X} \langle g, \Sigma_{YY}g\rangle_{\mathscr{H}_Y} \\
&\le \|\Sigma_{XX}\| \|f\|^2 \|\Sigma_{YY}\| \|g\|^2.
\end{aligned}$$

Taking square root on both sides, we have

$$|\langle f, \Sigma_{XY}g\rangle_{\mathscr{H}_X}| \le \|\Sigma_{XX}\|^{1/2} \|f\| \|\Sigma_{YY}\|^{1/2} \|g\|.$$

Let $f = \Sigma_{XY}g$. Then the above inequality becomes

$$\begin{aligned}
\|\Sigma_{XY}g\|^2_{\mathscr{H}_X} &\le \|\Sigma_{XX}\|^{1/2} \|\Sigma_{XY}g\| \|\Sigma_{YY}\|^{1/2} \|g\|_{\mathscr{H}_Y} \\
&\le \|\Sigma_{XX}\|^{1/2} \|\Sigma_{XY}\| \|\Sigma_{YY}\|^{1/2} \|g\|^2_{\mathscr{H}_Y}.
\end{aligned}$$

It follows that

$$\sup_{g \ne 0} \left\{ \|\Sigma_{XY}g\|^2_{\mathscr{H}_X} / \|g\|^2_{\mathscr{H}_Y} \right\} \le \|\Sigma_{XX}\|^{1/2} \|\Sigma_{XY}\| \|\Sigma_{YY}\|^{1/2}.$$

By definition, the left-hand side is simply $\|\Sigma_{XY}\|^2$.   □

When the operator Σ_{XX} is used, its domain is often modified to discard constant functions. Note that, if $f \in \ker(\Sigma_{XX})$, then

$$\langle f, \Sigma_{XX}f\rangle_{\mathscr{H}_X} = \mathrm{var}[f(X)] = 0.$$

That is, $f(X) = 0$ almost surely P_X. Evidently, constant functions are of no importance as long as statistical dependence is of interest. For this reason we can redefine the domain of Σ_{XX} to be $\ker(\Sigma_{XX})^{\perp}$. Because Σ_{XX} is a self adjoint operator, $\ker(\Sigma_{XX})^{\perp}$ is the same as $\overline{\mathrm{ran}}(\Sigma_{XX})$. As shown in Li and Song (2017), this set can be written down explicitly in our context as follows.

Lemma 12.2 *Under Assumption 12.1, we have*

$$\overline{\mathrm{ran}}(\Sigma_{XX}) = \overline{\mathrm{span}}\{\kappa_X(\cdot, x) - \mu_X : x \in \mathscr{H}_X\}.$$

PROOF. A member f of $\mathcal{H}_X$ is orthogonal to the subspace on the right-hand side if and only if $f \perp \kappa_X(\cdot, x) - \mu_x$ for all $x \in \Omega_X$. That is, $\langle f, \kappa_X(\cdot, x) - \mu_x \rangle_{\mathcal{H}_X} = 0$. However, this happens iff $f(x) = Ef(X)$ for all $x \in \mathcal{H}_X$, which is equivalent to $\mathrm{var}(f(X)) = 0$, or $\Sigma_{XX} f = 0$. Thus

$$\overline{\mathrm{span}}\{\kappa_X(\cdot, x) - \mu_x : x \in \mathcal{H}_X\}^{\perp} = \ker(\Sigma_{XX}),$$

which implies the desired equality because Σ_{XX} is self adjoint. □

Henceforth, we reset $\mathcal{H}_X$ to $\overline{\mathrm{ran}}(\Sigma_{XX})$. Note that, by Proposition 12.5, $\mathrm{ran}(\Sigma_{XY}) \in \overline{\mathrm{ran}}(\Sigma_{XX})$. So the redefined $\mathcal{H}_X$ still contains the range of the operator Σ_{XY}. Nevertheless, the redefinition means we need to restrict the domain of the original Σ_{XY} to $\overline{\mathrm{ran}}(\Sigma_{XX})$. Furthermore, resetting $\mathcal{H}_X$ to $\overline{\mathrm{ran}}(\Sigma_{XX})$ amounts to making the following assumption.

Assumption 12.2 $\ker(\Sigma_{XX}) = \{0\}$.

Note that, this assumption also implies that Σ_{XX} is invertible, though Σ_{XX}^{-1} is an unbounded operator. It can be shown that Σ_{XX} is a Hilbert-Schmidt operator (see Fukumizu et al. (2007)), whose eigenvalues converge to 0, which means Σ_{XX}^{-1} is unbounded. An unbounded operator is discontinuous, and cannot be estimated accurately by statistical functionals, which rely on continuity. Fortunately, in all our developments, Σ_{XX}^{-1} does not appear alone. It always appears in the form $\Sigma_{XX}^{-1} A$ for some operator A. We can impose assumptions on A so that $\Sigma_{XX}^{-1} A$ is bounded or even compact.

12.3 Coordinate Mapping

A linear operator is an abstract function independent of any basis. However, to implement it numerically we do need to start with a basis — or bases, if two Hilbert spaces are involved — and represent the operator as coordinates relative to the basis. In Section 1.9 we have already outlined some basic rules for coordinate representation. In this section we introduce the notion of coordinate mapping and through it to study further properties of coordinate representation.

As in Section 1.9, let $\mathcal{H}_1$, $\mathcal{H}_2$, and $\mathcal{H}_3$ be finite-dimensional Hilbert spaces, and let $\mathcal{B}_1$, $\mathcal{B}_2$, and $\mathcal{B}_3$ be their respective spanning systems. Let $G_{\mathcal{B}_i}$ be the Gram matrix of the basis $G_{\mathcal{B}_i}$, $i = 1, 2, 3$. We first consider the coordinate representation of a special linear operator — the tensor product, which appears in the covariance operators. If $f \in \mathcal{H}_1$ and $g \in \mathcal{H}_2$, then, for any $i = 1, \ldots, m_1$,

$$[(g \otimes f) b_i^{(1)}]_{\mathcal{B}_2} = \langle f, b_i^{(1)} \rangle_{\mathcal{H}_1} [g]_{\mathcal{B}_2} = [g]_{\mathcal{B}_2} [f]_{\mathcal{B}_1}^{\mathsf{T}} G_{\mathcal{B}_1} [b_i^{(1)}]_{\mathcal{B}_1}.$$

Since $[b_i^{(1)}]_{\mathcal{B}_1} = e_i$, we have

$$\begin{aligned}
{}_{\mathcal{B}_2}[g \otimes f]_{\mathcal{B}_1} &= \left([(g \otimes f) b_1^{(1)}]_{\mathcal{B}_2}, \ldots, [(g \otimes f) b_{m_1}^{(1)}]_{\mathcal{B}_2} \right) \\
&= [g]_{\mathcal{B}_2} [f]_{\mathcal{B}_1}^{\mathsf{T}} G_{\mathcal{B}_1} (e_1, \ldots, e_{m_1}) \\
&= [g]_{\mathcal{B}_2} [f]_{\mathcal{B}_1}^{\mathsf{T}} G_{\mathcal{B}_1}.
\end{aligned}$$

To summarize the properties about coordinating representation more systematically, let us consider the mappings:

$$\begin{aligned}
[\cdot]_{\mathcal{B}_1} : \quad & \mathcal{H}_1 \to \mathbb{R}^{m_1}, \quad f \mapsto [f]_{\mathcal{B}_1} \\
{}_{\mathcal{B}_2}[\cdot]_{\mathcal{B}_1} : \quad & \mathcal{B}(\mathcal{H}_1, \mathcal{H}_2) \to \mathbb{R}^{m_2 \times m_1}, \quad A \mapsto {}_{\mathcal{B}_2}[A]_{\mathcal{B}_1}.
\end{aligned} \tag{12.9}$$

We call these mappings coordinate mappings. It is easy to see that these are linear mappings. The following lemma, which is a part of Lemma 1 of Solea and Li (2016), summarizes some properties of the coordinate mappings.

Lemma 12.3 *Suppose $\mathscr{H}_1$, $\mathscr{H}_2$, $\mathscr{H}_3$ are finite-dimensional Hilbert spaces with spanning systems $\mathscr{B}_1, \mathscr{B}_2, \mathscr{B}_3$, and G_1 is the Gram matrix of $\mathscr{B}_1$. Then the following properties hold for the coordinate mappings in (12.9):*

1. *(evaluation)*   *If $h \in \mathscr{H}_1$ and $T \in \mathscr{B}(\mathscr{H}_1, \mathscr{H}_2)$, then $[Th]_{\mathscr{B}_2} = \left({}_{\mathscr{B}_2}[T]_{\mathscr{B}_1}\right)[h]_{\mathscr{B}_1}$;*
2. *(linearity)*   *If $f_1, f_2 \in \mathscr{H}_1$, $\alpha_1, \alpha_2 \in \mathbb{R}$, then*

$$[\alpha_1 f_1 + \alpha_2 f_2]_{\mathscr{B}_1} = \alpha_1 [f_1]_{\mathscr{B}_1} + \alpha_2 [f_2]_{\mathscr{B}_1};$$

   If $T_1, T_2 \in \mathscr{B}(\mathscr{H}_1, \mathscr{H}_2)$ and α_1, α_2, then

$${}_{\mathscr{B}_2}[\alpha_1 T_1 + \alpha_2 T_2]_{\mathscr{B}_1} = \alpha_1 \left({}_{\mathscr{B}_2}[T_1]_{\mathscr{B}_1}\right) + \alpha_2 \left({}_{\mathscr{B}_2}[T_2]_{\mathscr{B}_1}\right);$$

3. *(composition)*   *If $T_1 \in \mathscr{B}(\mathscr{H}_1, \mathscr{H}_2)$, $T_2 \in \mathscr{B}(\mathscr{H}_2, \mathscr{H}_3)$, then*

$${}_{\mathscr{B}_3}[T_2 T_1]_{\mathscr{B}_1} = \left({}_{\mathscr{B}_3}[T_2]_{\mathscr{B}_2}\right)\left({}_{\mathscr{B}_2}[T_1]_{\mathscr{B}_1}\right);$$

4. *(inner product)*   *If $f_1, f_2 \in \mathscr{H}_1$, then $\langle f_1, f_2 \rangle_{\mathscr{H}_1} = ([f_1]_{\mathscr{B}_1})^{\mathsf{T}} G_1 ([f_2]_{\mathscr{B}_1})$;*
5. *(tensor product)*   *If $f_1 \in \mathscr{H}_1$, $f_2 \in \mathscr{H}_2$, then ${}_{\mathscr{B}_2}[f_2 \otimes f_1]_{\mathscr{B}_1} = [f_2]_{\mathscr{B}_2} [f_1]_{\mathscr{B}_1}^{\mathsf{T}} G_1$.*

12.4   Coordinate of Covariance Operators

Let $X_1, \ldots, X_n$ be an i.i.d. sample of X. Let κ_x be a positive definite kernel. Let $\mathscr{H}_x$ be the linear space spanned by the functions $\{\kappa(\cdot, X_i) : i = 1, \ldots, n\}$. We estimate Σ_{xx} by the sample version of (12.8) — that is, by replacing E by E_n whenever possible:

$$\begin{aligned}
\hat{\Sigma}_{XX} &= E_n \kappa(\cdot, X) \otimes \kappa(\cdot, X) - E_n \kappa(\cdot, X) \otimes E_n \kappa(\cdot, X) \\
&= E_n \{ (\kappa(\cdot, X) - E_n \kappa(\cdot, X)) \otimes (\kappa(\cdot, X) - E_n \kappa(\cdot, X)) \}.
\end{aligned} \tag{12.10}$$

The subspace $\overline{\mathrm{ran}}\,(\hat{\Sigma}_{XX})$ is spanned by the set

$$\mathscr{B}_X = \{ \kappa(\cdot, X_i) - E_n \kappa(\cdot, X) : i = 1, \ldots, n \} = \{ b_1^{(X)}, \ldots, b_n^{(X)} \}. \tag{12.11}$$

Note that the members of this spanning system are not linearly independent because they sum to the zero function. Let us find the coordinate representation of $\hat{\Sigma}_{XX}$ relative to this spanning system. By Lemma 12.3 (evaluation, linearity, and tensor product parts), we have

$$\begin{aligned}
[\hat{\Sigma}_{XX} b_i^{(X)}]_{\mathscr{B}_X} &= n^{-1} \left[\left(\sum_{k=1}^n b_k^{(X)} \otimes b_k^{(X)} \right) b_i^{(X)} \right]_{\mathscr{B}_X} \\
&= n^{-1} \left(\sum_{k=1}^n [b_k^{(X)}]_{\mathscr{B}_X} [b_k^{(X)}]_{\mathscr{B}_X}^{\mathsf{T}} G_{\mathscr{B}_X} \right) [b_i^{(X)}]_{\mathscr{B}_X}.
\end{aligned}$$

Because $b_i^{(X)}$ is simply the ith member of the spanning system $\mathscr{B}_X$, we have $[b_i^{(X)}]_{\mathscr{B}_X} = e_i$. Moreover,

$$\begin{aligned}
\langle b_i^{(X)}, b_j^{(X)} \rangle_{\mathscr{H}_X} &= \left\langle \kappa(\cdot, X_i) - n^{-1} \sum_{k=1}^n \kappa(\cdot, X_k), \kappa(\cdot, X_j) - n^{-1} \sum_{\ell=1}^n \kappa(\cdot, X_\ell) \right\rangle \\
&= \kappa(X_i, X_j) - n^{-1} \sum_{\ell=1}^n \kappa(X_i, X_\ell) - n^{-1} \sum_{k=1}^n \kappa(X_j, X_k) + n^{-2} \sum_{k=1}^n \sum_{\ell=1}^n \kappa(X_k, X_\ell).
\end{aligned}$$

Let K_X denote the $n \times n$ matrix whose (i,j)th entry is $\kappa(X_i, X_j)$, and let Q denote the projection matrix $I_n - 1_n 1_n^\top / n$. Then it is easy to check that the above is simply the (i,j)th entry of the matrix $Q K_X Q$. In other words, the Gram matrix of the set $\mathscr{B}_X$ is $G_{\mathscr{B}_X} = Q K_X Q$. Thus by (12.10) we have

$$[\hat{\Sigma}_{XX} b_i^{(X)}]_{\mathscr{B}_X} = n^{-1} \left(\sum_{k=1}^{n} e_k e_k^\top \right) Q K_X Q e_i = n^{-1} Q K_X Q e_i,$$

from which it follows that

$$\mathscr{B}_X [\hat{\Sigma}_{XX}]_{\mathscr{B}_X} = \left(\mathscr{B}_X [\hat{\Sigma}_{XX}]_{\mathscr{B}_X}, \ldots, \mathscr{B}_X [\hat{\Sigma}_{XX}]_{\mathscr{B}_X} \right)$$
$$= n^{-1} (Q K_X Q)(e_1, \ldots, e_n) = n^{-1} Q K_X Q = n^{-1} G_{\mathscr{B}_X}.$$

The similar development can be carried out on Y. That is, $\mathscr{H}_Y$ is the space spanned by

$$\mathscr{B}_Y = \{ \kappa_Y(\cdot, Y_i) - E_n \kappa(\cdot, Y) : i = 1, \ldots, n \} \equiv \{ b_i^{(Y)} : i = 1, \ldots, n \}. \qquad (12.12)$$

And the coordinate of $\hat{\Sigma}_{YY}$ with respect to $\mathscr{B}_Y$ is

$$\mathscr{B}_Y [\hat{\Sigma}_{YY}]_{\mathscr{B}_Y} = n^{-1} Q K_Y Q = n^{-1} G_{\mathscr{B}_Y}.$$

Finally, consider the covariance operator $\hat{\Sigma}_{YX}$, which is defined by

$$n^{-1} \sum_{i=1}^{n} b_i^{(Y)} \otimes b_i^{(X)}.$$

By definition, the coordinate of this operator relative to the spanning system $(\mathscr{B}_Y, \mathscr{B}_X)$ is

$$\left(\left[\left(n^{-1} \sum_{k=1}^{n} b_k^{(Y)} \otimes b_k^{(X)} \right) b_1^{(X)} \right]_{\mathscr{B}_Y}, \ldots, \left[\left(n^{-1} \sum_{k=1}^{n} b_k^{(Y)} \otimes b_k^{(X)} \right) b_n^{(X)} \right]_{\mathscr{B}_Y} \right).$$

By Lemma 12.3, the jth column of this matrix is

$$\left[\left(n^{-1} \sum_{k=1}^{n} b_k^{(Y)} \otimes b_k^{(X)} \right) b_j^{(X)} \right]_{\mathscr{B}_Y} = n^{-1} \left(\sum_{k=1}^{n} ([b_k^{(Y)}]_{\mathscr{B}_Y})([b_k^{(X)}]_{\mathscr{B}_X})^\top \right) G_{\mathscr{B}_X} [b_j^{(X)}]_{\mathscr{B}_X}$$
$$= n^{-1} \left(\sum_{k=1}^{n} e_k e_k^\top \right) G_{\mathscr{B}_X} e_j = n^{-1} G_{\mathscr{B}_X} e_j.$$

It follows that $\mathscr{B}_Y [\hat{\Sigma}_{YX}]_{\mathscr{B}_X} = n^{-1} Q K_X Q = n^{-1} G_{\mathscr{B}_X}$. By the same argument we can show that $\mathscr{B}_X [\hat{\Sigma}_{XY}]_{\mathscr{B}_Y} = n^{-1} Q K_Y Q = n^{-1} G_{\mathscr{B}_Y}$. We summarize the above calculations into the following theorem.

Theorem 12.1 *If $\mathscr{B}_X$ and $\mathscr{B}_Y$ are defined as (12.11) and (12.12), respectively, then we have the following coordinate representations*

$$\mathscr{B}_X [\hat{\Sigma}_{XX}]_{\mathscr{B}_X} = n^{-1} G_{\mathscr{B}_X}, \quad \mathscr{B}_Y [\hat{\Sigma}_{YX}]_{\mathscr{B}_X} = n^{-1} G_{\mathscr{B}_X},$$
$$\mathscr{B}_X [\hat{\Sigma}_{XY}]_{\mathscr{B}_Y} = n^{-1} G_{\mathscr{B}_Y}, \quad \mathscr{B}_Y [\hat{\Sigma}_{YY}]_{\mathscr{B}_Y} = n^{-1} G_{\mathscr{B}_Y}.$$

12.5   Kernel Principal Component Analysis

Since the extension of Sufficient Dimension Reduction from the linear case to the nonlinear case is somewhat parallel to the extension of PCA to kernel PCA (see Scholkopt et al., 1998), it is illuminating to give an overview of the latter extension, and express the numerical procedure systematically using the coordinate representation. This gives the flavor of the developments to come.

Recall that the classical principal component analysis consists of maximizing the variance

$$L(\alpha) = \text{var}(\alpha^\mathsf{T} X)$$

successively. That is, in the first step, we minimize $L(\alpha)$ among all $\alpha \in \mathbb{R}^p$, $\|\alpha\| = 1$. Denote this minimizer as α_1 and $L(\alpha_1)$ as λ_1. At the kth step, minimize $L(\alpha)$ subject to $\alpha \perp \text{span}(\alpha_1, \ldots, \alpha_{k-1})$ and $\|\alpha\| = 1$. Denote the minimizer as α_k and $L(\alpha_k) = \lambda_k$. This gives us a set of real numbers $\{\lambda_i : i = 1, \ldots, p\}$ with $\lambda_1 \geq \cdots \geq \lambda_p$, and an orthonormal basis $\{\alpha_1, \ldots, \alpha_p\}$ of $\mathbb{R}^p$. Furthermore, (λ_i, α_i) are eigenvalue-eigenvector pairs of the positive semi-definite matrix $\text{var}(\alpha^\mathsf{T} X)$. The random variable $\alpha_i^\mathsf{T} X$ is the ith principal component at the population level.

The kernel (or nonlinear) principal component analysis is developed by carrying out the similar scheme for the variance operator Σ_{XX}. Let $\mathscr{H}$ be a Hilbert space and $A : \mathscr{H} \to \mathscr{H}$ be a bounded and self adjoint linear operator. A member of $\mathscr{H}$ is an eigenfunction of A if $\|f\|_{\mathscr{H}} = 1$ and there is a real number λ such that $Af = \lambda f$. If this is the case, then we say (λ, f) is an eigen pair of A. If A is a compact operator, then it can always be written as the following decomposition:

$$A = \sum_{i=1}^{\infty} \lambda_i (f_i \otimes f_i) \tag{12.13}$$

where (λ_i, f_i) are eigen pairs of A, $\{f_i\}$ is an orthonormal set in $\mathscr{H}$, and the convergence is in terms of the operator norm. See, for example, Conway (1990), Theorem 5.1. We assume, without loss of generality, that λ_i are ordered such that $\lambda_1 \geq \lambda_2 \cdots$. This expansion is called the spectral decomposition of A. Like in the classical setting, the ith eigenvalue λ_i is the maximizer of $\langle f, \Sigma_{XX} f \rangle_{\mathscr{H}}$ among the set

$$\{f \in \mathscr{H} : \|f\|_{\mathscr{H}} = 1, f \perp \{f_1, \ldots, f_{i-1}\}\}.$$

As mentioned earlier, the operators Σ_{XX} and Σ_{YY}, if they are defined, are Hilbert-Schmidt operators, which is a subclass of compact operators. Hence they all possess spectral decomposition (12.13). By making analogy to the classical PCA, we maximize

$$\text{var}[f(X)] = \text{var}[\langle f, \kappa(\cdot, X) \rangle_{\mathscr{H}}] = \langle f, \Sigma_{XX} f \rangle_{\mathscr{H}}$$

successively to obtain $f_1, f_2, \ldots$, and refer to the random variable $\langle f_i, \kappa(\cdot, X) \rangle_{\mathscr{H}} = f_i(X)$ as the ith kernel principal component of X. Note that, by the argument of the last paragraph, $f_1, f_2, \ldots$ are precisely the eigenfunctions of Σ_{XX}. The kernel principal component is thus simply the i eigenfunction evaluated at the observed samples. It is the function in $\mathscr{H}$ that captures the most variation in the data left from the previous principal components.

At the sample level, we set $\mathscr{H}_X$ and $\mathscr{H}_Y$ to be the subspaces spanned by the spanning systems $\mathscr{B}_X$ and $\mathscr{B}_Y$ in (12.11) and (12.12), respectively. We replace Σ_{XX} by $\hat{\Sigma}_{XX}$. The inner products involved in the eigenvalue problem are

$$\langle f, \hat{\Sigma}_{XX} f \rangle_{\mathscr{H}_X} = [f]_{\mathscr{B}_X}^\mathsf{T} G_{\mathscr{B}_X} (\mathscr{B}_X [\hat{\Sigma}_{XX}]_{\mathscr{B}_X}) [f]_{\mathscr{B}_X} = [f]_{\mathscr{B}_X}^\mathsf{T} G_{\mathscr{B}_X}^2 [f]_{\mathscr{B}_X}$$

$$\langle f, f \rangle_{\mathscr{B}_X} = [f]_{\mathscr{B}_X}^\mathsf{T} G_{\mathscr{B}_X} [f]_{\mathscr{B}_X}$$

Thus, we would like to successively maximize

$$[f]_{\mathscr{B}_X}^\mathsf{T} G_{\mathscr{B}_X}^2 [f]_{\mathscr{B}_X} \quad \text{subject to} \quad [f]_{\mathscr{B}_X}^\mathsf{T} G_{\mathscr{B}_X} [f]_{\mathscr{B}_X} = 1.$$

This is a generalized eigen problem $\mathrm{GEV}(G_{\mathscr{B}_X}^2, G_{\mathscr{B}_X})$, whose solution is described in Chapter 1. That is, let $v_1, v_2, \ldots, v_p$ be the eigenvector of G_X, then $u_i = G_{\mathscr{B}_X}^{\dagger 1/2} v_i$ are the eigenvectors of the generalized eigen problem. Let $\kappa_X(\cdot, X_{1:n})$ denote the vector $(\kappa_X(\cdot, X_1), \ldots, \kappa_X(\cdot, X_n))^\mathsf{T}$. Then the ith kernel principal component is defined by

$$G_{\mathscr{B}_X} u_i = G_{\mathscr{B}_X}^{1/2} v_i.$$

If we use the Gaussian radial basis function (12.2) as our kernel, then we need to determine the tuning constant γ. We use the formula given in Li and Solea (2017), as displayed in the algorithm below.

Algorithm 12.1 Kernel PCA

---

1. Standardize $X_1, \ldots, X_n$ marginally as before.
2. Choose a kernel function. If the Gaussian radial basis function is chosen, then choose the tuning parameter γ using

$$1/\sqrt{\gamma} = \binom{n}{2}^{-1} \sum_{i<j} \|X_i - X_j\|.$$

Compute the kernel matrix K_X using (11.4).
3. Compute the eigenvectors $v_1, \ldots, v_p$ of QK_XQ. The ith kernel principal component is n-vector $(QK_XQ)^{1/2} v_i$.

In the following R-codes for kernel PCA, the input x is the $n \times p$ matrix of $X_1, \ldots, X_n$; complexity is the coefficient to expand or shrink γ: a larger complexity means a more wiggly kernel function. The default value (as recommended above) is 1. The R-codes also include tr, which compute the trace of a matrix; gramx, which computes the matrix K_X; standmat, which standardize each column of a matrix.

R-codes for kernel PCA

```
kpca=function(x,complexity){
x=standmat(x);p=dim(x)[2 ];n=dim(x)[1]
one=rep(1,n);Q=diag(n)-one%*%t(one)/n
Kx=gramx(x,complexity);Gx=Q%*%Kx%*%Q
v=eigen(Gx)$vectors;u=mppower(Gx,-1/2,10^(-8))%*%v
return(Gx%*%u)}

tr=function(a) return(sum(diag(a)))

gramx=function(x,complexity){
n=dim(x)[1]
k2=x%*%t(x);k1=t(matrix(diag(k2),n,n));k3=t(k1);k=k1-2*k2+k3
sigma=(sum(sqrt(k))-tr(sqrt(k)))/(2*choose(n,2));gamma=1/(sigma^2)
return(exp(-complexity*gamma*(k1-2*k2+k3)))}
```

```
standmat=function(x) return(t((t(x)-apply(x,2,mean))/apply(x,2,sd)))
```

We applied the linear and kernel PCA to the training set of the pen digit data, and present in Figure 12.1 the perspective plots of the first three predictors obtained by the two methods. Note that, unlike the perspective plots shown in Section 6.9, these PCA and kernel PCA are based completely on the predictor X, without using any labeling information. Nevertheless, both methods achieve a reasonable degree of separation, with greater clarity by kernel PCA. Also note that, since the labeling variable Y is not used, it is not necessary to plot the predictor on the testing set; so the plots are based entirely on the training set.

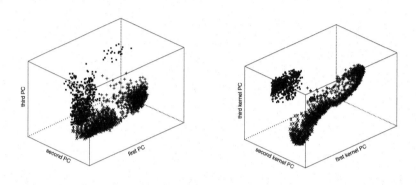

Figure 12.1 *First three linear principal components and first three kernel principal components for the pen digit data.*

12.6   Sufficient and Central σ-Field for Nonlinear SDR

Our goal is to replace the classical SDR statement $Y \perp\!\!\!\perp X | \beta^{\mathsf{T}} X$ by $Y \perp\!\!\!\perp X | f(X)$ for an arbitrary (and perhaps vector-valued) function f. Since conditional independence is only affected by the σ-field generated by the function being conditioned on — in particular, we can replace $f(X)$ by any one-to-one function of it without affecting the conditional independence — it is conceptually easier to start with a definition of nonlinear sufficient dimension reduction generally in terms of σ-field. We will resume the use of the notation $f(X)$ at a later stage. Let $\sigma(X)$ be the σ field generated by X.

Definition 12.3 *A sub σ-field $\mathscr{G}$ of $\sigma(X)$ is sufficient dimension reduction σ-field if*

$$Y \perp\!\!\!\perp X | \mathscr{G}.$$

Obviously, there are many sub-σ-fields that satisfy this relation; for example, $Y \perp\!\!\!\perp X | \sigma(X)$ always holds because, given $\sigma(X), X$ is nonrandom, and hence is independent of Y. This raises a question: is there a smallest σ-field that satisfies the above relation? If so, we should target the smallest one so as to achieve the greatest data reduction. Interestingly, the answer to this question seems to be easier than the parallel answer in the setting of linear sufficient dimension reduction. The next theorem gives such a sufficient condition. Its proof echoes Bahadur's proof in Bahadur (1954). In the following, we let Π_y denote the conditional probability measure $P_{X|Y}(\cdot|y)$, and let $\mathbb{P}$ denote the family probability measures $\{\Pi_y : y \in \Omega_Y\}$.

Theorem 12.2 *If $\mathbb{P}$ is dominated by a σ-finite measure, then there is a unique sub σ-field $\mathscr{G}^*$ of $\sigma(X)$ such that*

1. $Y \perp\!\!\!\perp X | \mathscr{G}^*$,

2. *if $\mathscr{G}$ is a sub σ-field of $\sigma(X)$ such that $Y \perp\!\!\!\perp X | \mathscr{G}$, then $\mathscr{G}^* \subseteq \mathscr{G}$.*

PROOF. Since $\mathbb{P}$ is dominated by a σ-finite measure, there is a countable subset $\mathbb{Q} = \{Q_k : k = 1, 2, \ldots\} \subseteq \mathbb{P}$ such that $\mathbb{Q} \equiv \mathbb{P}$, where the notation $\equiv$ means the two families of measures dominate each other. Let Q_0 be the mixture of $\{Q_k\}$ with positive mixing probabilities; that is, let $Q_0 = \sum_{k=1}^{\infty} c_k Q_k$ where $\{c_k : k = 1, 2, \ldots\}$ is a sequence of positive numbers that sum to 1. Then the singleton $\{Q_0\}$ dominates and is dominated by $\mathbb{Q}$, and hence also $\mathbb{P}$. Or, in symbols, $\{Q_0\} \equiv \mathbb{Q} \equiv \mathbb{P}$. Let $\pi_y = d\Pi_y / dQ_0$ and $\mathscr{G}$ be a sub σ-field of $\sigma(X)$. We claim that the following statements are equivalent:

1. $Y \perp\!\!\!\perp X | \mathscr{G}$;

2. π_y is essentially measurable with respect to $\mathscr{G}$ for all $y \in \Omega_Y$ modulo Q_0.

To prove $1 \Rightarrow 2$, we first note that, for any integrable function $h(X)$,

$$E_{Q_0}[h(x)\pi_y(x)] = \int h(x)\pi_y(x)dQ_0 = \int h(x)d\Pi_y = E_{\Pi_y}[h(X)]. \tag{12.14}$$

Now let $B \in \mathscr{F}_X$. Then

$$\begin{aligned} E_{Q_0}[\pi_y(X)I_B(X)] &= E_{\Pi_y}[I_B(X)] \\ &= E_{\Pi_y}[E_{\Pi_y}(I_B(X)|\mathscr{G})] \\ &= E_{Q_0}[E_{\Pi_y}(I_B(X)|\mathscr{G})\pi_y(X)] \\ &= E_{Q_0}[\Pi_y(B|\mathscr{G})\pi_y(X)], \end{aligned} \tag{12.15}$$

where the first and third equalities follow from (12.14). Since $Y \perp\!\!\!\perp X | \mathscr{G}$, $\Pi_y(B|\mathscr{G})$ is the same for all $y \in \Omega_Y$; that is $\Pi_y(B|\mathscr{G}) = \Pi_{y'}(B|\mathscr{G})$ for all $y' \in \Omega_Y$. In particular, $\Pi_y(B|\mathscr{G}) = Q_k(B|\mathscr{G})$ for all k, which implies $\Pi_y(B|\mathscr{G}) = Q_0(B|\mathscr{G})$. Hence we can rewrite the right-hand side of the above equalities as

$$\begin{aligned} E_{Q_0}[Q_0(B|\mathscr{G})\pi_y(X)] &= E_{Q_0}[E_{Q_0}(I_B(X)|\mathscr{G})\pi_y(X)] \\ &= E_{Q_0}[I_B(X)E_{Q_0}(\pi_y(X)|\mathscr{G})], \end{aligned} \tag{12.16}$$

where, for the second equality, we used Proposition 2.2. Comparing (12.15) and (12.16) we see that, for all $B \in \mathscr{F}_X$,

$$E_{Q_0}[\pi_y(X)I_B(X)] = E_{Q_0}[I_B(X)E_{Q_0}(\pi_y(X)|\mathscr{G})],$$

which implies $\pi_y(X) = E_{Q_0}[\pi_y(X)|\mathscr{G}]$ a.s. Q_0.

To prove $2 \Rightarrow 1$, let $A \in \mathscr{G}$. Then

$$\begin{aligned} E_{\Pi_y}[E_{Q_0}(I_B(X)|\mathscr{G})I_A(X)] &= E_{Q_0}[E_{Q_0}(I_B(X)|\mathscr{G})I_A(X)\pi_y(X)] \\ &= E_{Q_0}[I_B(X)I_A(X)E_{Q_0}(\pi_y(X)|\mathscr{G})], \end{aligned}$$

where the first equality follows from (12.14), and the second follows from Proposition 2.2. By assertion 2, $E_{Q_0}[\pi_y(X)|\mathscr{G}] = \pi_y(X)$. Hence the right-hand side becomes

$$E_{Q_0}[I_B(X)I_A(X)\pi_y(X)] = E_{\Pi_y}[I_B(X)I_A(X)] = \Pi_y(X \in A \cap B),$$

where the first equality follows from (12.14). Thus, by the definition of conditional probability, $E_{Q_0}(I_B(X)|\mathscr{G}) = Q_0(B|\mathscr{G})$ is the conditional probability $\Pi_y(B|\mathscr{G})$, which means that $\Pi_y(B|\mathscr{G})$ does not depend on y, which is assertion 1.

Now let $\mathscr{G}^*$ be the intersection of all SDR σ-fields $\mathscr{G}$; that is, let

$$\mathscr{G}^* = \cup\{\mathscr{G} : \mathscr{G} \in \mathfrak{A}\},$$

where

$$\mathfrak{A} = \{\mathscr{G} : \mathscr{G} \text{ is a sub } \sigma\text{-field of } \sigma(X) \text{ and } Y \perp\!\!\!\perp X|\mathscr{G}\}. \tag{12.17}$$

Then $\mathscr{G}^*$ is itself a σ-field. Since π_y is essentially measurable with respect to all SDR σ-fields for all $y \in \Omega_Y$, it is also essentially measurable with respect to $\mathscr{G}^*$ for all $y \in \Omega_Y$. Consequently, $\mathscr{G}^*$ is itself an SDR σ-field, which, by the proved equivalence of assertions 1 and 2, implies $Y \perp\!\!\!\perp X|\mathscr{G}^*$. That is, $\mathscr{G}^*$ is the smallest SDR σ-field. Finally, if $\mathscr{G}^{**}$ is another smallest SDR σ-field, then we know $\mathscr{G}^* \subseteq \mathscr{G}^{**}$ and $\mathscr{G}^{**} \subseteq \mathscr{G}^*$, implying the uniqueness of $\mathscr{G}^*$.   □

The above result justifies the following definition of the central σ-field, the target of nonlinear sufficient dimension reduction.

Definition 12.4 *If $\{\Pi_y : y \in \Omega_Y\}$ is dominated by a σ-finite measure, then the σ-field $\cap_{\mathscr{G} \in \mathfrak{A}} \mathscr{G}$ is called the central dimension reduction σ-field, or the central σ-field, where $\mathfrak{A}$ is as defined in (12.17).*

We denote the central σ-field by $\mathscr{G}_{Y|X}$. By construction, the central σ-field is the minimal sufficient σ-field.

12.7   Complete Sub σ-Field for Nonlinear SDR

It turns out that the central σ-field $\mathscr{G}_{Y|X}$ can be recovered fully by a method called the Generalized Sliced Inverse Regression developed in the next chapter, provided that the central σ-field is also in some sense complete. We first give a definition of completeness that suits our purpose, which bears close resemblance to the completeness in classical statistical inference (see, for example, Lehmann and Casella (1998b)).

Definition 12.5 *A sub σ-field of $\sigma(X)$ is complete if, for any $\mathscr{G}$-measurable f,*

$$E[f(X)|\mathscr{G}] = constant \ almost \ surely \ \Rightarrow \ E[f(X)|Y] = constant \ almost \ surely.$$

Recall that completeness is a rather restrictive assumption in classical statistical inference: indeed, complete and sufficient statistics exist only in very special cases, such as exponential family and uniform distribution. However, sufficient and complete SDR σ-field is not a restrict requirement. To demonstrate this point, in the next two propositions we give two examples of forward and inverse regression models, which cover a wide range of statistical applications, for which complete and sufficient sub-σ-field exist.

Proposition 12.7 *If there is a function $h \in \mathscr{H}_X^q$ such that*

$$Y = h(X) + \varepsilon, \tag{12.18}$$

where $\varepsilon \perp\!\!\!\perp X$ and $E(\varepsilon) = 0$, then $\sigma[h(X)]$ is a complete and sufficient dimension reduction σ-field for Y versus X.

PROOF. Suppose m is measurable $\mathscr{G}$ and $E[m(X)|Y] = $ constant a.s. P. We can assume, without loss of generality, that this constant is 0, for we can always reset m to $m - E[m(X)]$. Then there is a measurable function $g : \mathbb{R}^q \to \mathbb{R}$ such that $m = g \circ h$. Let $U = h(X)$. Then $E[g(U)|Y] = 0$ a.s. P. Hence,

$$E[g(U)e^{it^\top Y}] = E\{E[g(U)|Y]e^{it^\top Y}\} = 0,$$

for all $t \in \mathbb{R}^q$, where $i = \sqrt{-1}$. Because $U \perp\!\!\!\perp \varepsilon$, the above implies

$$E[g(U)e^{it^\top U}] = E[g(U)e^{it^\top U}e^{it^\top \varepsilon}]/E(e^{it^\top \varepsilon})$$
$$= E[g(U)e^{it^\top Y}]/E(e^{it^\top \varepsilon}) = 0.$$

By the uniqueness of inverse Fourier transformation we see that $g(U) = 0$ a.s. P, which implies $m(X) = (g \circ h)(X) = 0$ a.s. P. □

Model (12.18) covers many useful regression settings in Statistics and Econometrics, such as the single index and the multiple index models (Ichimura (1993), Hardle et al. (1993), Yin et al. (2008)), are special cases of (12.18). Thus, complete and sufficient σ-fields exist for all these models.

The next proposition considers a type of inverse regression model, where X is transformed into two components, the response Y and an independent error. Inverse regressions of this type are used in Cook (2007), Cook and Forzani (2009), and Cook et al. (2010) in the context of linear SDR.

Proposition 12.8 *Suppose $q < p$, Ω_Y has a nonempty interior, and P_Y is dominated by the Lebesgue measure on $\mathbb{R}^q$. Suppose there exist functions $g \in \mathscr{H}_X^q$ and $h \in \mathscr{H}_X^{p-q}$ such that*

1. $g(X) = Y + \varepsilon$, where $Y \perp\!\!\!\perp \varepsilon$, and $\varepsilon \sim N(0, \Sigma)$;

2. $\sigma(g(X), h(X)) = \sigma(X)$;

3. $h(X) \perp\!\!\!\perp (Y, g(X))$;

4. the induced measure $P_X \circ g^{-1}$ is dominated by the Lebesgue measure on $\mathbb{R}^q$.

Then $\sigma[g(X)]$ is a complete sufficient dimension reduction σ-field for Y versus X nonlinear SDR.

PROOF. By Theorem 2.1, condition 3 implies $Y \perp\!\!\!\perp h(X)|g(X)$. Hence, by condition 2, $Y \perp\!\!\!\perp X|g(X)$. That is, $\sigma[g(X)]$ is a sufficient σ-field. Let $u(X)$ be a measurable function of $\mathscr{G}$ such that $E[u(X)|\mathscr{G}]$ is a constant almost surely P. Let $U = g(X)$. Then $u = v \circ g$ for some measurable function $v : \mathbb{R}^q \to \mathbb{R}$ and $E[v(U)|Y] = $ constant almost surely P. Assume, without loss of generality, $E[v(U)|Y] = 0$ almost surely P. Because $Y \perp\!\!\!\perp \varepsilon$, this implies $P_Y(\{y : E[v(y + \varepsilon)] = 0\}) = 1$; that is,

$$\int_{\mathbb{R}^q} v(t) \frac{1}{(2\pi)^{q/2}|\Sigma|^{1/2}} e^{-(t-y)^\top \Sigma^{-1}(t-y)/2} dt = 0$$

a.s. P_Y. This implies

$$\int v(t)e^{-t^\mathsf{T}\Sigma^{-1}t/2}e^{y^\mathsf{T}\Sigma^{-1}t}dt = 0 \Rightarrow \int v(\Sigma s)e^{-s^\mathsf{T}\Sigma s/2}e^{y^\mathsf{T}s}ds = 0$$

a.s. P_Y, where $s = \Sigma^{-1}t$. Since Ω_Y contains an open set in $\mathbb{R}^q$ and the above function of y is analytic, by the analytic continuation theorem, the above function is 0 everywhere on $\mathbb{R}^q$. Hence, by the uniqueness of inverse Laplace transformation, we have

$$v(\Sigma s)e^{-s^\mathsf{T}\Sigma s/2} = 0 \text{ almost surely } \lambda,$$

where λ is the Lebesgue measure on $\mathbb{R}^q$. Hence $v(\Sigma s) = 0$ almost surely λ, or equivalently $v(t) = 0$ almost surely λ. Since

$$\int_{v\circ g(x)\neq 0} dP_X = \int_{v(t)\neq 0} dP_X\circ g^{-1},$$

by condition 4, $P_X\circ g^{-1} \ll \lambda$, the above integral is 0, implying $v\circ g(X) = 0$ almost surely P.   □

From the above two propositions we see that completeness in the sense we defined it is widely applicable, particularly when the dependence of Y on X is through the conditional mean $E(Y|X)$.

12.8   Converting σ-Fields to Function Classes for Estimation

Since a sub σ-field is a rather abstract object to estimate, we convert it into a subspace, which can then be estimated by the range of a linear operator. This again is analogous to the linear SDR, where the object of estimation is a subspace, which is estimated by the columns space of a candidate matrix. Let $\mathscr{H}_X$ be a Hilbert space of functions defined on Ω_X and take values in $\mathbb{R}$. We assume that there exist functions $f_1, \dots, f_d \in \mathscr{H}_X$ such that

$$Y \perp\!\!\!\perp X | f_1(X), \dots, f_d(X).$$

The central σ-field is $\mathscr{G}_{Y|X} = \sigma[f_1(X), \dots, f_d(X)]$. For any sub-$\sigma$-field $\mathscr{G}$ of $\sigma(X)$, let $\mathscr{H}_X(\mathscr{G})$ be the subspace of $\mathscr{H}_X$ spanned by the functions in $\mathscr{H}_X$ that are measurable with respect to $\mathscr{G}$. That is,

$$\mathscr{H}_X(\mathscr{G}) = \overline{\text{span}}\{f \in \mathscr{H}_X : f \text{ is measurable } \mathscr{G}\}.$$

Definition 12.6 *We say that $\mathscr{H}_X(\mathscr{G})$ is a sufficient or complete dimension reduction class for Y versus X if $\mathscr{G}$ is sufficient or complete σ-field for Y versus X. We call $\mathscr{H}_X(\mathscr{G}_{Y|X})$ the central sufficient dimension reduction class, or the central class, and denote it by $\mathfrak{S}_{Y|X}$.*

We used the word "class" instead of "space" to distinguish these objects from their counterparts in linear SDR. Note that there is also a conceptual difference: in linear SDR, the central subspace is the space spanned by the coefficient vectors; that is, $\text{span}(\beta)$ where β appears in $\beta^\mathsf{T}X$. In the current context, we cannot in general separate f from X. Thus we define the central class as the space of random variables rather than the space of parameters. Our goal in nonlinear SDR is to estimate the central class — or a basis of it.

We now outline a general strategy and related concepts of estimation for nonlinear SDR, which will be fully developed in the next chapter. Let $\mathfrak{F}$ be the class of all probability measures on $\Omega_X \times \Omega_Y$, let F_0 be the true distribution of (X, Y), and let F_n be the empirical distribution based on an i.i.d. sample $(X_1, Y_1), \ldots, (X_n, Y_n)$. Let $\mathcal{M}$ be another Hilbert space; for example, $\mathcal{M}$ can be $\mathcal{H}_X$ itself, or $\mathcal{H}_Y$, a Hilbert space of functions defined on Ω_Y and taking values in $\mathbb{R}$. Consider the following statistical functional

$$T : \mathfrak{F} \to \mathcal{B}(\mathcal{M}, \mathcal{H}_X).$$

Thus, for each distribution $F \in \mathfrak{F}$, $T(F)$ is a bounded linear operator from $\mathcal{M}$ to $\mathcal{H}_X$. Adopting the nomenclature in linear SDR, we call $T(F_n)$ an estimate of $\mathfrak{S}_{Y|X}$, and $T(F_0)$ a candidate operator.

Definition 12.7 *If $\overline{\mathrm{ran}}\,[T(F_0)] \subseteq \mathfrak{S}_{Y|X}$, then we say $T(F_n)$ is an unbiased estimate of the central class; if $\overline{\mathrm{ran}}\,[T(F_0)] = \mathfrak{S}_{Y|X}$, then we say that $T(F_n)$ is an exhaustive estimate of the central class.*

Although our theoretical formulation of nonlinear SDR does not require $\mathcal{H}_X$ to be an RKHS — indeed one of our estimation procedures, GSAVE, uses the $L_2(P_X)$-inner product rather than the RKHS inner product — RKHS offers most direct connection with the linear SDR. Suppose now that $\mathcal{H}_X$ is an RKHS generated by a positive kernel κ_X. Then by the reproducing property our nonlinear SDR problem can be stated as

$$Y \perp\!\!\!\perp X \,|\, \langle f_1, \kappa_X(\cdot, X) \rangle_{\mathcal{H}_X}, \ldots, \langle f_d, \kappa_X(\cdot, X) \rangle_{\mathcal{H}_X}. \tag{12.19}$$

This form is analogous to the linear SDR problem

$$Y \perp\!\!\!\perp X \,|\, \beta_1^\mathsf{T} X, \ldots, \beta_d^\mathsf{T} X. \tag{12.20}$$

The only difference between (12.19) and (12.20) is that the inner product $\beta_p^\mathsf{T} X$ is replaced by $\langle f, \kappa_X(\cdot, X) \rangle_{\mathcal{H}_X}$. This framework is the one we use in developing the first estimator in the next chapter — the generalized sliced inverse regression estimator.

Chapter 13

Generalized Sliced Inverse Regression

13.1	Regression Operator	212
13.2	Generalized Sliced Inverse Regression	213
13.3	Exhaustiveness and Completeness	215
13.4	Relative Universality	216
13.5	Implementation of GSIR	217
13.6	Precursors and Variations of GSIR	220
13.7	Generalized Cross Validation for Tuning ε_X and ε_Y	220
13.8	k-Fold Cross Validation for Tuning $\rho_X, \rho_Y, \varepsilon_X, \varepsilon_Y$	223
13.9	Simulation Studies	225
13.10	Applications	226
	13.10.1 Pen Digit Data	227
	13.10.2 Face Sculpture Data	228

In this chapter we describe the first estimator of the central class, which is based on the linear operator $\Sigma_{XX}^{-1}\Sigma_{XY}$. We call this operator the regression operator. It can be shown that, under very mild conditions, the range of this operator is always contained in the central class. Interestingly, the sufficient conditions for this to hold are much weaker than those in the linear SDR setting. In particular, the linear conditional mean is not required. Another important point is that, when the central class is complete, the range of the regression operator is, in fact, the central class itself. That is, the method is exhaustive under completeness. We will develop the sample estimate of this estimator through coordinate representation. In a special case, the range of this operator can be presented as the nonlinear eigenvalue problem

$$\text{maximize} \quad \text{var}[f(X)] \quad \text{subject to} \quad \|f\|_{\mathscr{H}_X} = 1,$$

which parallels the eigenvalue problem underlying the Sliced Inverse Regression; that is

$$\text{maximize} \quad \text{var}(\beta^{\mathsf{T}}X) \quad \text{subject to} \quad \|\beta\| = 1.$$

This motivates the name Generalized Sliced Inverse Regression.

13.1   Regression Operator

In order to define the operator $\Sigma_{XX}^{-1}\Sigma_{XY}$ we need to make some assumptions. First, as mentioned earlier, since constants are not important for conditional independence, we start with an RKHS $\mathcal{H}_X$, construct covariance operator Σ_{XX}, and then reset the space $\mathcal{H}_X$ to $\overline{\text{ran}}(\Sigma_{XX})$. We do the same for Y. Under this construction, Σ_{XX} and Σ_{YY} are invertible operators because $\ker(\Sigma_{XX}) = 0$ and $\ker(\Sigma_{YY}) = 0$. Because Σ_{XX} has domain $\overline{\text{ran}}(\Sigma_{XX})$ and range $\text{ran}(\Sigma_{XX})$, its inverse Σ_{XX}^{-1} has domain $\text{ran}(\Sigma_{XX})$ and range $\overline{\text{ran}}(\Sigma_{XX})$. Thus, a precondition for $\Sigma_{XX}^{-1}\Sigma_{XY}$ to be defined is

$$\text{ran}(\Sigma_{XY}) \subseteq \text{dom}(\Sigma_{XX}^{-1}) = \text{ran}(\Sigma_{XX}).$$

We record this condition (and its parallel for Y) in the following definition.

Assumption 13.1 $\text{ran}(\Sigma_{XY}) \subseteq \text{ran}(\Sigma_{XX})$, $\text{ran}(\Sigma_{YX}) \subseteq \text{ran}(\Sigma_{YY})$.

As we will see in the discussion following Assumption 13.2, the condition $\text{ran}(\Sigma_{YX}) \subseteq \text{ran}(\Sigma_{YY})$ is, in fact, an assumption on the smoothness of the relation between X and Y. The same can be said about the assumption $\text{ran}(\Sigma_{YX}) \subseteq \text{ran}(\Sigma_{YY})$. We also note that by Proposition 12.5, $\text{ran}(\Sigma_{YX}) \subseteq \overline{\text{ran}}(\Sigma_{YY})$ and $\text{ran}(\Sigma_{XY}) \subseteq \overline{\text{ran}}(\Sigma_{XX})$. Thus the above assumption is not very strong. Also, as we mentioned in Section 12.2, Σ_{XX} and Σ_{YY} are Hilbert Schmidt operators, and as a result their inverses are unbounded operators. Thus we make the following assumptions.

Assumption 13.2 *The operators $\Sigma_{YY}^{-1}\Sigma_{YX}$ and $\Sigma_{XX}^{-1}\Sigma_{XY}$ are compact.*

We now use $\Sigma_{YY}^{-1}\Sigma_{YX}$ to illustrate the meaning of this assumption. Intuitively, since the sequence of eigenvalues of Σ_{YY} goes to 0, for $\Sigma_{YY}^{-1}\Sigma_{YX}$ to be a compact operator, the operator Σ_{YX} must send all incoming functions into the eigenspaces of Σ_{YY} with relatively large eigenvalues. That is, Σ_{YX} must send functions to the low-frequency range of the eigenspaces of Σ_{YY}. So the above assumption is akin to a smoothness condition.

The operator $\Sigma_{YY}^{-1}\Sigma_{YX}$ is called the Regression Operator in Lee et al. (2016); see also Li (2017). It was used earlier in, for example, Fukumizu et al. (2007) and Fukumizu et al. (2009). The reason for calling it the "Regression Operator" is that it transforms a function of X to its conditional expectation given Y, modulo constants, as shown in the next lemma.

Lemma 13.1 *Suppose $\mathcal{H}_X$ is dense in $L_2(P_Y)$ modulo constants. Then, for any $f \in \mathcal{H}_X$,*

$$\Sigma_{YY}^{-1}\Sigma_{YX}f = E[f(X)|Y] + constant.$$

PROOF. It suffices to show that, for any $g \in L_2(P_Y)$, we have

$$\text{cov}[f(X) - (\Sigma_{YY}^{-1}\Sigma_{YX}f)(Y), g(Y)] = 0, \qquad (13.1)$$

where $(\Sigma_{YY}^{-1}\Sigma_{YX}f)(Y)$ means the function $\Sigma_{YY}^{-1}\Sigma_{YX}f$ evaluated at Y. Because $\mathcal{H}_Y$ is dense in $L_2(P_Y)$ modulo constants, there is a sequence $\{g_n\} \subseteq \mathcal{H}_Y$ such that $\text{var}\{g(Y) - g_n(Y)\} \to 0$. Using g_n we perform the following decomposition

$$\begin{aligned}
&\text{cov}\{(\Sigma_{YY}^{-1}\Sigma_{YX}f)(Y), g(Y)\} \\
&= \text{cov}\{(\Sigma_{YY}^{-1}\Sigma_{YX}f)(Y), g_n(Y)\} + \text{cov}\{(\Sigma_{YY}^{-1}\Sigma_{YX}f)(Y), g(Y) - g_n(Y)\}.
\end{aligned} \qquad (13.2)$$

The first term on the right-hand side is

$$\begin{aligned}
\langle \Sigma_{YY}^{-1}\Sigma_{YX}f, \Sigma_{YY}g_n \rangle_{\mathcal{H}_Y} &= \langle \Sigma_{YX}f, g_n \rangle_{\mathcal{H}_Y} \\
&= \text{cov}\{f(X), g_n(Y)\} \\
&= \text{cov}\{f(X), g(Y)\} + \text{cov}\{f(X), g_n(Y) - g(Y)\}.
\end{aligned}$$

Hence the equality (13.2) can be rewritten as

$$\text{cov}\{(\Sigma_{YY}^{-1}\Sigma_{YX}f)(Y), g(Y)\}$$
$$= \text{cov}\{f(X), g(Y)\} + \text{cov}\{f(X) - (\Sigma_{YY}^{-1}\Sigma_{YX}f)(Y), g_n(Y) - g(Y)\}.$$

Because $\text{var}\{g_n(Y) - g(Y)\} \to 0$, by the Cauchy-Schwarz inequality, the second term on the right tends to 0 as $n \to \infty$. Therefore

$$\text{cov}\{(\Sigma_{YY}^{-1}\Sigma_{YX}f)(Y), g(Y)\} = \text{cov}\{f(X), g(Y)\},$$

which implies (13.1). □

13.2 Generalized Sliced Inverse Regression

In this section we show that the range of the regression operator is always contained in central class $\mathfrak{S}_{Y|X}$, so that the sample estimate of the regression operator provides an unbiased estimator of the central class. We first prove a lemma, which is a direct consequence of Proposition 2.2.

Lemma 13.2 *If U and V are random variables and $\mathscr{F}$ is a σ-field, then*

$$\text{cov}[E(U|\mathscr{F}), V] = \text{cov}[U, E(V|\mathscr{F})] = \text{cov}[E(U|\mathscr{F}), E(V|\mathscr{F})]. \quad (13.3)$$

PROOF. By the definition of covariance, we have

$$\text{cov}[E(U|\mathscr{F}), V] = E[E(U|\mathscr{F})V^{\mathsf{T}}] - E[E(U|\mathscr{F})]EV^{\mathsf{T}}$$
$$= E[UE(V^{\mathsf{T}}|\mathscr{F})] - E(U)E[E(V^{\mathsf{T}}|\mathscr{F})] = \text{cov}[U, E(V|\mathscr{F})],$$

where the second equality follows from Proposition 2.2. The second equality follows from

$$\text{cov}[U, E(V|\mathscr{F})] = \text{cov}\{U, E[E(V|\mathscr{F})|\mathscr{F}]\}.$$

By the first equality in (13.3), the right-hand side is $\text{cov}\{E(U|\mathscr{F}), E(V|\mathscr{F})\}$. □

In the following we use $L_2(P_X|\mathscr{G}_{Y|X})$ to denote the collection of $\mathscr{G}_{Y|X}$-measurable functions in $L_2(P_X)$. We now prove the main theorem of this section.

Theorem 13.1 Suppose Assumptions 12.1, 12.2, 13.1, and 13.2 hold, and $\mathfrak{S}_{Y|X}$ is dense in $L_2(P_X|\mathscr{G}_{Y|X})$ modulo constants. Then

$$\overline{\text{ran}}\,(\Sigma_{XX}^{-1}\Sigma_{XY}) \subseteq \mathfrak{S}_{Y|X}.$$

PROOF. We first show that

$$\overline{\text{ran}}\,(\Sigma_{XY}) \subseteq \Sigma_{XX}\mathfrak{S}_{Y|X}, \quad (13.4)$$

which is equivalent to $(\Sigma_{XX}\mathfrak{S}_{Y|X})^{\perp} \subseteq \overline{\text{ran}}\,(\Sigma_{XY})^{\perp}$. Since $\overline{\text{ran}}\,(\Sigma_{XY})^{\perp} = \ker(\Sigma_{YX})$, it suffices to show that $(\Sigma_{XX}\mathfrak{S}_{Y|X})^{\perp} \subseteq \ker(\Sigma_{YX})$. Let $f \in (\Sigma_{XX}\mathfrak{S}_{Y|X})^{\perp}$. Then, for all $g \in \mathfrak{S}_{Y|X}$,

$$\langle f, \Sigma_{XX}g \rangle_{\mathscr{H}_X} = \text{cov}\{f(X), g(X)\} = 0.$$

Because g is measurable with respect to $\mathscr{G}_{Y|X}$, we have $g(X) = E[g(X)|\mathscr{G}_{Y|X}]$. By Lemma 13.2,

$$\text{cov}\{f(X), E[g(X)|\mathscr{G}_{Y|X}]\} = \text{cov}\{E[f(X)|\mathscr{G}_{Y|X}], g(X)\}.$$

Since $\mathfrak{S}_{Y|X}$ is dense in $L_2(P_X|\mathscr{G}_{Y|X})$ modulo constants, there exists a sequence $\{f_n\} \subseteq \mathfrak{S}_{Y|X}$ such that $\mathrm{var}[f_n(X) - f(X)] \to 0$. Then

$$\mathrm{cov}\{E[f(X)|\mathscr{G}_{Y|X}], f_n(X)\} = 0. \tag{13.5}$$

But on the other hand,

$$\begin{aligned}\mathrm{cov}\{E[f(X)|\mathscr{G}_{Y|X}], f_n(X)\} &\to \mathrm{cov}\{E[f(X)|\mathscr{G}_{Y|X}], f(X)\} \\ &= \mathrm{cov}\{E[f(X)|\mathscr{G}_{Y|X}], E[f(X)|\mathscr{G}_{Y|X}]\}.\end{aligned} \tag{13.6}$$

Combining (13.5) and (13.6) we see that

$$\mathrm{var}\{E[f(X)|\mathscr{G}_{Y|X}]\} = 0.$$

This implies that $E[f(X)|\mathscr{G}_{Y|X}] = $ constant almost surely. Since $\mathscr{G}_{Y|X}$ is sufficient, we have $E[f(X)|\mathscr{G}_{Y|X}] = E[f(X)|Y, \mathscr{G}_{Y|X}]$. So $E[f(X)|Y, \mathscr{G}_{Y|X}] = $ constant almost surely. Consequently, $E[f(X)|Y] = $ constant almost surely. That means $\Sigma_{YY}^{-1}\Sigma_{YX}f = $ constant. But the only constant in $\mathscr{H}_Y$ is 0. So $\Sigma_{YY}^{-1}\Sigma_{YX}f = 0$, which implies $\Sigma_{YX}f = \Sigma_{YY}0 = 0$. Thus we have proved (13.4).

By (13.4) we have

$$\mathrm{ran}\,(\Sigma_{XY}) \subseteq \Sigma_{XX}\mathfrak{S}_{Y|X},$$

which implies $\Sigma_{XX}^{-1}\mathrm{ran}\,(\Sigma_{XY}) \subseteq \mathfrak{S}_{Y|X}$. Note that

$$\begin{aligned}\Sigma_{XX}^{-1}\mathrm{ran}\,(\Sigma_{XY}) &= \{\Sigma_{XX}^{-1}f : f = \Sigma_{XY}g, g \in \mathscr{H}_Y\} \\ &= \{\Sigma_{XX}^{-1}\Sigma_{XY}g : g \in \mathscr{H}_Y\} = \mathrm{ran}\,(\Sigma_{XX}^{-1}\Sigma_{XY}).\end{aligned}$$

Finally, because $\mathfrak{S}_{Y|X}$ is closed, we have $\overline{\mathrm{ran}}\,(\Sigma_{XX}^{-1}\Sigma_{XY}) \subseteq \mathfrak{S}_{Y|X}$. $\qquad\square$

From this we see that, for any invertible operator $A : \mathscr{H}_Y \to \mathscr{H}_Y$ we have

$$\overline{\mathrm{ran}}\,(\Sigma_{XX}^{-1}\Sigma_{XY}A\Sigma_{YX}\Sigma_{XX}^{-1}) \subseteq \mathfrak{S}_{Y|X}. \tag{13.7}$$

The space (13.7) can be recovered by performing the generalized eigenvalue problem $\mathrm{GEC}(\Sigma_{XY}A\Sigma_{YX}, \Sigma_{XX})$. That is, we solve the following problem:

$$\begin{aligned}&\text{maximize} \quad \langle f, \Sigma_{XY}A\Sigma_{XY}f\rangle_{\mathscr{H}_X} \\ &\text{subject to} \quad \langle f, \Sigma_{XX}f\rangle_{\mathscr{H}_X} = 1, f \perp \mathscr{S}_{k-1}.\end{aligned} \tag{13.8}$$

where $\mathscr{S}_k = \mathrm{span}(f_1, \ldots, f_{k-1})$ and $f_1, \ldots, f_{k-1}$ are the solutions to this constrained maximization problem in the previous steps.

If we take $A = \Sigma_{YY}^{-1}$, then this procedure is a nonlinear parallel of SIR in the following sense. In this case we need to recover the subspace

$$\overline{\mathrm{ran}}\,(\Sigma_{XX}^{-1}\Sigma_{XY}\Sigma_{YY}^{-1}\Sigma_{YX}\Sigma_{XX}^{-1}).$$

The generalized eigenvalue problem $\mathrm{GEV}(\Sigma_{XY}\Sigma_{YY}^{-1}\Sigma_{YX}, \Sigma_{XX})$ can be restated in terms of the successive maximization problem: at the kth step

$$\begin{aligned}&\text{maximize} \quad \langle f, \Sigma_{XY}\Sigma_{YY}^{-1}\Sigma_{YX}f\rangle_{\mathscr{H}_X} \\ &\text{subject to} \quad \langle f, \Sigma_{XX}f\rangle_{\mathscr{H}_X} = 1, f \perp \mathscr{S}_{k-1}.\end{aligned} \tag{13.9}$$

However, by Lemma 13.1,

$$\langle f, \Sigma_{XY}\Sigma_{YY}^{-1}\Sigma_{YX}f\rangle_{\mathscr{H}_X} = \langle \Sigma_{YY}^{-1}\Sigma_{YX}f, \Sigma_{YY}\Sigma_{YY}^{-1}\Sigma_{YX}f\rangle_{\mathscr{H}_X}$$
$$= \text{var}[E(f(X)|Y)]$$
$$= \text{var}\{E[\langle f, \kappa_X(\cdot,X)\rangle_{\mathscr{H}_X}|Y]\}.$$

Also, the inner product $\langle f, \Sigma_{XX}f\rangle_{\mathscr{H}_X}$ can be written as $\text{var}[\langle f, \kappa_X(\cdot,X)\rangle_{\mathscr{H}_X}]$. Thus, the maximization problem (13.9) becomes

$$\text{maximize} \quad \text{var}\{E[\langle f, \kappa_X(\cdot,X)\rangle_{\mathscr{H}_X}|Y]\}$$
$$\text{subject to} \quad \text{var}[\langle f, \kappa_X(\cdot,X)\rangle_{\mathscr{H}_X}] = 1, \ f \perp \mathscr{S}_{k-1}.$$

Contrasting this to the sequential maximization underlying SIR:

$$\text{maximize} \quad \text{var}[E(\beta^\mathsf{T}X|Y)]$$
$$\text{subject to} \quad \text{var}(\beta^\mathsf{T}X) = 1, \ f \perp \mathscr{S}'_{k-1},$$

where $\mathscr{S}'_{k-1}$ is the subspace in $\mathbb{R}^p$ spanned by the solutions in the previous steps, we see that the two problems have exactly the same structure except that the Euclidean space $\mathbb{R}^p$ is replaced by the reproducing kernel Hilbert space $\mathscr{H}_X$, and that the Euclidean inner product $\beta^\mathsf{T}X$ is replaced by the $\mathscr{H}_X$-inner product $\langle f, \kappa_X(\cdot,X)\rangle_{\mathscr{H}_X}$. For this reason, we call the method based on the sample estimate of the range of the regression operator $\Sigma_{XX}^{-1}\Sigma_{XY}$ the Generalized Sliced Inverse Regression, or GSIR.

13.3 Exhaustiveness and Completeness

In this section we show that, if $\mathfrak{S}$ is a sufficient and complete dimension reduction class, then the closure of the range of the regression operator is the same as $\mathfrak{S}$. This is significant in three ways. First, it means that the GSIR estimate is exhaustive. The second is that a complete and sufficient nonlinear SDR class is the central class; that is, sufficiency and completeness imply minimal sufficiency, which is parallel to a well known result in classical statistical inference (see Lehmann (1981)). The third is that such a result does not exist in the linear SDR setting. Furthermore, as we have shown in Section 12.7, completeness in our sense applies to a wide range of nonparametric models, which means the exhaustive of GSIR also applies widely.

Theorem 13.2 *Suppose the assumptions in Theorem 13.1 hold and, in addition, $\mathfrak{S}_{Y|X}$ is complete. Then*

$$\overline{\text{ran}}(\Sigma_{XX}^{-1}\Sigma_{XY}) = \mathfrak{S}_{Y|X}.$$

PROOF. Since, by Theorem 13.1, $\overline{\text{ran}}(\Sigma_{XY}) \subseteq \Sigma_{XX}\mathfrak{S}_{Y|X}$, we only need to show $\Sigma_{XX}\mathfrak{S}_{Y|X} \subseteq \overline{\text{ran}}(\Sigma_{XY})$, or equivalently, $\ker(\Sigma_{YX}) \subseteq (\Sigma_{XX}\mathfrak{S}_{Y|X})^\perp$. Let $f \in \ker(\Sigma_{YX})$. Then $\Sigma_{YX}f = 0$, which implies that $\Sigma_{YY}^{-1}\Sigma_{YX}f = 0$. Then, by Lemma 13.1, $E[f(X)|Y] = \text{constant}$, which implies $E\{E[f(X)|\mathscr{G}_{Y|X}]|Y\} = \text{constant}$. Because $E[f(X)|\mathscr{G}_{Y|X}]$ is measurable with respect to $\mathscr{G}_{Y|X}$, and $\mathscr{G}_{Y|X}$ is complete, we have $E[f(X)|\mathscr{G}_{Y|X}] = \text{constant}$. It follows that, for any $g \in \Sigma_{XX}\mathfrak{S}_{Y|X}$, we have

$$\text{cov}[f(X),g(X)] = \text{cov}\{f(X),E[g(X)|\mathscr{G}_{Y|X}]\} = \text{cov}\{E[f(X)|\mathscr{G}_{Y|X}],g(X)\} = 0.$$

That is, $f \in (\Sigma_{XX}\mathfrak{S}_{Y|X})^\perp$, as desired.   □

This result also implies that a complete and sufficient dimension reduction class is necessarily the central class.

Corollary 13.1 *If the assumptions in Theorem 13.1 hold, then a sufficient and complete dimension reduction class is the central class.*

PROOF. Let $\mathfrak{S}$ be a complete and sufficient dimension reduction class. Then, Theorem 13.1 tells us that $\overline{\mathrm{ran}}\,(\Sigma_{XX}^{-1}\Sigma_{XY}) \subseteq \mathfrak{S}_{Y|X}$, and Theorem 13.2 tells us that $\overline{\mathrm{ran}}\,(\Sigma_{XX}^{-1}\Sigma_{XY}) = \mathfrak{S}$. Therefore, $\mathfrak{S} \subseteq \mathfrak{S}_{Y|X}$. But we also know $\mathfrak{S}_{Y|X} \subseteq \mathfrak{S}$ because $\mathfrak{S}_{Y|X}$ is the intersection of all sufficient dimension reduction classes. $\qquad\square$

13.4   Relative Universality

A crucial condition in Theorems 13.1 and 13.2 is that $\mathfrak{S}_{Y|X}$ is dense in $L_2(P_X|\mathscr{G}_{Y|X})$ modulo constants. As discussed earlier, appropriate choices of κ_X, such as the choice of the Gauss radial basis function, can guarantee that $\mathscr{H}_X$ is dense in $L_2(P_X)$. Intuitively, this should imply that $\mathfrak{S}_{Y|X}$ is also dense in $L_2(P_X|\mathscr{G}_{Y|X})$. The next theorem shows that this is indeed the case. In fact, we are going to prove a more general result. Let $\mathscr{G}$ be a sub σ-field of $\sigma(X)$ generated by some functions in $\mathscr{H}_X$. Let $\mathscr{H}_X(\mathscr{G})$ be the closed linear span of the collection of all $\mathscr{G}$-measurable functions in $\mathscr{H}_X$. Let $L_2(P_X|\mathscr{G})$ be the collection of all $\mathscr{G}$-measurable functions in $L_2(P_X)$. Then $\mathscr{H}_X(\mathscr{G})$ is dense in $L_2(P_X|\mathscr{G})$ modulo constants. We call this property *relative universality with respect to a σ-field*. To our knowledge this result has not been proven in the current literature.

Theorem 13.3 *If $\mathscr{H}_X$ is dense in $L_2(P_X)$ modulo constants, then $\mathscr{H}_X(\mathscr{G})$ is dense in $L_2(P_X|\mathscr{G})$ modulo constants.*

PROOF. Since Σ_{XX} is defined, it is Hilbert-Schmidt, and therefore compact. Let $\mathscr{A} = \{h_n\}$ be the sequence of eigenfunctions of Σ_{XX}. Because $\mathscr{H}_X = \overline{\mathrm{ran}}\,(\Sigma_{XX})$, $\mathscr{A}$ is dense in $\mathscr{H}_X$ in terms of $\mathscr{H}_X$-topology. Let $\mathscr{A}_{\mathscr{G}} = \{g_n\}$ be a subset of $\mathscr{A}$ that are measurable with respect to $\mathscr{G}$, and let $\mathscr{M}_{\mathscr{G}}$ be the closed linear span of $\mathscr{A}_{\mathscr{G}}$. Then the closed linear span of $\mathscr{A} \setminus \mathscr{A}_{\mathscr{G}}$ is the orthogonal complement of $\mathscr{M}_{\mathscr{G}}$ in $\mathscr{H}_X$. Denote this as $\mathscr{M}_{\mathscr{G}}^{\perp}$. Note that for $g^{(1)} \in \mathscr{M}_{\mathscr{G}}$ and $g^{(2)} \in \mathscr{M}_{\mathscr{G}}^{\perp}$, we not only have $\langle g^{(1)}, g^{(2)} \rangle_{\mathscr{M}_{\mathscr{G}}} = 0$, but also $\mathrm{cov}[g^{(1)}(X), g^{(2)}(X)] = 0$. This is because $\mathscr{M}_{\mathscr{G}}$ and $\mathscr{M}_{\mathscr{G}}^{\perp}$ are spanned by eigenfunctions of Σ_{XX}. In other words, $\mathscr{M}_{\mathscr{G}}$ and $\mathscr{M}_{\mathscr{G}}^{\perp}$ are orthogonal not only in $\mathscr{H}_X$-geometry, but also in $L_2(P_X)$-geometry.

Now let $f \in L_2(P_X)$ be a measurable function of $\mathscr{G}$; that is, $f \in L_2(p_X|\mathscr{G})$. Our goal is to show that there is a sequence $\{f_n\}$ in $\mathscr{H}_X(\mathscr{G})$ such that $\mathrm{var}[f_n(X) - f(X)] \to 0$. Because $\mathscr{M}_{\mathscr{G}}$ is a subset of $\mathscr{H}_X(\mathscr{G})$, it suffices to show that there is a sequence $\{r_n\}$ in $\mathscr{M}_{\mathscr{G}}$ such that

$$\mathrm{var}[r_n(X) - f(X)] \to 0. \tag{13.10}$$

Because $\mathscr{H}_X$ is dense in $L_2(P_X)$ modulo constants, there is a sequence $\{s_n\}$ in $\mathscr{H}_X$ such that $\mathrm{var}[s_n(X) - f(X)] \to 0$. This implies that $\{s_n\}$ is a var-Cauchy sequence, by which we mean for any $\varepsilon > 0$ there exists n_0 such that $\mathrm{var}[s_m(X) - s_n(X)] < \varepsilon$ for all $n, m > n_0$. Write s_n as $s_n^{(1)} + s_n^{(2)}$, where $s_n^{(1)} \in \mathscr{M}_{\mathscr{G}}$ and $s_n^{(2)} \in \mathscr{M}_{\mathscr{G}}^{\perp}$. Because

$$\mathrm{var}[s_n(X) - s_m(X)] = \mathrm{var}[s_n^{(1)}(X) - s_m^{(1)}(X)] + \mathrm{var}[s_n^{(2)}(X) - s_m^{(2)}(X)],$$

the sequences $\{s_n^{(1)}\}$ and $\{s_n^{(2)}\}$ are themselves var-Cauchy. Therefore, there exist $s^{(1)}, s^{(2)} \in L_2(P_X)$ such that

$$\mathrm{var}[s_n^{(1)}(X) - s^{(1)}(X)] \to 0, \quad \mathrm{var}[s_n^{(2)}(X) - s^{(2)}(X)] \to 0.$$

Consequently,

$$\begin{aligned}
&\operatorname{var}\{[s_n^{(1)}(X) + s_n^{(2)}(X)] - [s^{(1)}(X) + s^{(2)}(X)]\} \\
&= \operatorname{var}\{[s_n^{(1)}(X) - s^{(1)}(X)] + [s_n^{(2)}(X) - s^{(2)}(X)]\} \\
&= \operatorname{var}[s_n^{(1)}(X) - s^{(1)}(X)] + \operatorname{var}[s_n^{(2)}(X) - s^{(2)}(X)] \\
&\quad + 2\operatorname{cov}[s_n^{(1)}(X) - s^{(1)}(X), s_n^{(2)}(X) - s^{(2)}(X)] \to 0.
\end{aligned}$$

It follows that $s^{(1)} + s^{(2)} = f + \text{constant}$, and hence that

$$\operatorname{var}[s^{(1)}(X)|\mathscr{G}] + \operatorname{var}[s^{(2)}(X)|\mathscr{G}] = \operatorname{var}[f(X)|\mathscr{G}].$$

Since $\operatorname{var}[s^{(1)}(X)|\mathscr{G}] = 0$, $\operatorname{var}[f(X)|\mathscr{G}] = 0$, we have $\operatorname{var}[s^{(2)}(X)|\mathscr{G}] = 0$. Because $\mathscr{A} \setminus \mathscr{A}_{\mathscr{G}}$ contains only functions that are not measurable with respect to $\mathscr{G}$, the only $s^{(2)}$ that satisfies the above equality is 0 (no constants other than 0 are in $\mathscr{H}_X$). Consequently, $\operatorname{var}[s_n^{(2)}(X)] \to 0$. By the triangular inequality,

$$\{\operatorname{var}[s_n^{(1)}(X) - f(X)]\}^{1/2} \leq \{\operatorname{var}[s_n(X) - f(X)]\}^{1/2} + \{\operatorname{var}[s_n^{(2)}(X)]\}^{1/2}.$$

Hence $\lim_n \operatorname{var}[s_n^{(1)}(X) - f(X)] = 0$, which proves (13.10).   □

13.5   Implementation of GSIR

As in the case for kernel PCA, at the sample level, the spaces $\mathscr{H}_X$ and $\mathscr{H}_Y$ are spanned, respectively, by $\mathscr{B}_X$ and $\mathscr{B}_Y$ in (12.11) and (12.12). Let G_X denote $QK_X Q$. The first inner product in (13.8) is

$$\langle f, \hat{\Sigma}_{XY} A \hat{\Sigma}_{YX} f \rangle_{\mathscr{H}_X} = [f]^\mathsf{T} G_X (_{\mathscr{B}_X}[\hat{\Sigma}_{XY}]_{\mathscr{B}_Y})(_{\mathscr{B}_Y}[A]_{\mathscr{B}_Y})(_{\mathscr{B}_Y}[\hat{\Sigma}_{YX}]_{\mathscr{B}_X})[f]_{\mathscr{B}_X}. \tag{13.11}$$

The second inner product in (13.9) is

$$[f]_{\mathscr{B}_X} G_X \Sigma_{XX} [f]_{\mathscr{B}_X} = [f]_{\mathscr{B}_X} G_X^2 [f]_{\mathscr{B}_X}.$$

We recommend two choices of A. The first choice is the identity operator $A = I$, in which case $_{\mathscr{B}_Y}[A]_{\mathscr{B}_Y}$ is simply Q. Thus our generalized eigenvalue problem becomes

$$\text{maximize} \quad [f]_{\mathscr{B}_X}^\mathsf{T} G_X G_Y G_X [f]_{\mathscr{B}_X} \quad \text{subject to} \quad [f]_{\mathscr{B}_X} G_X^2 [f]_{\mathscr{B}_X} = 1.$$

Let $v = G_X[f]_{\mathscr{B}_X}$. Solving this equation, we obtain $[f]_{\mathscr{B}_X} = G_X^\dagger v$. To control the level of overfitting we use the Tychonoff-regularized inverse $(G_X + \eta_X I_n)^{-1}$ to replace the Moore-Panrose inverse, where η_X is a tuning constant. To endow η_X with appropriate scale, we let $\eta_X = \varepsilon_X \lambda_{\max}(G_X)$, where $\lambda_{\max}(\cdots)$ is the maximum eigenvalue of a matrix. As a result, we have

$$[f]_{\mathscr{B}_X} = (G_X + \varepsilon_X \lambda_{\max}(G_X) I_n)^{-1} v.$$

As will be discussed in the next section, the factor ε_X will be determined by a cross-validation criterion. Thus v is the eigenvector of the following matrix

$$\Lambda_{\text{GSIR}}^{(1)} = (G_X + \eta_X I_n)^{-1} G_X G_Y G_X (G_X + \eta_X I_n)^{-1}. \tag{13.12}$$

The advantage of this choice is that we do not need to choose the tuning parameter ε_Y. This choice also tends to work better (than the second choice below) for smaller sample sizes.

The second choice is $A = \hat{\Sigma}_{YY}^{-1}$, in which case the coordinate $_{\mathscr{B}_Y}[A]_{\mathscr{B}_Y}$ is $G_Y^\dagger$. Again we replace this with its regularized version $(G_Y + \eta_Y I_n)^{-1}$, where $\eta_Y = \varepsilon_Y \lambda_{\max}(G_Y)$. Thus we now have the following generalized eigenvalue problem:

$$\text{maximize} \quad [f]_{\mathscr{B}_X}^\mathsf{T} G_X G_Y (G_Y + \eta_Y I_n)^{-1} G_X [f]_{\mathscr{B}_X} \quad \text{subject to} \quad [f]_{\mathscr{B}_X} G_X^2 [f]_{\mathscr{B}_X} = 1.$$

Let v be the eigenvector of the matrix

$$\Lambda_{\text{GSIR}}^{(2)} = (G_X + \eta_X I_n)^{-1} G_X G_Y (G_Y + \eta_Y I_n)^{-1} G_X (G_X + \eta_X I_n)^{-1}. \tag{13.13}$$

Then the coordinate $[f]_{\mathscr{B}_X}$ is $(G_X + \varepsilon_X \lambda_{\max}(G_X) I_n)^{-1} v$. We summarize the steps for implementing GSIR at the sample level in the following algorithm.

Algorithm 13.1 GSIR

---

1. Standardize $X_1, \ldots, X_n$ marginally and, if Y is a random vector, standardize $Y_1, \ldots, Y_n$ marginally.

2. Choose kernel functions κ_X and κ_Y. If the Gaussian radial basis function is chosen, then choose the tuning parameters γ_X and γ_Y using as described in Section 13.7. Also, choose ε_X and ε_Y as described in Section 13.7.

3. Compute the first d eigenvectors $v_1, \ldots, v_d$ of the matrix $\Lambda_{\text{GSIR}}^{(1)}$ in (13.12) or $\Lambda_{\text{GSIR}}^{(2)}$ in (13.13). The coordinate $[f_i]_{\mathscr{B}_X}$ of the ith eigenfunction of the generalized eigenvalue problem (13.9) is estimated by $(G_X + \varepsilon_X I_n)^{-1} v_i$, $i = 1, \ldots, d$.

4. The sufficient predictors are

$$v_i^\mathsf{T} (G_X + \varepsilon_X I_n)^{-1} b_{1:n}^{(X)}(X_j), \quad i = 1, \ldots, d, \ j = 1, \ldots, n,$$

where $b_{1:n}^{(X)}$ is $(b_1^{(X)}, \ldots, b_n^{(X)})^\mathsf{T}$, with $b_1^{(X)}, \ldots, b_n^{(X)}$ being defined in (12.11).

When Y is categorical — say $\Omega_Y = \{a_1, \ldots, a_m\}$, where $a_1, \ldots, a_m$ are integers representing the labels of groups — the kernel κ_Y should be the discrete kernel. That is,

$$\kappa_Y(a_i, a_j) = \delta_{ij},$$

where δ_{ij} is the Kronecker δ-function. Furthermore, if the number of classes is few compared with the sample size, we recommend that the Tychonoff regularized matrix inverse $(G_Y + \eta_Y I_n)^{-1}$ be replaced by the Moore-Penrose inverse $G_Y^\dagger$. When the number of categorical is large, we recommend using the Tychonoff regularized inverse with η_Y determined by generalized cross validation, as in the continuous case.

The following is a set of R-codes to implement GSIR. The R-code `gsir` computes the predictors of GSIR, which can either be those evaluated at the training set or the testing set. Among the input variables, `x` is the $n \times p$ matrix of the predictors; `x.new` can be set to `x` or the predictor matrix for the testing set; `y` is the response variable; `ytype` is a variable indicating the type of response: it can be "`categorical`" or "`scalar`", and when it is `scalar`, Y can be either real-valued or vector-valued; `atype` is a variable indicating the type of matrix A in (13.11): it can be "`identity`" or "`Gyinv`", the first indicating $A = I$ and the second indicating $A = \hat{\Sigma}_{YY}^{-1}$; `complex.x` and `complex.y` are the ρ_X and ρ_Y in (13.16), which measure the complexity of $\mathscr{H}_X$ and $\mathscr{H}_Y$; r is the number of leading GSIR predictors in the output. The R-code `gram.gauss` computes the gram matrix based on the Gaussian radial basis kernel. Different from the code `gramx` in the R-codes for PCA, `gram.gauss` computes the inner products between two sets of variables, which can be the training set and the test set. When the two sets are the same, it gives the same result as gramx. The code gram.dis computes the discrete kernel. The R-code onorm computes the operator norm (the largest eigenvalue) of a symmetric matrix. The R-code sym symmetrizes a matrix, in the situations where a matrix is theoretically symmetric but is numerically asymmetric due to numerical error.

R-codes for GSIR

```
gsir=function(x,x.new,y,ytype,atype,ex,ey,complex.x,complex.y,r){
n=dim(x)[1];p=dim(x)[2];Q=diag(n)-rep(1,n)%*%t(rep(1,n))/n
Kx=gram.gauss(x,x,complex.x)
if(ytype=="scalar") Ky=gram.gauss(y,y,complex.y)
if(ytype=="categorical") Ky=gram.dis(y)
Gx=Q%*%Kx%*%Q;Gy=Q%*%Ky%*%Q
Gxinv=matpower(sym(Gx+ex*onorm(Gx)*diag(n)),-1)
if(ytype=="categorical") Gyinv=mppower(sym(Gy),-1,1e-9)
if(ytype=="scalar") Gyinv=matpower(sym(Gy+ey*onorm(Gy)*diag(n)),-1)
a1=Gxinv%*%Gx
if(atype=="identity") a2=Gy
if(atype=="Gyinv") a2=Gy%*%Gyinv
gsir=a1%*%a2%*%t(a1)
v=eigen(sym(gsir))$vectors[,1:r]
Kx.new=gram.gauss(x,x.new,complex.x)
pred.new=t(t(v)%*%Gxinv%*%Q%*%Kx.new)
return(pred.new)
}


gram.gauss=function(x,x.new,complexity){
x=as.matrix(x);x.new=as.matrix(x.new)
n=dim(x)[1];m=dim(x.new)[1]
k2=x%*%t(x);k1=t(matrix(diag(k2),n,n));k3=t(k1);k=k1-2*k2+k3
sigma=sum(sqrt(k))/(2*choose(n,2));gamma=complexity/(2*sigma^2)
k.new.1=matrix(diag(x%*%t(x)),n,m)
k.new.2=x%*%t(x.new)
k.new.3=matrix(diag(x.new%*%t(x.new)),m,n)
return(exp(-gamma*(k.new.1-2*k.new.2+t(k.new.3))))
}


gram.dis=function(y){
n=length(y);yy=matrix(y,n,n);diff=yy-t(yy);vecker=rep(0,n^2)
vecker[c(diff)==0]=1;vecker[c(diff)!=0]=0
return(matrix(vecker,n,n))}


onorm=function(a) return(eigen(round((a+t(a))/2,8))$values[1])


sym=function(a) return(round((a+t(a))/2,9))
```

13.6   Precursors and Variations of GSIR

While the theory of nonlinear Sufficient Dimension Reduction was formally laid out by Lee et al. (2013), various methods similar to GSIR was proposed by earlier authors without articulating a general framework of nonlinear sufficient dimension reduction. For example, Wu (2008) introduced the Kernel Sliced Inverse Regression, which applies the idea of Sliced Inverse Regression to a nonlinear feature mapping $\Phi(X)$ of X. That is, we first divide the sample of Y into several slices, and then take the slice mean of $\Phi(X)$ to form a candidate matrix, on which spectral decomposition is then performed. The feature $\Phi(X)$ can be expressed equivalently in a kernel form. This method was further developed by Yeh et al. (2009). A similar procedure was also proposed in the Ph.D. thesis of Yu Wang (Wang, 2008), a former student of the author of the book. These methods can be expressed as a special case of GSIR, by taking a discrete kernel for $\kappa_Y(y_1, y_2)$; that is,

$$\kappa_Y(y_1, y_2) = \begin{cases} 1 & \text{if } y_1 \text{ and } y_2 \text{ belong to the same slice} \\ 0 & \text{if they belong to different slices} \end{cases}$$

Even though KSIR is similar to GSIR numerically, the proposed theory behind KSIR is quite different: it was assume that $E[\Phi(X)|\beta^\mathsf{T}\Phi(X)]$ is linear in $\Phi(X)$, as we assumed for linear SDR. Our more general theory does not need this assumption.

Another line of earlier development is the method called Kernel Canonical Correlation Analysis (KCCA) developed by Akaho (2001), Bach and Jordan (2002), and Fukumizu et al. (2007), which is a kernel generalization of Canonical Correlation Analysis (CCA). The CCA solves the following Singular-Value Decomposition problem:

$$\text{maximize} \quad \text{cov}^2[\beta^\mathsf{T}X, \gamma^\mathsf{T}X]$$
$$\text{subject to} \quad \text{var}(\beta^\mathsf{T}X) = 1, \ \text{var}(\gamma^\mathsf{T}X) = 1.$$

The above authors extended this problem using RKHS. Their argument leads to the following procedure in our notation: let α be the leading eigenvector of the matrix

$$(G_X + \eta_X I_n)^{-1} G_X G_Y (G_Y + \eta_Y I_n)^{-2} G_Y G_X (G_X + \eta_X I_n)^{-1}.$$

Then the canonical correlation component is $\alpha^\mathsf{T}(G_X + \eta_X I_n)^{-1} b_{1:n}^{(X)}(X_i)$. We can see that the above matrix is quite similar to $\Lambda_{\text{GSIR}}^{(2)}$ in (13.13).

The GSIR developed in Lee et al. (2013) was slightly different from the GSIR presented here in that the former uses the $L_2(F_n)$-inner product instead of the RKHS-inner product. Their development leads to the following solution. Let $v_1, \ldots, v_d$ be the first d-eigenvectors of

$$(G_X + \eta_X I_n)^{-3/2} G_X^{3/2} (G_Y + \eta_Y I_n)^{-1} G_Y^2 (G_Y + \eta_Y I_n)^{-1} G_X^{3/2} (G_X + \eta_X I_n)^{-3/2}.$$

The sufficient predictors are

$$v_i^\mathsf{T}(G_X + \eta_X I_n)^{-1} b_{1:n}(X_j), \quad i = 1, \ldots, d, \ j = 1, \ldots, n.$$

13.7   Generalized Cross Validation for Tuning ε_X and ε_Y

We use the Generalized Cross Validation to choose the tuning constants ε_X and ε_Y in the algorithm. Generalized Cross Validation was introduced by Golub et al. (1979) as a general method for choosing the tuning parameter for ridge regression in linear models. It takes the following general form

$$V(\varepsilon) = n^{-1} \|[I_n - A(\varepsilon)]\mathbb{Y}\|^2 / \{n^{-1}\text{tr}[I_n - A(\varepsilon)]\}^2 \tag{13.14}$$

where $\mathbb{Y}$ is the vector $(Y_1, \ldots, Y_n)^\mathsf{T}$, and $A(\varepsilon)$ is the matrix $\mathbb{X}(\mathbb{X}^\mathsf{T}\mathbb{X} + \varepsilon I_n)^{-1}\mathbb{X}^\mathsf{T}$, in which $\mathbb{X}$ is the matrix $(X_1, \ldots, X_n)^\mathsf{T}$. The idea behind this criterion is the following. The term $A(\varepsilon)\mathbb{Y}$ approximates the conditional expectation $E(Y|X)$ under the linear model. The constant ε here determines the degree of complexity. The smaller the ε, the closer $A(\varepsilon)$ is to the identity matrix, and the smaller the estimation error $[I_n - A(\varepsilon)]\mathbb{Y}$. This roughly corresponds to using more parameters in the regression model. As ε increases, $A(\varepsilon)$ becomes further apart from the identity matrix. In fact, as ε increases, $A(\varepsilon)\mathbb{Y}$ converges to the zero vector, which means we are not fitting the model at all. If we only want to minimize the prediction error, the smaller the ε, the better. But this causes overfitting. For example, if $p \geq n$ and $\varepsilon \approx 0$, then we are essentially using $\mathbb{Y}$ to predict $\mathbb{Y}$. The denominator penalizes the smallness of the distance between $A(\varepsilon)$ and the identity matrix I_n, where the smallness represents model complexity.

We use the same idea to select ε_X and ε_Y. However, note that in nonlinear Sufficient Dimension Reduction, we are not fitting a single response but rather a family of functions of the response variable. That is, we would like to fit the functions $\kappa_Y(\cdot, Y_i), i = 1, \ldots, n$ using the family of functions $\kappa_X(\cdot, X_i), i = 1, \ldots, n$. First consider a single function $\kappa_Y(\cdot, Y_i)$. The conditional expectation $E[\kappa_Y(Y, y_i)|X]$ is approximated at the sample level by the regression operator

$$\hat{M}_{XX}^{-1}\hat{M}_{XY}\kappa(\cdot, y_i) \tag{13.15}$$

where M_{XX} and M_{XY} are the second moment operators

$$\hat{M}_{XX} = E_n[\kappa_X(\cdot, X) \otimes \kappa_X(\cdot, X)], \quad \hat{M}_{XY} = E_n[\kappa_X(\cdot, X) \otimes \kappa_Y(\cdot, Y)].$$

Now consider the Hilbert spaces $\mathscr{H}_X$ and $\mathscr{H}_Y$ spanned by $\{\kappa_X(\cdot, X_i) : i = 1, \ldots, n\}$ and $\{\kappa_Y(\cdot, Y_j) : j = 1, \ldots, n\}$. Here, we use the uncentered space because, for prediction, constants do matter. Let $\mathscr{C}_X$ be the basis $\{\kappa(\cdot, X_i) : i = 1, \ldots, n\}$ and $\mathscr{C}_Y$ be the basis $\{\kappa_Y(\cdot, Y_i) : i = 1, \ldots, n\}$. By the calculations that lead to Theorem 12.1, it can be shown that

$$_{\mathscr{C}_X}[\hat{M}_{XX}]_{\mathscr{C}_X} = K_X, \quad _{\mathscr{C}_X}[\hat{M}_{XY}]_{\mathscr{C}_X} = K_Y.$$

Hence, with Tychonoff regularization, the coordinate of (13.15) is

$$[\hat{M}_{XX}^{-1}\hat{M}_{XY}\kappa_Y(\cdot, y_i)]_{\mathscr{C}_X} = (K_X + \varepsilon_X \lambda_{\max}(K_X)I_n)^{-1}K_Y e_i$$

where we have used the fact $[\kappa_Y(\cdot, Y_i)]_{\mathscr{C}_Y} = e_i$. In other words, if (X', Y') is a draw from the distribution of (X, Y), then the random variable $\kappa_Y(Y', Y_i)$ is approximated by

$$\kappa_X(X', X_{1:n})^\mathsf{T}(K_X + \varepsilon_X \lambda_{\max}(K_X)I_n)^{-1}K_Y e_i$$

where $\kappa_X(\cdot, X_{1:n})$ stands for the vector $(\kappa_X(\cdot, X_1), \ldots, \kappa_X(\cdot, X_n))^\mathsf{T}$. In particular, $\kappa_Y(Y_j, Y_i)$ is approximated by

$$\hat{\kappa}_{Y,ij} = \kappa_X(X_j, X_{1:n})^\mathsf{T}(K_X + \varepsilon_X \lambda_{\max}(K_X)I_n)^{-1}K_Y e_i,$$

which is simply the (j, i)th entry of the matrix

$$\hat{K}_Y = K_X(K_X + \varepsilon_X \lambda_{\max}(K_X)I_n)^{-1}K_Y.$$

The sum of squared error for estimating the function $\kappa_Y(\cdot, X_i)$ is

$$\sum_{i=1}^{n} (\kappa_Y(Y_j, Y_i) - \hat{\kappa}_{Y,ji})^2.$$

The aggregated error for estimating $\kappa_Y(\cdot, 1), \ldots, \kappa_Y(\cdot, n)$ is therefore

$$\sum_{i=1}^{n} \sum_{j=1}^{n} (\kappa_Y(Y_j, Y_i) - \hat{\kappa}_{Y,ij})^2,$$

which is simply the Frobenius norm $\|K_Y - \hat{K}_Y\|_F^2$. Following the construction of Golub et al. (1979), we come up with the natural GCV criterion

$$\text{GCV}_X(\varepsilon_X) = \frac{\|K_Y - K_X(K_X + \varepsilon_X \lambda_{\max}(K_X)I_n)^{-1}K_Y\|_F^2}{\{\text{tr}[I_n - K_X(K_X + \varepsilon_X \lambda_{\max}(K_X)I_n)^{-1}]\}^2}.$$

Similar to the GCV criterion in (13.14), the numerator is the prediction error; the denominator is to prevent $K_X(K_X + \varepsilon_X \lambda_{\max}(K_X)I_n)]$ from getting too close to the identity matrix. The GCV criterion for ε_Y is exactly the same, except that the roles of X and Y are switched. That is,

$$\text{GCV}_Y(\varepsilon_Y) = \frac{\|K_X - K_Y(K_Y + \varepsilon_Y \lambda_{\max}(K_Y)I_n)^{-1}K_X\|_F^2}{\{\text{tr}[I_n - K_X(K_X + \varepsilon_Y \lambda_{\max}(K_Y)I_n)^{-1}]\}^2}.$$

We minimize this criterion over a relative coarse grid, such as $\{10^{-6}, \ldots, 10^{-1}\}$, to find the optimal tuning constant.

Since it is unclear how to generalize the GCV criterion to the current setting where the tuning parameters do not resemble those in the ridge regression, we choose the parameters γ_X and γ_Y as some fixed, but reasonable, quantities. Note that the tuning constants $\varepsilon_X, \varepsilon_Y$ and the tuning constants γ_X, γ_Y are not independent. They both control the complexity of the functions in $\mathscr{H}_X$ and $\mathscr{H}_Y$. That is, the larger γ_X, γ_Y are, the more wiggly the functions in $\mathscr{H}_X$ and $\mathscr{H}_Y$ are; the smaller $\varepsilon_X, \varepsilon_Y$, the more wiggly these functions are. For this reason, the general principal for choosing γ_X, γ_Y is that they shouldn't be too small, because once the functions are smoothed by γ_X, γ_Y, they cannot be unsmoothed by $\varepsilon_X, \varepsilon_Y$. On the other hand, if γ_X, γ_Y undersmoothed the functions (meaning they are too small), then functions can be further smoothed by $\varepsilon_X, \varepsilon_Y$. From our experiences the following choices seem to work well:

$$\gamma_X = \frac{\rho_X}{2\sigma_X^2}, \quad \gamma_Y = \frac{\rho_Y}{2\sigma_Y^2}, \tag{13.16}$$

where

$$\sigma_X^2 = \binom{n}{2}^{-1} \sum_{i<j} \|X_i - X_j\|^2, \quad \sigma_Y^2 = \binom{n}{2}^{-1} \sum_{i<j} \|Y_i - Y_j\|^2, \quad \rho_X, \rho_Y \in (1/10, 10).$$

The following is the R-code to evaluate $\text{GCV}_X(\varepsilon_X)$ or $\text{GCV}_Y(\varepsilon_Y)$ for given ε_X and ε_Y. The input variables are the same as those in gsir except eps, which is the value of ε_X or ε_Y, and the variable which, which can be ex or ey, indicating whether $\text{GCV}_X(\varepsilon_X)$ or $\text{GCV}_Y(\varepsilon_Y)$ is desired.

R-code for GCV

```
gcv=function(x,y,eps,which,ytype,complex.x,complex.y){
p=dim(x)[2];n=dim(x)[1]
Kx=gram.gauss(x,x,complex.x)
if(ytype=="scalar") Ky=gram.gauss(y,y,complex.y)
if(ytype=="categorical") Ky=gram.dis(y)
if(which=="ey") {G1=Kx;G2=Ky}
if(which=="ex") {G1=Ky;G2=Kx}
G2inv=matpower(G2+eps*onorm(G2)*diag(n),-1)
nu=sum((G1-G2%*%G2inv%*%G1)^2)
tr=function(a) return(sum(diag(a)))
de=(1-tr(G2inv%*%G2)/n)^2
return(nu/de)
}
```

13.8   *k*-Fold Cross Validation for Tuning $\rho_X, \rho_Y, \varepsilon_X, \varepsilon_Y$

The advantage of GCV is that it is fast, because no repeated removal of samples is performed. However, it is unclear how to generalize Cross-Validation to accommodate a tuning parameter that does not resemble the tuning parameter in the ridge regression setting. On the other hand, Cross-Validation (CV, see, for example, Stone (1977)) is a flexible criterion that applies to virtually any tuning parameters, at the expense of increased computing time. Nevertheless, expense of computing time of CV can be mitigated by using a *k*-fold Cross Validation rather than leave-one-out Cross Validation.

The *k*-fold Cross Validation, in our setting, proceeds as follows. Divide the sample $(X_1, Y_1), \ldots, (X_n, Y_n)$ into *k* sub-samples, of roughly equal sizes, say $D_1, \ldots, D_k$. Let *D* denote the full data set. For each $i = 1, \ldots, k$, use $D \setminus D_i = D_i^c$ as the training set, and D_i as the testing set. Use the functions in the training set to develop the predictor, and substitute into the predictor the *X* in the testing set to predict the response in the testing set. Compute the prediction error for each *i* and aggregate the error for all rotations $i = 1, \ldots, k$, which yields an overall cross-validation error. This overall error depends on the tuning parameters, and is then minimized over a grid of the tuning parameters to obtain the optimal tuning parameters.

It remains to develop the specific form of the cross-validation error, for which we need some notations. For a subset *C* of *D*, let $X(C)$ be the set of X_i such that (X_i, Y_i) belongs to *C*; that is,

$$X(C) = \{X_i : (X_i, Y_i) \in C\}.$$

For example, for the full data set *D*, $X(D)$ would be $X_{1:n} = (X_1, \ldots, X_n)$. Also, if $C, E \subseteq D$, then $\kappa_X(X(C), X(E))$ denotes the matrix

$$\{\kappa_X(x, x') : x \in X(C), x' \in X(E)\},$$

which is a $\text{card}(C) \times \text{card}(E)$ matrix. Define $Y(C)$, $\kappa_Y(Y(C), Y(E))$ in the same way.

Next, for a subset *C* of *D*, let $\mathscr{H}_X(C)$ be the reproducing kernel Hilbert space generated by the kernel matrix $\kappa_X(X(C), X(C))$, as described earlier. Define $\mathscr{H}_Y(C)$ similarly. For each rotation $i = 1, \ldots, k$, we predict the functions in $\mathscr{H}_Y(D^i)$ using the regression operator developed from the two spaces $\mathscr{H}_X(D_i^c)$ and $\mathscr{H}_X(D_i)$. Following through the development in the last section, it is not difficult to see that the squared error of this estimation is

$$\|K_Y(Y(D_i), Y(D_i^c)) - \hat{K}_Y(Y(D_i), Y(D_i^c))\|_F^2,$$

where $\|\cdot\|_F$ is the Frobenius norm, and

$$\hat{K}_Y(Y(D_i), Y(D_i^c)) = K_X(X(D_i), X(D_i^c))$$
$$\{K_X[X(D_i^c), X(D_i^c)] + \varepsilon_X \lambda_{\max}[K_X(X(D_i^c), X(D_i^c))]I_{\text{card}(D_i^c)}\}^{-1}$$
$$K_Y[X(D_i^c), X(D_i^c)].$$

The aggregated cross-validation error is, then,

$$\mathrm{CV}(\rho_X, \varepsilon_X) = \sum_{i=1}^{k} \|K_Y(Y(D_i), Y(D_i^c)) - \hat{K}_Y(Y(D_i), Y(D_i^c))\|_F^2.$$

The optimal tuning parameter is obtained by minimizing the above criterion over a grid of (ρ_X, ε_X). The grid we used in the simulation is

$$\{1/5, 1/4, \ldots, 1, 2, \cdots 5\} \times \{10^{-6}, \ldots, 10^{-1}\}.$$

The two sets of parameters (ρ_X, ε_X) and (ρ_Y, ε_X) are tuned separately, not jointly. However, in determining ρ_X, ε_X, for example, we do need to assign a value to ρ_Y, as it determines the Hilbert spaces $\mathcal{H}_Y(D_i)$ and $\mathcal{H}_Y(D_i^c)$. This value is not of critical importance, because we simply need a set of functions of Y to predict. Thus, when tuning (ρ_X, ε_X), we set ρ_Y to be 1; when tuning (ρ_Y, ε_Y), we set ρ_X to be 1.

In the following R-code for evaluating k-fold cross validation criterion $\mathrm{CV}(\rho_Y, \varepsilon_Y)$, the inputs x and y are the predictor matrix and the response vector or matrix; k is the number of folds in the k-fold CV. complex.x and ex are the tuning parameters ρ_X and ε_X. This code can also be used to evaluate $\mathrm{CV}(\rho_X, \varepsilon_X)$ by simply reversing the roles of X and Y. The code calls four functions, gram.gauss, gram.dis, matpower, and onorm, which were provided previously.

R-code for k-fold Cross-Validation

```
cv.kfold=function(x,y,k,complex.x,ex){
x=as.matrix(x);y=as.matrix(y);n=dim(x)[1]
ind=numeric();for(i in 1:(k-1)) ind=c(ind,floor(n/k))
ind[k]=n-floor(n*(k-1)/k)
cv.out=0
for(i in 1:k){
if(i<k) groupi=((i-1)*floor(n/k)+1):(i*floor(n/k))
if(i==k) groupi=((k-1)*floor(n/k)+1):n
groupic=(1:n)[-groupi]
x.tra=as.matrix(x[groupic,]);y.tra=as.matrix(y[groupic,])
x.tes=as.matrix(x[groupi,]);y.tes=as.matrix(y[groupi,])
Kx=gram.gauss(x.tra,x.tra,complex.x)
Kx.tes=gram.gauss(x.tra,x.tes,complex.x)
Ky=gram.gauss(y.tra,y.tra,1)
Ky.tes=gram.gauss(y.tra,y.tes,1)
```

```
cvi=sum((t(Ky.tes)-t(Kx.tes)%*%
    matpower(Kx+ex*onorm(Kx)*diag(dim(y.tra)[1]),-1)%*%Ky)^2)
cv.out=cv.out+cvi
}
return(cv.out)
}
```

13.9   Simulation Studies

In this section we investigate the performance of GSIR in three simulated models, as given below

$$
\begin{cases}
\text{I}: & Y = (X_1^2 + X_2^2)^{1/2} \log(X_1^2 + X_2^2)^{1/2} + \varepsilon \\
\text{II}: & Y = X_1/(1 + e^{X_2}) + \varepsilon \\
\text{III}: & Y = \sin(\pi(X_1 + X_2)/10) + \varepsilon
\end{cases}
, \quad \varepsilon \perp\!\!\!\perp X, \ \varepsilon \sim N(0, 0.25), \ p = 10.
$$

For the distribution of the 10-dimensional predictor X, we consider three scenarios: (A) independent Gaussian predictors, (B) independent non-Gaussian predictors, and (C) correlated Gaussian predictors. Specifically,

$$
\begin{cases}
\text{A}: & X \sim N(0, I_p) \\
\text{B}: & X \sim (1/2)N(-1_p, I_p) + (1/2)N(1_p, I_p) \\
\text{C}: & X \sim N(0, 0.6I_p + 0.41_p 1_p^\mathsf{T})
\end{cases}
$$

where, in Scenario B, the notation $(1/2)N(-1_p, I_p) + (1/2)N(1_p, I_p)$ is to be interpreted as the mixture distribution of $N(-1_p, I_p)$ and $N(1_p, I_p)$ with mixing probabilities $(1/2, 1/2)$. Note that the central σ-fields for the three models I, II, and III are generated by $X_1^2 + X_2^2$, $X_1/(1 + e^{X_2})$, and $\sin(\pi(X_1 + X_2)/10)$, respectively. These models, as well as the distributional scenarios, are taken from Lee et al. (2013).

Under the framework of linear Sufficient Dimension Reduction, the dimension of the central subspaces for models I and II are 2, with sufficient linear predictors X_1, X_2; the central subspace for model I has dimension 1, with linear sufficient predictor $X_1 + X_2$. In comparison, in the framework of nonlinear Sufficient Dimension Reduction, each central class is generated by a single nonlinear or linear function of X. To gain some intuition about the working of GSIR, we first present the result in one sample of Model I and scenario A, with sample size $n = 200$. In Figure 13.1, upper panel, we show the scatter plot of the true predictor $(X_1^2 + X_2^2)^{1/2} \log(X_1^2 + X_2^2)^{1/2}$ versus the first GSIR predictor. We can see that the method works very well, successfully capturing the nonlinear predictor hidden in 10 dimensions. The lower left panel is the response versus the true nonlinear predictor, which is closely resembled by the plot on the lower left panel, the scatter plot of Y versus the first GSIR predictor.

To further study the performance of GSIR we need a numerical measurement of its effectiveness. Unlike in the linear SDR setting, we cannot use the difference between subspaces, such as criterion described in Section 11.9, as the measure of effectiveness, because the target of estimation is not a subspace but the predictors themselves. For this reason, we assess the quality of an estimated sufficient predictor by its closeness to the true sufficient predictor or, when the true sufficient predictor is unknown, its closeness to the response. Since the nonlinear SDR can only provide an estimate of monotone functions of the predictors, we use Spearman's correlation, which is not affected by monotone transformations, as the measure of closeness.

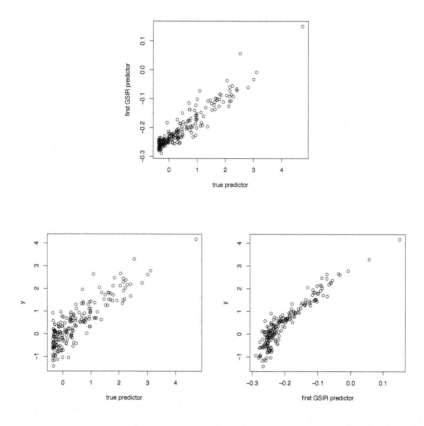

Figure 13.1 *GSIR for Model I, Scenario A.*

For each combinations of sample sizes $n = 100, 200$, dimensions $p = 10, 20$, models I, II, III, and scenarios A, B, and C, we generate 100 samples. We applied the two versions of GSIR — $\Lambda^{(1)}_{GSIR}$ and $\Lambda^{(2)}_{GSIR}$ — to the 100 sample, and compute the means of the Spearman's correlation between the true and the estimated predictors. The Gauss radial basis function is used for κ_X and κ_Y. The tuning parameters $\rho_X, \varepsilon_X, \rho_Y, \varepsilon_Y$ are chosen by 10-fold Cross-Validation. Since the optimal tuning parameters change little from sample to sample, for each of the 72 combinations, we tune the parameters using the first sample, and use the resulting $\rho_X, \varepsilon_X, \rho_Y, \varepsilon_Y$ for the rest of the 100 samples. The means of the Spearman's correlations are shown in Table 13.1. Overall, both versions of GSIR work very well, with Spearman's correlation ranges between $0.8 \sim 1$ in most of the cases. The two versions perform similarly, with $\Lambda^{(1)}_{GSIR}$ more stable for smaller n and larger p, and $\Lambda^{(2)}_{GSIR}$ slightly more accurate when for larger n and smaller p.

As we mentioned, there are some minor differences between the GSIR presented here and that in Lee et al. (2013). In addition, the tuning methods are also different. Nevertheless, the results presented here are quite similar to those presented in Table 1 of Lee et al. (2013).

Table 13.1 *Performance of GSIR (with options $\Lambda^{(1)}_{\text{GSIR}}$ and $\Lambda^{(2)}_{\text{GSIR}}$) under Models I, II, III. Scenarios A, B, C, and four different combinations of sample sizes and dimensions*

(n, p)	model	$\Lambda^{(1)}_{\text{GSIR}}$			$\Lambda^{(2)}_{\text{GSIR}}$		
		A	B	C	A	B	C
(100, 10)	I	0.84	0.91	0.84	0.46	0.95	0.85
	II	0.86	0.91	0.91	0.82	0.92	0.90
	III	0.89	0.97	0.94	0.88	0.97	0.94
(100, 20)	I	0.58	0.87	0.55	0.66	0.86	0.67
	II	0.79	0.90	0.87	0.84	0.90	0.87
	III	0.81	0.94	0.91	0.81	0.94	0.90
(200, 10)	I	0.88	0.95	0.89	0.89	0.96	0.90
	II	0.86	0.95	0.91	0.89	0.94	0.88
	III	0.84	0.98	0.94	0.84	0.97	0.95
(200, 20)	I	0.80	0.94	0.81	0.40	0.94	0.80
	II	0.81	0.92	0.88	0.86	0.94	0.91
	III	0.79	0.95	0.94	0.88	0.95	0.92

13.10   Applications

13.10.1   Pen Digit Data

We now apply GSIR to the pen digit data. We use the discrete kernel for the response, and the Gaussian radial basis function as the kernel K_X. For the tuning parameters, we choose γ_X to be (13.16) with ρ_X chosen to be 1. The tuning parameter ε_X is chosen by the Generalized Cross Validation, as described in the last section, which turned out to be 10^{-6}. Since the response Y is categorical, there are no tuning parameters γ_Y and ε_Y. As in linear Sufficient Dimension Reduction, when the response is categorical with number of classes equal to k, GSIR can provide at most $k - 1$ sufficient predictors. This is because the matrix G_X has rank $k - 1$.

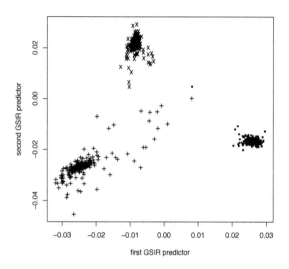

Figure 13.2 *First two GSIR predictors for the pen digit data.*

We plot the first two GSIR predictors in Figure 13.2, where the plot symbol "+" represents digit 0, "×" represents digit 6, and "·" represents digit 9. As in the previous applications to this data set, the predictors are developed from the training set, of sample size 2219, and the sufficient predictors are evaluated at the testing set, of sample size 1035. As we can see the three groups are very well separated. Comparing this plot with the left panel of Figure 6.6 of Chapter 6, which is the first to SIR predictors for this data set, we see that the nonlinear method seem to offer clearer separation.

13.10.2   Face Sculpture Data

This is a well known data set used in Tenenbaum et al. (2000). It contains 698 images of a face sculpture photographed at different horizontal angles, vertical angles, and light directions. Each image has 64×64 pixels. These can be regarded as a sample of 698 vector-valued predictors each having dimension 4096. The response Y is of dimension 3: the first component of Y is a number that measures the direction of the light source; the second component is the horizontal angle of the face; the third component is the vertical angle of the face.

To give a comprehensive view of the data set, in Figure 13.3 we put these images into a 3-d perspective plot, where the three variables, light, horizontal angle, and vertical angle are represented by the three axes of the perspective plot. Thus, along the long edge of the box, the light direction is gradually changed from one direction to another; along the short (inward) edge of the box, the faces change their horizontal orientations; along the vertical edge of the box, the faces change their vertical orientations. Since the images are photos from the same face sculpture, they are completely determined by the three numbers. The goal of nonlinear sufficient dimension reduction is to try to extract these three numbers (or a one-to-one function of these three numbers) by just observing the 698 sample faces. That is, we need to reduce the dimension from 4096 to 3.

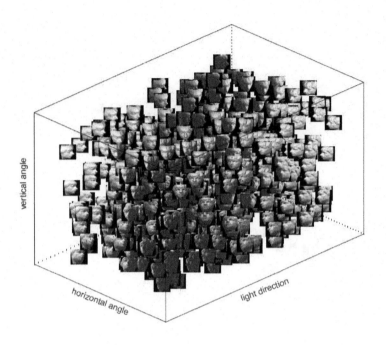

Figure 13.3 *Face data representation.*

We applied GSIR to this data set. The kernels κ_X and κ_Y are both chosen to be the Gaussian radial basis function. The tuning parameters γ_X and γ_Y are chosen by (13.16) with $\rho_X = \rho_Y = 1/5$. The tuning parameters ε_X and ε_Y are chosen by the Generalized Cross Validation described in the last section, which results in the optimal choices $\varepsilon_X = \varepsilon_Y = 10^{-6}$. We split the data set randomly into a training set, of sample size 558, and a testing set, of sample size 140. The training set is roughly 80% of the full data set. We developed a 3-dimensional predictor from the training set, and evaluate it at the testing set. The perspective plot of images is shown in Figure 13.4, where the predictors were assigned to the three axes. We can see that the images are roughly aligned with those in Figure 13.3. The very possibility of such an alignment already reveals that all three variables in the true response have been successfully captured. Indeed, like the perspective image plot for the true responses, the direction of light source changes along the third GSIR direction, which is represented by the long edge of the box; the faces changes their horizontal angles along the first GSIR direction, which is represented by the short, inward edge; the faces change their vertical angles along the second GSIR direction, which is represented by the vertical axis.

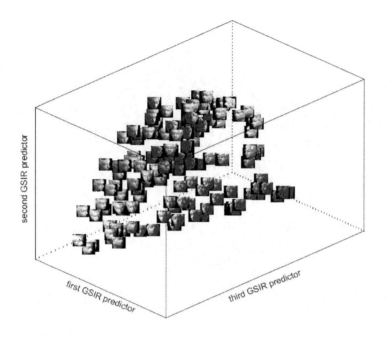

Figure 13.4 *First three GSIR predictors as evaluated at the testing set.*

To further investigate the relation between the true responses and the first three predictors, we show in Figure 13.5 the scatter plot of the first GSIR predictor versus the horizontal angle (upper panel), the scatter plot of the third GSIR predictor versus the light direction (lower-left panel), and the scatter plot of the second GSIR predictor versus the vertical angle (lower-right panel). These plots show strong linear relationship, with absolute values of correlations 0.94, 0.89, and 0.58. The multiple correlation between the two sets of three variables is 2.28 (the maximum value of the multiple correlation in this case is 3).

It is interesting to note that, although our theory only guarantees, at the population level, that nonlinear SDR recovers the σ-field of the underlying variables, the three sufficient GSIR predictors correspond nicely with the three true response variables. Scatter plots (note presented here) show that there is no discernable dependence between other variables among the two sets. The correlation matrix between (horizontal angle, light direction, vertical angle) and (GSIR1, GSIR3, GSIR2) is

$$\begin{pmatrix} -0.946 & 0.268 & 0.014 \\ 0.316 & 0.893 & -0.214 \\ -0.035 & 0.063 & 0.577 \end{pmatrix},$$

where the off-diagonal elements are quite small, indicating no significant statistical dependence. Whether this is merely a happy coincidence or there is a deeper reason behind it deserves further investigation.

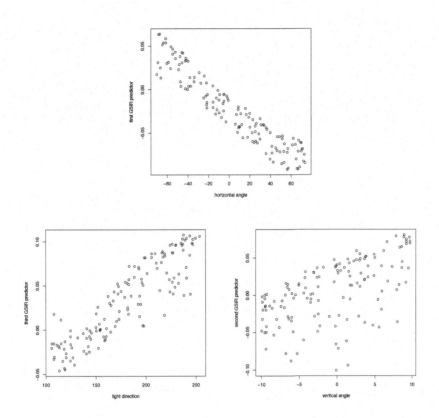

Figure 13.5 *Relation between the true responses and the first three GSIR predictors in the face data.*

Chapter 14

Generalized Sliced Average Variance Estimator

14.1 Generalized Sliced Average Variance Estimation   233
14.2 Relation with GSIR   237
14.3 Implementation of GSAVE   239
14.4 Simulation Studies and an Application   248
14.5 Relation between Linear and Nonlinear SDR   251

In this section we describe the second estimator of the central class — the Generalized Sliced Average Variance Estimator, or GSAVE, which is a generalization of SAVE in the linear SDR setting, introduced by Lee et al. (2013). It can recover a larger portion of the central subspace than GSIR, when the central subspace is not complete. It is based on what we call the heteroscedastic conditional covariance operator, which is a different operator than the conditional variance operator developed prior to Lee et al. (2013). Because the heteroscedastic conditional operator involves the conditional mean such as $E[f(X)g(X)|Y]$, it is convenient to construct the estimator using the $L_2(F_n)$ inner product, in a space that contains a constant. For this reason, the geometry used in this chapter is different from that used in Chapter 13.

14.1 Generalized Sliced Average Variance Estimation

As shown in Theorem 13.2, GSIR is unbiased and exhaustive when the central class is sufficient and complete. However, when the central class is incomplete, there is no guarantee that it will recover the central class fully. This was the original motivation for developing GSAVE in Lee et al. (2013).

Recall that the SAVE in the linear SDR setting relies on the conditional variance matrix $\text{var}(X|Y)$. In GSAVE, the conditional variance matrix is replaced by a conditional variance operator. Let us then first define the conditional variance operator. To define this operator we need to involve conditional expectation of the form $E[f(X)g(X)|Y]$ for arbitrary f and g. Since $f(X)g(X)$ is not centered even if both $g(X)$ and $f(X)$ are centered, it is more convenient and more intuitive to develop GSAVE using a Hilbert space that contains a constant function, which is not always true for an RKHS (for example, it is not true if the kernel is the Gaussian radial basis function). This problem is circumvented in Lee et al. (2013) by using the space

$$\mathscr{H}_Y = \overline{\mathrm{span}} \left(\{1\} \cup \{ \kappa_Y(\cdot, y) : y \in \Omega_Y \} \right)$$

coupled with the $L_2(P_Y)$-inner product; that is, $\langle f, g \rangle_{\mathscr{H}_X} = E[f(X)g(X)]$. This is the framework we adopt in this chapter. We define $\mathscr{H}_X$ and its inner product similarly. For this construction to be possible we need the following assumption.

Assumption 14.1 *For each* $x \in \Omega_X$, $y \in \Omega_Y$, $E[\kappa_X^2(X,x)] < \infty$ *and* $E[\kappa_Y^2(Y,y)] < \infty$.

This assumption is easily satisfied by many kernels, including the Gauss radial basis function. Next, let us introduce the regression operator. Consider the bilinear form

$$\mathscr{H}_X \times \mathscr{H}_Y \to \mathbb{R}, \quad (f,g) \mapsto E[f(X)g(Y)].$$

By the Cauchy-Schwarz inequality, this is a bounded bilinear form — in fact, the bound of the bilinear form is no greater than 1. Thus there is a bounded linear operator in $A \in \mathscr{B}(\mathscr{H}_X, \mathscr{H}_Y)$ such that

$$\langle g, Af \rangle_{\mathscr{H}_Y} = E[f(X)g(Y)].$$

We call this operator the cross-moment operator, and write it as M_{XY}. By construction, we see that M_{YX} is simply the adjoint operator of M_{XY}. This is the same construction as that in Section 12.2 except that the inner product has changed. Interestingly, it is easy to see that the operator M_{XX} induced by the bilinear form $(f,g) \mapsto E[f(X)g(X)]$ is simply the identity operator. Not surprisingly, then, the regression operator in this setting coincides with the cross-moment operator, as the next theorem shows.

Theorem 14.1 *If* $\mathscr{H}_Y$ *is dense in* $L_2(P_Y)$, *then, for any* $f \in \mathscr{H}_X$,

$$(M_{YX}f)(y) = E[f(X)|Y = y].$$

PROOF. Let $f \in \mathscr{H}_X$. By the definition of conditional expectation, $M_{YX}f$ is the condition expectation of $f(X)$ given Y if, for any $g \in L_2(P_Y)$,

$$E\{[f(X) - (M_{XY}f)(Y)]g(Y)\} = 0. \tag{14.1}$$

Because $\mathscr{H}_Y$ is dense in $L_2(P_Y)$, there is a sequence $\{g_n\} \subseteq \mathscr{H}_Y$ such that $\|g_n - g\|_{\mathscr{H}_Y} \to 0$. For any g_n in the sequence we have

$$E[f(X)g_n(Y)] = \langle f, M_{XY}g \rangle_{\mathscr{H}_X} = \langle M_{YX}f, g_n \rangle_{\mathscr{H}_Y} = E[(M_{YX}f)(Y)g_n(Y)].$$

Hence equality (14.1) holds for g_n. That is,

$$E\{[f(X) - (M_{YX}f)(Y)]g_n(Y)\} = 0.$$

Now take $\lim_{n \to \infty}$ on both sides to complete the proof. □

This theorem shows that M_{YX} sends any member $f \in \mathscr{H}_X$ to $\mathscr{H}_Y$. As a byproduct, it also shows that if $f \in \mathscr{H}_X$, then $E[f(X)|Y]$ is a member of $\mathscr{H}_Y$. Let us now define a random linear operator that is a generalization of conditional variance matrix $\mathrm{var}(X|Y)$, which is, in general, a random matrix.

Before generalizing the conditional variance, we first generalize the first two conditional moments. For each fixed $y \in \Omega_Y$, consider the linear functional

$$L_y : \mathscr{H}_X \to \mathbb{R}, \quad f \mapsto E[f(X)|Y = y].$$

Under the assumption that this is a bounded linear functional, there is a member $f_y \in \mathscr{H}_X$ such that

$$\langle f_y, f \rangle_{\mathscr{H}_X} = E[f_y(X)f(X)] = E[f(X)|y].$$

The member f_y is called the conditional mean element of X given $Y = y$. We write the function f_y as $\mu_{X|Y}(y)$. We can define the second moment operators in a similar way. For each $y \in \Omega_Y$, consider the bilinear form

$$T_y : \mathscr{H}_X \times \mathscr{H}_X \to \mathbb{R}, \quad (f, g) \mapsto E[f(X)g(X)|Y = y].$$

If this bilinear form is bounded, then it induces a linear operator $A_y \in \mathscr{B}(\mathscr{H}_X, \mathscr{H}_X)$ such that

$$\langle f, A_y g \rangle_{\mathscr{H}_X} = E[f(X)g(X)|Y = y].$$

We call this operator the conditional cross moment operator, and denote it by $M_{XX|Y}(y)$. For these constructions to be possible, we make the following assumption.

Assumption 14.2 *For each $y \in \Omega_Y$, the linear functional L_y is a bounded linear functional; the bilinear form T_y is a bounded bilinear form.*

We now formally define the linear operator that generalizes the conditional variance matrix $\mathrm{var}(X|y)$.

Definition 14.1 *Under Assumption 14.2, the following operator*

$$V_{XX|Y}(y) = M_{XX|Y}(y) - \mu_{X|Y}(y) \otimes \mu_{X|Y}(y)$$

is called the heteroscedastic conditional variance operator.

Here we have deliberately avoided the more natural notation $\Sigma_{XX|Y}$, because the latter was used by Fukumizu et al. (2004) and Fukumizu et al. (2009) to denote another type of conditional variance operator, defined as $\Sigma_{XX|Y} = \Sigma_{XX} - \Sigma_{XY}\Sigma_{YY}^{-1}\Sigma_{YX}$. The latter is actually a generalization of $E[\mathrm{var}(X|Y)]$ instead of $\mathrm{var}(X|Y)$. To make a distinction, we refer to $V_{XX|Y}$ as the heteroscedastic conditional variance operator, and $\Sigma_{XX|Y}$ as the homoscedastic conditional variance operator.

By construction, for each $y \in \Omega_Y$, $f, g \in \mathscr{H}_X$, we have

$$\langle f, V_{XX|Y}(y)g \rangle_{\mathscr{H}_X} = \langle f, M_{XX|Y}(y)g \rangle_{\mathscr{H}_X} - \langle f, \mu_{X|Y}(y) \otimes \mu_{X|Y}(y)g \rangle_{\mathscr{H}_X}.$$

The first term on the right-hand side is, by construction, $E[f(X)g(Y)|Y = y]$. The second term is

$$\begin{aligned}
\langle f, \mu_{X|Y}(y) \otimes \mu_{X|Y}(y)g \rangle_{\mathscr{H}_X} &= \langle f, \mu_{X|Y}(y)\langle \mu_{X|Y}(y), g \rangle_{\mathscr{H}_X} \rangle_{\mathscr{H}_X} \\
&= \langle f, \mu_{X|Y}(y) \rangle_{\mathscr{H}_X} \langle \mu_{X|Y}(y), g \rangle_{\mathscr{H}_X} \\
&= E[f(X)|Y = y]E[g(X)|Y = y].
\end{aligned}$$

It follows that

$$\langle f, V_{xx|Y}(y)g \rangle_{\mathscr{H}_X} = \mathrm{cov}[f(X), g(X)|Y = y],$$

as to be expected.

Since the operator $V_{xx|Y}(y)$ depends on y, it is a random operator that maps from Ω_Y to $\mathscr{B}(\mathscr{H}_X, \mathscr{H}_X)$. We are going to take the expectation of the square of this random operator to define a generalization of SAVE. We define the expectation of a random operator using the same mechanism described in Section 12.2, except that the inner product is different.

Definition 14.2 *The GSAVE operator is*

$$\Lambda_{\mathrm{GSAVE}} = E\{[\Sigma_{XX} - V_{XX|Y}(Y)]^2\}.$$

It is then plausible that the solution to the generalized eigenvalue problem $\mathrm{GCV}(\Lambda_{\mathrm{GSAVE}}, \Sigma_{XX})$ would recover the central class $\mathfrak{S}_{Y|X}$ in the nonlinear setting. To show this, we need to make two more assumptions: the first is a nonlinear extension of the constant variance assumption (Assumption 5.1) in Section 5.2; the second is parallel to Assumption 13.2, which guarantees $\Sigma_{XX}^{-1}\Lambda_{\mathrm{GSAVE}}$ is a compact operator.

Assumption 14.3 *For any function $f \perp \mathfrak{S}_{Y|X}$, the heteroscedastic conditional variance operator $V_{XX|Y}(Y)$ is a nonrandom operator.*

Since, under this assumption, the heteroscedastic conditional variance operator $V_{XX|Y}(Y)$ coincides with the homoscedastic conditional variance operator $\Sigma_{XX|Y}$, we refer to this assumption as the homoscedastic conditional variance operator assumption, which echoes the "constant conditional variance" assumption in Section 5.2.

Assumption 14.4 Λ_{GSAVE} *is a compact operator.*

We now prove the unbiasedness of the estimator based on Λ_{GSAVE}.

Theorem 14.2 *If Assumptions 14.1 through 14.4 are satisfied, and $\mathfrak{S}_{Y|X}$ is dense in $L_2(P_X|\mathscr{G}_{Y|X})$ modulo constant, then*

$$\overline{\mathrm{ran}}(\Lambda_{\mathrm{GSAVE}}) \subseteq \mathfrak{S}_{Y|X}. \tag{14.2}$$

We should note that, in both Assumption 14.4 and Theorem 14.2, the operator Σ_{XX}^{-1} is absent, as compared with SAVE in the linear SDR setting in Chapter 5, or as compared with GSIR in the last chapter. This is because the marginal variance of any function $f(X)$ is absorbed by the $L_2(P_X)$-inner product. In fact, Σ_{XX} in this setting is just $I : \mathscr{H}_X \to \mathscr{H}_X$, the identity mapping.

PROOF OF THEOREM 14.2. Because Λ_{GSAVE} is a self adjoint operator, it suffices to show that

$$\mathfrak{S}_{Y|X}^{\perp} \subseteq \ker(\Lambda_{\mathrm{GSAVE}}). \tag{14.3}$$

Let then $f \perp \mathfrak{S}_{Y|X}$. We first show that $y \in \Omega_Y$,

$$\langle f, [V - V_{X|Y}(y)]f \rangle_{L_2(P_X)} = 0. \tag{14.4}$$

Since $Y \perp\!\!\!\perp X | \mathscr{G}_{Y|X}$,

$$\text{var}[f(X)|Y] = \text{var}[E(f(X)|\mathscr{G}_{Y|X})|Y] + E[\text{var}(f(X)|\mathscr{G}_{Y|X})|Y]. \tag{14.5}$$

Since $f \perp \mathfrak{S}_{Y|X}$, for any $g \in \mathscr{S}_{Y|X}$, we have

$$\text{cov}[f(X), g(X)] = 0.$$

Because $\mathscr{G}_{Y|X}$ is dense in $L_2(P_X | \mathscr{G}_{Y|X})$ modulo constants, the above implies

$$\text{cov}\{f(X), E[f(X)|\mathscr{G}_{Y|X}]\} = \text{var}\{E[f(X)|\mathscr{G}_{Y|X}]\} = 0,$$

where the first equality follows from Lemma 13.2. Thus $E[f(X)|\mathscr{G}_{Y|X}]$ is almost surely constant, and consequently the first term on the right-hand side of (14.5) is 0. Since, by Assumption 14.3, $\text{var}[f(X)|\mathscr{G}_{Y|X}]$ is nonrandom, the second term on the right-hand side of (14.5) is $\text{var}[f(X)|\mathscr{G}_{Y|X}]$. Hence

$$\text{var}[f(X)|Y] = \text{var}[f(X)|\mathscr{G}_{Y|X}].$$

Apply the same argument to obtain

$$\text{var}[f(X)] = \text{var}\{E[f(X)|\mathscr{G}_{Y|X}]\} + E\{\text{var}[f(X)|\mathscr{G}_{Y|X}]\} = \text{var}[f(X)|\mathscr{G}_{Y|X}].$$

Comparing the above two displayed equations we see that $\text{var}[f(X)|Y] = \text{var}[f(X)]$, which implies (14.4).

Since $V - V_{X|Y}(y)$ is self-adjoint, (14.4) implies $f \in \ker[V_{X|Y}(y)]$. Consequently,

$$\langle f, [V - V_{X|Y}(y)]^2 f \rangle_{L_2(P_X)} = 0.$$

Taking unconditional expectation on both sides of the above operator equation, and using the defining relation (12.6) of the expectation of a random operator, we have

$$E\left(\langle f, [V - V_{X|Y}(y)]^2 f \rangle_{L_2(P_X)} \right) = \langle f, E([V - V_{X|Y}(y)]^2) f \rangle_{L_2(P_X)}$$
$$= \langle f, \Lambda_{\text{GSAVE}} f \rangle_{L_2(P_X)}$$

Hence $f \in \ker(\Lambda_{\text{GSAVE}})$, as desired. $\square$

14.2   Relation with GSIR

Recall that, as shown in Theorem 5.2 in the linear Sufficient Dimension Reduction setting, SAVE recovers a larger portion of the central subspace when SIR is not exhaustive. That is, $\mathscr{S}_{\text{SIR}} \subseteq \mathscr{S}_{\text{SAVE}}$. This relation was first noticed in Cook and Critchley (2000). A parallel result also holds in the nonlinear SDR setting. It would have been quite easy to prove this generalization had we stayed in the same geometry. However, now that we have defined GSIR and GSAVE in different geometries, we need to first convert GSIR into the current $L_2(P_X)$-geometry first.

Recall that, in the $L_2(P_X)$-geometry, the regression operator becomes M_{XY}. Let 1_X be the function $f(x) \equiv 1$ in $\mathscr{H}_X$ and 1_Y be the function $g(y) \equiv 1$ in $\mathscr{H}_X$. Then the covariance operator between X and Y is defined by

$$\Sigma_{XY} = M_{XY} - 1_X \otimes 1_Y.$$

In fact, it is easy to check that, for any $f, g \in \mathscr{H}_X$,

$$\langle f, \Sigma_{XY} g \rangle_{\mathscr{H}_X} = \mathrm{cov}[f(X), g(Y)].$$

The following lemma is a parallel to the combination of Theorems 13.1 and 13.2; its proof is parallel to those of the two theorems, and is omitted.

Lemma 14.1 *If Assumptions 14.1 through 14.4 hold and $\mathfrak{S}_{Y|X}$ is dense in $L_2(P_X|\mathscr{G}_{Y|X})$, then* $\overline{\mathrm{ran}}(\Sigma_{XY}) \subseteq \mathfrak{S}_{Y|X}$. *If, moreover, the central class is complete, then* $\overline{\mathrm{ran}}(\Sigma_{XY}) = \mathfrak{S}_{Y|X}$.

Let $\mathfrak{S}_{\mathrm{GSIR}}$ be the subspace $\overline{\mathrm{ran}}(\Sigma_{XY})$ and let $\mathfrak{S}_{\mathrm{GSAVE}}$ be the subspace $\overline{\mathrm{ran}}(\Lambda_{\mathrm{GSAVE}})$. We now describe the relation between GSIR and GSAVE.

Theorem 14.3 *Under Assumptions 14.1, 14.2, and 14.4, we have*

$$\mathfrak{S}_{\mathrm{GSIR}} \subseteq \mathfrak{S}_{\mathrm{GSAVE}}.$$

PROOF. This statement is equivalent to $\overline{\mathrm{ran}}(\Sigma_{XY}) \subseteq \overline{\mathrm{ran}}(\Lambda_{\mathrm{GSAVE}})$, which, since Λ_{GSAVE} is self adjoint and the adjoint operator of Σ_{XY} is Σ_{YX}, is equivalent to $\ker(\Lambda_{\mathrm{GSAVE}}) \subseteq \ker(\Sigma_{XY})$. For any $f \in \ker(\Lambda_{\mathrm{GSAVE}})$,

$$\langle f, \Lambda_{\mathrm{GSAVE}} f \rangle_{\mathscr{H}_X} = E\{ \langle f, [\Sigma_{XX} - V_{XX|Y}(Y)]^2 f \rangle_{\mathscr{H}_X} \} = 0.$$

Because $\langle f, [\Sigma_{XX} - V_{XX|Y}(Y)]^2 f \rangle_{\mathscr{H}_X} \geq 0$, we have

$$\langle f, [\Sigma_{XX} - V_{XX|Y}(Y)]^2 f \rangle_{\mathscr{H}_X} = 0$$

almost surely P. Therefore $[\Sigma_{XX} - V_{XX|Y}(Y)]f = 0$ almost surely P. Consequently,

$$E\{ \langle f, [\Sigma_{XX} - V_{XX|Y}(Y)]f \rangle_{\mathscr{H}_X} \} = 0.$$

By the definitions of Σ_{XX} and $V_{XX|Y}$, the above equality can be rewritten as

$$\mathrm{var}[f(X)] - E[\mathrm{var}(f(X)|Y)] = \mathrm{var}[E(f(X)|Y)] = 0.$$

Hence $E[f(X)|Y]$ is a constant almost surely, which implies $E\{f(X) - E[f(X)]|Y\} = 0$ almost surely. By Lemma 14.1, this means

$$M_{YX} f - E[f(X)]M_{YX} 1_X = 0. \tag{14.6}$$

Since $M_{YX} 1_X = E(1|Y) = 1_Y$, we have

$$M_{YX} E[f(X)] = 1_Y E[f(X)] = 1_Y \langle 1_X, f \rangle_{\mathscr{H}_X} = (1_Y \otimes 1_X) f.$$

Substituting this into (14.6), we see that

$$\Sigma_{YX} f = M_{YX} f - (1_Y \otimes 1_X) f = E[f(X) - E(f(X))|Y] = 0,$$

which means $f \in \ker(\Sigma_{YX})$. $\qquad\square$

Interestingly, this theorem does not require Assumption 14.3. In other words, the relation $\mathfrak{S}_{\mathrm{GSIR}} \subseteq \mathfrak{S}_{\mathrm{GSAVE}}$ holds regardless of whether $\mathfrak{S}_{\mathrm{GSAVE}}$ is a subspace of the central class. However, if Assumption 14.3 does hold, then we have

$$\mathfrak{S}_{\mathrm{GSIR}} \subseteq \mathfrak{S}_{\mathrm{GSAVE}} \subseteq \mathfrak{S}_{Y|X}.$$

If, furthermore, the central class is complete, then

$$\mathfrak{S}_{\mathrm{GSIR}} = \mathfrak{S}_{\mathrm{GSAVE}} = \mathfrak{S}_{Y|X}.$$

14.3 Implementation of GSAVE

We first derive the coordinate representation of the operator $V_{XX|Y}(y)$. Let

$$\mathcal{H}_X = \mathrm{span}\{1_X, \kappa_X(\cdot, X_1), \ldots, \kappa_X(\cdot, X_n)\},$$
$$\mathcal{H}_Y = \mathrm{span}\{1_Y, \kappa_Y(\cdot, Y_1), \ldots, \kappa_Y(\cdot, Y_n)\}. \tag{14.7}$$

Let

$$\mathscr{C}_X = \{1_X, \kappa_X(\cdot, X_1), \ldots, \kappa_X(\cdot, X_n)\}, \quad \mathscr{C}_Y = \{1_Y, \kappa_Y(\cdot, Y_1), \ldots, \kappa_Y(\cdot, Y_n)\}$$

be the spanning systems of the spaces in (14.7). Let $\ell_X(x)$ and $\ell_Y(y)$ be the functions

$$x \mapsto (1, \kappa_X(x, X_1), \ldots, \kappa_X(x, X_n))^\mathsf{T}, \quad y \mapsto (1, \kappa_X(y, X_1), \ldots, \kappa_X(y, X_n))^\mathsf{T},$$

respectively. Let ℓ_X^i be the ith component of ℓ_X. Let L_X and L_Y be the $(n+1) \times n$ matrices

$$(\ell_X(X_1), \ldots, \ell_X(X_n)), \quad (\ell_Y(Y_1), \ldots, \ell_Y(Y_n)),$$

respectively. In these notations, the Gram matrix of the basis $\mathscr{C}_X$ is specified by

$$\langle \ell_X^i, \ell_X^j \rangle_{\mathcal{H}_X} = n^{-1} \sum_{a=1}^{n} \ell_X^i(X_a) \ell_X^j(X_a),$$

which is simply the (i, j)th entry of the matrix $n^{-1} L_X L_X^\mathsf{T}$. Similarly, the Gram matrix for $\mathscr{C}_Y$ is $n^{-1} L_Y L_Y^\mathsf{T}$. These imply that the inner products in $\mathcal{H}_X$ and $\mathcal{H}_Y$ can be expressed in the following coordinate forms.

Lemma 14.2 *If f and g are members of $\mathcal{H}_X$, and r, s are members of $\mathcal{H}_Y$, then*

$$\langle f, g \rangle_{\mathcal{H}_X} = [f]_{\mathscr{C}_X}^\mathsf{T} \left(n^{-1} L_X L_X^\mathsf{T} \right) [g]_{\mathscr{C}_X}, \quad \langle r, s \rangle_{\mathcal{H}_Y} = [r]_{\mathscr{C}_Y}^\mathsf{T} \left(n^{-1} L_Y L_Y^\mathsf{T} \right) [s]_{\mathscr{C}_Y}.$$

Since we are only going to deal with centered quantities unconditionally, we will use the centered version of $\mathcal{H}_X$:

$$\mathcal{H}_X^0 = \mathrm{span}\{\kappa_X(\cdot, X_1) - E_n \kappa_X(X, X_n), \ldots, \kappa_X(\cdot, X_n) - E_n \kappa_X(X, X_n)\}.$$

This is simply $\mathcal{H}_X \ominus \mathrm{span}(1_X)$, the orthogonal complement of the space spanned by 1_X. The next lemma gives the form of the inner product between members of $\mathcal{H}_X^0$. For a function on Ω_X, we use $f(X_{1:n})$ to denote the n-dimensional vector $(f(X_1), \ldots, f(X_n))^\mathsf{T}$.

Lemma 14.3 *If f and g are members of $\mathcal{H}_X^0$, then*

$$\langle f, g \rangle_{\mathscr{C}_X} = [f]_{\mathscr{C}_X}^\mathsf{T} \left(n^{-1} L_X Q L_X^\mathsf{T} \right) [g]_{\mathscr{C}_X},$$

where $Q = I_n - 1_n 1_n^\mathsf{T}/n$.

PROOF. First, we note that, if $f \in \mathcal{H}_X^0$, then $E_n f(X) = n^{-1} f(X_{1:n})^\mathsf{T} 1_n = 0$. Consequently,

$$f(X_{1:n}) = Q f(X_{1:n}).$$

Because $f(X_{1:n}) = L_X^\mathsf{T}[f]_{\mathscr{C}_X}$, the above equality implies $f(X_{1:n}) = QL_X^\mathsf{T}[f]_{\mathscr{C}_X}$. By the definition of inner product in $\mathscr{H}_X$,

$$\langle f, g \rangle_{\mathscr{H}_X} = n^{-1} f(X_{1:n})^\mathsf{T} g(X_{1:n}) = n^{-1} (QL_X^\mathsf{T}[f]_{\mathscr{C}_X})^\mathsf{T} (QL_X^\mathsf{T}[f]_{\mathscr{C}_X})$$
$$= n^{-1} [f]_{\mathscr{C}_X}^\mathsf{T} L_X Q L_X^\mathsf{T}[f]_{\mathscr{C}_X},$$

as desired. $\square$

As usual, we estimate the operators such as M_{XX} and M_{XY} by replacing the population-level expectation E by the sample average E_n. For example, $\hat{M}_{XY}$ is defined through the bilinear form

$$\mathscr{H}_X \times \mathscr{H}_Y \to \mathbb{R}, \quad (f,g) \mapsto E_n[f(X)g(Y)].$$

We define $\hat{\Sigma}_{XX}$ to be the operator from $\mathscr{H}_X^0$ to $\mathscr{H}_X^0$ induced by the bilinear form

$$\mathscr{H}_X^0 \times \mathscr{H}_X^0 \to \mathbb{R}, \quad (f,g) \mapsto E_n[f(X)g(X)] = \mathrm{cov}_n[f(X)g(X)].$$

The next lemma gives the coordinate of the operator $\hat{M}_{YX}$.

Lemma 14.4 *For any* $f \in \mathscr{H}_X$ *and* $g \in \mathscr{H}_Y$,

$$_{\mathscr{C}_X}[\hat{M}_{YX}]_{\mathscr{C}_X} = \left(n^{-1} L_Y L_Y \right)^\dagger \left(n^{-1} L_Y L_X^\mathsf{T} \right).$$

PROOF. Let $f \in \mathscr{H}_X$ and $g \in \mathscr{H}_Y$. By the definition of $\hat{M}_{YX}$,

$$\langle \hat{M}_{YX} f, g \rangle_{\mathscr{H}_Y} = E_n[f(X)g(Y)] = n^{-1} f(X_{1:n})^\mathsf{T} g(X_{1:n}).$$

Because $f(X_{1:n}) = QL_X^\mathsf{T}[f]_{\mathscr{C}_X}$ and $g(X_{1:n}) = L_Y^\mathsf{T}[g]_{\mathscr{C}_Y}$, the above equality can be rewritten as

$$\langle \hat{M}_{YX} f, g \rangle_{\mathscr{H}_Y} = [f]_{\mathscr{C}_X}^\mathsf{T} \left(n^{-1} L_X Q L_Y^\mathsf{T} \right) [g]_{\mathscr{C}_Y}. \tag{14.8}$$

Meanwhile, by Lemma 14.2,

$$\langle \hat{M}_{YX} f, g \rangle_{\mathscr{H}_Y} = n^{-1} [M_{YX} f]_{\mathscr{C}_Y}^\mathsf{T} L_Y L_Y^\mathsf{T} [g]_{\mathscr{C}_Y}$$
$$= n^{-1} [f]_{\mathscr{C}_Y}^\mathsf{T} \left(_{\mathscr{C}_Y}[M_{YX}]_{\mathscr{C}_X} \right)^\mathsf{T} L_Y L_Y^\mathsf{T} [g]_{\mathscr{C}_Y}. \tag{14.9}$$

Since the right-hand sides of (14.8) and (14.9) are equal for all $[f]_{\mathscr{C}_X} \in \mathbb{R}^{n+1}$ and $[g]_{\mathscr{C}_X} \in \mathbb{R}^{n+1}$, we have the desired equality. $\square$

To proceed further, we need to develop the sample estimate of the conditional expectations such as $E[f(X)|Y]$. Motivated by Theorem 14.1, we use $\hat{M}_{YX} f$ to estimate this conditional expectation, and denote the estimated conditional expectation as $E_n[f(X)|Y]$. Following this logic further, we define the conditional mean element, the sample version of $\mu_{X|Y}$, through the following equation

$$\langle f, \hat{\mu}_{X|Y}(y) \rangle_{\mathscr{H}_X} = E_n[f(X)|Y = y]. \tag{14.10}$$

Likewise, for $f \in \mathscr{H}_X$ and $g \in \mathscr{H}_X$, the conditional expectation $E[f(X)g(Y)|Y]$ estimated by $\hat{M}_{YX}(fg)$, and write the estimated conditional expectation as $E_n[f(X)g(X)|Y]$. We define the conditional second moment operator $\hat{M}_{XX|Y}(y)$ through the relation

$$\langle f, \hat{M}_{XX|Y}(y)g \rangle_{\mathscr{H}_X} = E_n[f(X)g(X)|Y = y]. \tag{14.11}$$

With these concepts and notations in mind, we now derived the coordinate of $\hat{V}_{XX|Y}(y)$ for each $y \in \Omega_X$. By Lemma 12.3, we have

$$\mathscr{C}_X[\hat{V}_{XX|Y}(y)]_{\mathscr{C}_X} = \mathscr{C}_X[\hat{M}_{XX|Y}(y)]_{\mathscr{C}_X} - \mathscr{C}_X[\hat{\mu}_{X|Y}(y) \otimes \hat{\mu}_{X|Y}(y)]_{\mathscr{C}_X}. \tag{14.12}$$

We should emphasize that we regard the operators $\hat{V}_{XX|Y}(y)$, $\hat{M}_{XX|Y}(y)$, and $\hat{\mu}_{X|Y}(y) \otimes \hat{\mu}_{X|Y}(y)$ as operators from $\mathscr{H}_X^0$ to $\mathscr{H}_X^0$, rather than from $\mathscr{H}_X$ to $\mathscr{H}_X$.

Lemma 14.5 *The coordinate of the operator $\hat{\mu}_{X|Y}(y) \otimes \hat{\mu}_{X|Y}(y)$ satisfies the following relation:*

$$\left(n^{-1} L_X Q L_X^{\mathsf{T}} \right) \left(\mathscr{C}_X[\hat{\mu}_{X|Y}(y) \otimes \hat{\mu}_{X|Y}(y)]_{\mathscr{C}_X} \right) = L_X Q \tau(y) \tau(y)^{\mathsf{T}} Q L_X^{\mathsf{T}}.$$

where $\tau(y) = L_Y^{\mathsf{T}} (L_Y L_Y^{\mathsf{T}})^{\dagger} \ell_Y(y)$.

PROOF. For any $f, g \in \mathscr{H}_X^0$,

$$
\begin{aligned}
\langle f, \hat{\mu}_{X|Y}(y) \otimes \hat{\mu}_{X|Y}(y) g \rangle_{\mathscr{H}_X} &= \langle f, \hat{\mu}_{X|Y}(y) \rangle_{\mathscr{H}_X} \langle \hat{\mu}_{X|Y}(y), g \rangle_{\mathscr{H}_X} \\
&= E_n[f(X)|y] E_n[g(X)|y] \\
&= (\hat{M}_{YX} f)(y) (\hat{M}_{YX} g)(y),
\end{aligned} \tag{14.13}
$$

where the first equality follows from the definition of tensor product; the second follows from (14.10); the third is simply the definition of the notation $E_n(\cdots|\cdots)$. By Lemma 12.3, the right-hand side of (14.13) can be rewritten as

$$
\left((\mathscr{C}_X[M_{YX}]_{\mathscr{C}_X})[f]_{\mathscr{C}_X} \right)^{\mathsf{T}} \ell_Y(y) \left((\mathscr{C}_X[M_{YX}]_{\mathscr{C}_X})[g]_{\mathscr{C}_X} \right)^{\mathsf{T}} \ell_Y(y)
$$
$$
= [f]_{\mathscr{H}_X}^{\mathsf{T}} \left((\mathscr{C}_X[M_{YX}]_{\mathscr{C}_X})^{\mathsf{T}} \ell_Y(y) \ell_Y(y)^{\mathsf{T}} (\mathscr{C}_X[M_{YX}]_{\mathscr{C}_X}) \right) [g]_{\mathscr{H}_X},
$$

where, by Lemma 14.4,

$$
\begin{aligned}
[f]_{\mathscr{C}_X}^{\mathsf{T}} (\mathscr{C}_X[M_{YX}]_{\mathscr{C}_X})^{\mathsf{T}} \ell_Y(y) &= [f]_{\mathscr{C}_X}^{\mathsf{T}} L_X L_Y^{\mathsf{T}} (L_Y L_Y^{\mathsf{T}})^{\dagger} \ell_Y(y) \\
&= [f]_{\mathscr{C}_X}^{\mathsf{T}} L_X Q L_Y^{\mathsf{T}} (L_Y L_Y^{\mathsf{T}})^{\dagger} \ell_Y(y) = [f]_{\mathscr{C}_X}^{\mathsf{T}} L_X Q \tau(y),
\end{aligned}
$$

where the second equality follows from the fact that $f \in \mathscr{H}_X^0$, which implies $LX^{\mathsf{T}}[f]_{\mathscr{C}_X} = QL_X^{\mathsf{T}}[f]_{\mathscr{C}_X}$. By the same argument we can show that

$$
\ell_Y(y)^{\mathsf{T}} (\mathscr{C}_X[M_{YX}]_{\mathscr{C}_X})[g]_{\mathscr{C}_X} = \tau(y)^{\mathsf{T}} Q L_X^{\mathsf{T}}[g]_{\mathscr{C}_X}.
$$

Hence

$$
[f]_{\mathscr{C}_X}^{\mathsf{T}} (\mathscr{C}_X[M_{YX}]_{\mathscr{C}_X})^{\mathsf{T}} \ell_Y(y) = [f]_{\mathscr{C}_X}^{\mathsf{T}} L_X Q \tau(y) \tau(y)^{\mathsf{T}} Q L_X^{\mathsf{T}}[g]_{\mathscr{C}_X}. \tag{14.14}
$$

On the other hand, we also have

$$
\begin{aligned}
\langle f, \hat{\mu}_{X|Y}(y) \otimes \hat{\mu}_{X|Y}(y) g \rangle_{\mathscr{H}_X} &\\
= [f]_{\mathscr{H}_X}^{\mathsf{T}} \left(n^{-1} L_X Q L_X^{\mathsf{T}} \right) \left(\mathscr{C}_X[\hat{\mu}_{X|Y}(y) \otimes \hat{\mu}_{X|Y}(y)]_{\mathscr{C}_X} \right) [g]_{\mathscr{H}_X}. &
\end{aligned} \tag{14.15}
$$

Since the right-hand sides of (14.14) and (14.15) are equal to each other for all $f, g \in \mathscr{H}_X^0$, we have the desired equality.   □

To proceed further we need to introduce some machineries in linear algebra. We first develop some properties of the Moore-Penrose inverse of symmetric matrices. Recall that, for a symmetric matrix U, a matrix V is its Moore-Penrose inverse if and only if the following equalities hold

$$UV = VU, \quad UVU = U, \quad VUV = V. \tag{14.16}$$

See, for example, Rao and Mitra (1972) and McCullagh (1987).

Lemma 14.6 *Suppose $A \in \mathbb{R}^{n \times m}$, $n \geq m$, is a matrix with full column rank. Let Q be the matrix $I_m - 1_m 1_m^\mathsf{T}/m$. Then*

1. $A^\mathsf{T}(AA^\mathsf{T})^\dagger A = I_m$;
2. $(AA^\mathsf{T})(AA^\mathsf{T})^\dagger A = A$;
3. $QA^\mathsf{T}(AQA^\mathsf{T})^\dagger AQ = Q$;
4. $(AQA^\mathsf{T})^\dagger AQ = (AA^\mathsf{T})^\dagger AQ$.

PROOF. *1.* We first show that

$$(AA^\mathsf{T})^\dagger = A(A^\mathsf{T}A)^{-2}A^\mathsf{T}. \tag{14.17}$$

For this to hold, we need to check the three equalities in (14.16) with $U = AA^\mathsf{T}$ and $V = A(A^\mathsf{T}A)^{-2}A^\mathsf{T}$, as is done below:

$$(AA^\mathsf{T})[A(A^\mathsf{T}A)^{-2}A^\mathsf{T}] = A(AA^\mathsf{T})^{-1}A^\mathsf{T} = [A(A^\mathsf{T}A)^{-2}A^\mathsf{T}](AA^\mathsf{T})$$
$$(AA^\mathsf{T})[A(A^\mathsf{T}A)^{-2}A^\mathsf{T}](AA^\mathsf{T}) = AA^\mathsf{T}$$
$$[A(A^\mathsf{T}A)^{-2}A^\mathsf{T}](AA^\mathsf{T})[A(A^\mathsf{T}A)^{-2}A^\mathsf{T}] = A(A^\mathsf{T}A)^{-2}A^\mathsf{T}.$$

Now substitute (14.17) into $A(A^\mathsf{T}A)^\dagger A^\mathsf{T}$ to prove the desired identity.

2. This follows from the first asserted equality:

$$(AA^\mathsf{T})(AA^\mathsf{T})^\dagger A = AA^\mathsf{T}(AA^\mathsf{T})^\dagger A = A.$$

3. We first show that

$$(AQA^\mathsf{T})^\dagger = A(A^\mathsf{T}A)^{-1}Q(A^\mathsf{T}A)^{-1}A^\mathsf{T}. \tag{14.18}$$

We check (14.16) with $U = AQA^\mathsf{T}$ and $V = A(A^\mathsf{T}A)^{-1}Q(A^\mathsf{T}A)^{-1}A^\mathsf{T}$ as follows

$$(AQA^\mathsf{T})[A(A^\mathsf{T}A)^{-1}Q(A^\mathsf{T}A)^{-1}A^\mathsf{T} = A(A^\mathsf{T}A)^{-1}Q(A^\mathsf{T}A)^{-1}A^\mathsf{T}$$
$$= [A(A^\mathsf{T}A)^{-1}Q(A^\mathsf{T}A)^{-1}A^\mathsf{T}(AQA^\mathsf{T})$$
$$[A(A^\mathsf{T}A)^{-1}Q(A^\mathsf{T}A)^{-1}](AQA^\mathsf{T})[A(A^\mathsf{T}A)^{-1}Q(A^\mathsf{T}A)^{-1}A^\mathsf{T} = A(A^\mathsf{T}A)^{-1}Q(A^\mathsf{T}A)^{-1}A^\mathsf{T}$$
$$(AQA^\mathsf{T})[A(A^\mathsf{T}A)^{-1}Q(A^\mathsf{T}A)^{-1}A^\mathsf{T}](AQA^\mathsf{T}) = AQA^\mathsf{T}.$$

Substitute (14.18) into $QA^\mathsf{T}(AQA^\mathsf{T})^\dagger AQ$ to prove this equality.

4. By (14.18) again,

$$(AQA^\mathsf{T})^\dagger AQ = A(A^\mathsf{T}A)^{-1}Q(A^\mathsf{T}A)^{-1}A^\mathsf{T}AQ = A(A^\mathsf{T}A)^{-1}AQ.$$

Meanwhile, by (14.17),

$$(AA^\mathsf{T})^\dagger AQ = A(A^\mathsf{T}A)^{-2}A^\mathsf{T}AQ = A(A^\mathsf{T}A)^{-1}AQ.$$

Hence both sides of the equality in statement *4* equal the same matrix. $\qquad\square$

Next, we develop a property of the Hadamard product between matrices of the same dimensions, which is simply the elementwise product. That is, if $A = \{a_{ij}\}$ and $B = \{b_{ij}\}$ are matrices of the same dimensions, then their Hadamard product, denoted by $A \odot B$, is the matrix $\{a_{ij}b_{ij}\}$. The following property of the Hadamard product can be verified by simple calculation.

Lemma 14.7 *If a,b,c are vectors in $\mathbb{R}^m$, then $(a \odot b)^\mathsf{T} c = a^\mathsf{T}\mathrm{diag}(c)b$.*

We now derive the coordinate representation of $\hat{M}_{XX|Y}(y)$.

Lemma 14.8 *The coordinate of the operator $\hat{M}_{XX|Y}(y)$ satisfies*

$$\left(n^{-1}L_X Q L_X^\mathsf{T}\right)\left({}_{\mathscr{C}_X}[\hat{M}_{XX|Y}(y)]_{\mathscr{C}_X}\right) = L_X Q \mathrm{diag}[\tau(y)]Q L_X^\mathsf{T}.$$

PROOF. For each $f \in \mathscr{H}_X^0$ and $g \in \mathscr{H}_X^0$, we have

$$\begin{aligned}
\langle f, \hat{M}_{XX|Y}g\rangle_{\mathscr{H}_X} &= E_n[f(X)g(X)|y]\\
&= [\hat{M}_{YX}(fg)](y) \qquad\qquad (14.19)\\
&= [fg]_{\mathscr{C}_X}^\mathsf{T} L_X L_Y^\mathsf{T}(L_Y L_Y^\mathsf{T})^\dagger \ell_Y(y),
\end{aligned}$$

where the first equality follows from (14.11), the second from the definition of $E_n(\cdots|\cdots)$, and the third from Lemma 14.4. Now let us compute $[fg]_{\mathscr{H}_X}$. Since, for any $x \in \Omega_X$,

$$(fg)(x) = [fg]_{\mathscr{C}_X}^\mathsf{T} \ell_X(x),$$

we have

$$(fg)(X_{1:n}) = L_X^\mathsf{T}[fg]_{\mathscr{C}_X} \;\Rightarrow\; [fg]_{\mathscr{C}_X} = (L_X L_X^\mathsf{T})^\dagger L_X(fg)(X_{1:n}).$$

The vector $(fg)(X_{1:n})$ on the right-hand side can be expressed in terms of the coordinates of f and g as follows

$$(fg)(X_{1:n}) = f(X_{1:n}) \odot g(X_{1:n}) = (QL_X^\mathsf{T}[f]_{\mathscr{C}_X}) \odot (QL_X^\mathsf{T}[g]_{\mathscr{C}_X}),$$

so that the coordinate of fg with respect to $\mathscr{C}_X$ can be expressed as

$$[fg]_{\mathscr{C}_X} = (L_X L_X^\mathsf{T})^\dagger L_X\left((QL_X^\mathsf{T}[f]_{\mathscr{C}_X}) \odot (QL_X^\mathsf{T}[g]_{\mathscr{C}_X})\right).$$

Substitute this into the right-hand side of (14.19) to obtain

$$\begin{aligned}
&\langle f, \hat{M}_{XX|Y}(y)g\rangle_{\mathscr{H}_X}\\
&= \{(L_X L_X^\mathsf{T})^\dagger L_X[(QL_X^\mathsf{T}[f]_{\mathscr{C}_X}) \odot (QL_X^\mathsf{T}[g]_{\mathscr{C}_X})]\}^\mathsf{T} L_X L_Y^\mathsf{T}(L_Y L_Y^\mathsf{T})^\dagger \ell_Y(y)\\
&= \{[(QL_X^\mathsf{T}[f]_{\mathscr{C}_X}) \odot (QL_X^\mathsf{T}[g]_{\mathscr{C}_X})]\}^\mathsf{T} L_X^\mathsf{T}(L_X L_X^\mathsf{T})^\dagger L_X L_Y^\mathsf{T}(L_Y L_Y^\mathsf{T})^\dagger \ell_Y(y) \qquad (14.20)\\
&= \{[(QL_X^\mathsf{T}[f]_{\mathscr{C}_X}) \odot (QL_X^\mathsf{T}[g]_{\mathscr{C}_X})]\}^\mathsf{T} L_Y^\mathsf{T}(L_Y L_Y^\mathsf{T})^\dagger \ell_Y(y)\\
&= [f]_{\mathscr{C}_X}^\mathsf{T}(L_X Q \mathrm{diag}[\tau(y)]Q L_X^\mathsf{T})[g]_{\mathscr{C}_X},
\end{aligned}$$

where the third equality follows from Lemma 14.6, part 1, and the last equality follows from Lemma 14.7.

In the meantime,

$$\langle f, \hat{M}_{xx|Y}(y)g \rangle_{\mathscr{H}_X} = [f]_{\mathscr{C}_X}^{\mathsf{T}} \left(n^{-1} L_X Q L_X^{\mathsf{T}} \right) \left({}_{\mathscr{C}_X}[\hat{M}_{xx|Y}(y)]_{\mathscr{C}_X} \right) [g]_{\mathscr{C}_X}. \tag{14.21}$$

Since the right-hand sides of (14.20 and 14.21) are equal for all $f, g \in \mathscr{H}_X^0$, we have the desired equality. □

Having obtained the coordinate representations of $\hat{\mu}_{X|Y}(y) \otimes \hat{\mu}_{X|Y}(y)$ and $\hat{M}_{xx|Y}(y)$, it is then easy to derive the coordinate representation of $\hat{V}_{xx|Y}(y)$.

Corollary 14.1 *The coordinate of the operator $\hat{V}_{xx|Y}(y)$ is*

$$\left(n^{-1} L_X Q L_X^{\mathsf{T}} \right) \left({}_{\mathscr{C}_X}[\hat{V}_{xx|Y}(y)]_{\mathscr{C}_X} \right) = L_X Q \left(\operatorname{diag}[\tau(y)] - \tau(y)\tau(y)^{\mathsf{T}} \right) Q L_X^{\mathsf{T}}.$$

Next, we derive the coordinate representation of $\hat{\Sigma}_{xx}$.

Lemma 14.9 *The coordinate of the operator $\hat{\Sigma}_{xx}$ is*

$$\left({}_{\mathscr{C}_X}[\hat{\Sigma}_{xx}]_{\mathscr{C}_X} \right) = \left(L_X Q L_X^{\mathsf{T}} \right)^{\dagger} \left(L_X Q L_X^{\mathsf{T}} \right).$$

PROOF. For any $f, g \in \mathscr{H}_X^0$,

$$\langle f, \hat{\Sigma}_{xx}g \rangle_{\mathscr{H}_X} = E_n[f(X)g(X)] = \langle f, g \rangle_{\mathscr{H}_X},$$

where right-hand side is, by Lemma 14.2,

$$[f]_{\mathscr{C}_X}^{\mathsf{T}} \left(n^{-1} L_X Q L_X^{\mathsf{T}} \right) [g]_{\mathscr{C}_X}.$$

In the meantime, we also have

$$\langle f, \hat{\Sigma}_{xx}g \rangle_{\mathscr{H}_X} = [f]_{\mathscr{C}_X}^{\mathsf{T}} \left(n^{-1} L_X Q L_X^{\mathsf{T}} \right) \left({}_{\mathscr{C}_X}[\hat{\Sigma}_{xx}]_{\mathscr{C}_X} \right) [g]_{\mathscr{C}_X}.$$

Hence we have the desired equality. □

Corollary 14.1 and Lemma 14.9 immediately lead to the coordinate representation of $\hat{\Sigma}_{xx} - \hat{V}_{xx|Y}(y)$, which is given by the next corollary. In the following, we use $\Lambda(y)$ to denote the matrix $\operatorname{diag}[\tau(y)] - \tau(y)\tau(y)^{\mathsf{T}}$.

Corollary 14.2 *We have the following coordinate representation for the operator $\hat{\Sigma}_{xx} - \hat{V}_{xx|Y}(y)$:*

$$_{\mathscr{C}_X}[\hat{\Sigma}_{xx} - \hat{V}_{xx|Y}(y)]_{\mathscr{C}_X} = \left(n^{-1} L_X Q L_X^{\mathsf{T}} \right)^{\dagger} L_X Q(Q/n - \Lambda(Y)) Q L_X^{\mathsf{T}}.$$

Finally, we use the definition of the expectation of a random operator to derive the coordinate representation of GSAVE operator $\hat{\Lambda}_{\mathrm{GSAVE}}$.

Theorem 14.4 *The coordinate of the operator $\hat{\Lambda}_{\mathrm{GSAVE}}$ is*

$$_{\mathscr{C}_X}[\hat{\Lambda}_{\mathrm{GSAVE}}]_{\mathscr{C}_X} = \left(n^{-1} L_X Q L_X^{\mathsf{T}} \right)^{\dagger} \left(L_X \sum_{a=1}^{n} \left(Q/n - Q\Lambda(Y_a)Q \right)^2 L_X^{\mathsf{T}} \right).$$

PROOF. Let f, g be arbitrary members of $\mathscr{H}_X^0$. By the definition of the expectation of a random operator, as applied to the empirical measure, we have

$$\langle f, E_n\{[\Sigma_{XX} - V_{XX|Y}(Y)]^2\}g\rangle_{\mathscr{H}_X} = E_n\langle f, [\Sigma_{XX} - V_{XX|Y}(Y)]^2 g\rangle_{\mathscr{H}_X}. \quad (14.22)$$

By Lemma 14.2, the inner product on the right-hand side can be expressed as

$$\langle f, [\Sigma_{XX} - V_{XX|Y}(Y)]^2 g\rangle_{\mathscr{H}_X} = [f]_{\mathscr{C}_X}^{\mathsf{T}} \left(n^{-1}L_X Q L_X^{\mathsf{T}}\right) \left(\mathscr{C}_X[\Sigma_{XX} - V_{XX|Y}(Y)]_{\mathscr{C}_X}\right)^2 [g]_{\mathscr{C}_X}.$$

Expand the squared term on the right using Corollary 14.2 to obtain

$$\left(\mathscr{C}_X[\Sigma_{XX} - V_{XX|Y}(Y)]_{\mathscr{C}_X}\right)^2$$

$$= \left(n^{-1}L_X Q L_X^{\mathsf{T}}\right)^{\dagger} L_X Q\left(Q/n - \Lambda(Y)\right)Q L_X^{\mathsf{T}} \left(n^{-1}L_X Q L_X^{\mathsf{T}}\right)^{\dagger} L_X Q\left(Q/n - \Lambda(Y)\right)Q L_X^{\mathsf{T}}$$

$$= n\left(n^{-1}L_X Q L_X^{\mathsf{T}}\right)^{\dagger} L_X\left(Q/n - \Lambda(Y)\right)\left(Q/n - \Lambda(Y)\right)L_X^{\mathsf{T}}$$

$$= n\left(n^{-1}L_X Q L_X^{\mathsf{T}}\right)^{\dagger} L_X\left(Q/n - Q\Lambda(Y)Q\right)^2 L_X^{\mathsf{T}},$$

where the second equality follows from part 3 of Lemma 14.6. Using the above relation we derive the following inner product:

$$\langle f, [\Sigma_{XX} - V_{XX|Y}(Y)]^2 g\rangle_{\mathscr{H}_X}$$

$$= [f]_{\mathscr{C}_X}^{\mathsf{T}} \left(n^{-1}L_X Q L_X^{\mathsf{T}}\right) n\left(n^{-1}L_X Q L_X^{\mathsf{T}}\right)^{\dagger} L_X\left(Q/n - Q\Lambda(Y)Q\right)^2 L_X^{\mathsf{T}}[g]_{\mathscr{C}_X}$$

$$= n[f]_{\mathscr{C}_X}^{\mathsf{T}} L_X\left(Q/n - Q\Lambda(Y)Q\right)^2 L_X^{\mathsf{T}}[g]_{\mathscr{C}_X},$$

where the second equality follows again from part 3 of Lemma 14.6. Substituting the above relation into the right-hand side of (14.22), we find

$$\langle f, \hat{\Lambda}_{\mathrm{GSAVE}}\, g\rangle_{\mathscr{C}_X} = [f]_{\mathscr{C}_X}^{\mathsf{T}} L_X \sum_{a=1}^{n}\left(Q/n - Q\Lambda(Y_a)Q\right)^2 L_X^{\mathsf{T}}[g]_{\mathscr{C}_X}.$$

In the meantime, by Lemma 14.2, we have

$$\langle f, \hat{\Lambda}_{\mathrm{GSAVE}}\, g\rangle_{\mathscr{C}_X} = [f]_{\mathscr{C}_X}^{\mathsf{T}} \left(n^{-1}\sum_{a=1}^{n} L_X Q L_X^{\mathsf{T}}\right)\left(\mathscr{C}_X[\hat{\Lambda}_{\mathrm{GSAVE}}]_{\mathscr{C}_X}\right)[g]_{\mathscr{C}_X}.$$

Comparing the above two equalities, we see that the asserted equality is true. $\qquad\square$

   With the coordinate $\hat{\Lambda}_{\mathrm{GSAVE}}$ available, we now derive the numerical procedure for estimating the central class using GSAVE. Theorem 14.3 suggests that an unbiased estimate of the central class can be obtained by solving the generalized eigenvalue problem $\mathrm{GEV}(\hat{\Lambda}_{\mathrm{GSAVE}}, \hat{\Sigma}_{XX})$. That is, we maximize

$$\langle f, \hat{\Lambda}_{\mathrm{GSAVE}} f\rangle_{\mathscr{H}_X} \quad \text{subject to} \quad \langle f, \hat{\Sigma}_{XX} f\rangle_{\mathscr{H}_Y} = 1. \quad (14.23)$$

In terms of coordinate representation, the first inner product is

$$[f]_{\mathscr{C}_X}^{\mathsf{T}} L_X \sum_{a=1}^{n}(Q/n - Q\hat{\Lambda}(X_a)Q)^2 L_X^{\mathsf{T}}[f]_{\mathscr{C}_X},$$

The second inner product in (14.23) is, by Lemma 14.9,

$$[f]_{\mathscr{G}_X}^{\mathsf{T}} \left(L_X Q L_X^{\mathsf{T}} \right) \left(_{\mathscr{G}_X} [\hat{\Sigma}_{XX}]_{\mathscr{G}_X} \right) [f]_{\mathscr{G}_X} = [f]_{\mathscr{G}_X}^{\mathsf{T}} (L_X Q L_X^{\mathsf{T}}) [f]_{\mathscr{G}_X},$$

where we have ignored the proportionality constant n^{-1}, which does not affect the result. To solve this generalized eigenvalue problem, we let

$$v = (L_X Q L_X^{\mathsf{T}})^{1/2} [f]_{\mathscr{G}_X}.$$

Solving this equation for $[f]_{\mathscr{G}_X}$ with Tychonoff regularization, we have

$$[f]_{\mathscr{G}_X} = (L_X Q L_X^{\mathsf{T}} + \varepsilon_X \lambda_{\max}(L_X Q L_X^{\mathsf{T}}) I_{n+1})^{-1/2} v$$

Abbreviating the number $\varepsilon_X \lambda_{\max}(L_X Q L_X^{\mathsf{T}})$ by η_X, the problem becomes a standard eigenvalue problem:

$$\text{maximize} \quad v^{\mathsf{T}} (L_X Q L_X^{\mathsf{T}} + \eta_X I_{n+1})^{-1/2} L_X \sum_{a=1}^{n} (Q/n - Q\Lambda(Y_a)Q)^2 L_X^{\mathsf{T}} (L_X Q L_X^{\mathsf{T}} + \eta_X I_{n+1})^{-1/2} v$$

$$\text{subject to} \quad v^{\mathsf{T}} v = 1.$$

For the matrix inverse involved in $\hat{\Lambda}(y)$, we also use Tychonoff regularization. Specifically, recall that $\Lambda(y) = \text{diag}(\tau(y)) - \tau(y)\tau(y)^{\mathsf{T}}$, where $\tau(y) = L_Y^{\mathsf{T}} (L_Y^{\mathsf{T}} L_Y)^{\dagger} \ell_Y(y)$. We replace the inverse Moore-Penrose inverse $(L_Y^{\mathsf{T}} L_Y)^{\dagger}$ by

$$(L_Y L_Y^{\mathsf{T}} + \varepsilon_Y \lambda_{\max}(L_X L_X^{\mathsf{T}}) I_{n+1})^{-1} \equiv (L_Y L_Y^{\mathsf{T}} + \eta_Y I_{n+1})^{-1}.$$

Another point to note is that the computation of the term

$$\sum_{a=1}^{n} (Q/n - Q\Lambda(Y_a)Q)^2$$

in the above eigenvalue problem — if done by an R-loop — can be quite time consuming when n is large, because each summand is an $n \times n$ matrix. However, this can be done much more quickly by converting the sum into matrix operations. First, rewrite this sum as

$$Q/n - (2/n)Q \sum_{a=1}^{n} \Lambda(Y_a)Q + Q \left(\sum_{a=1}^{n} \Lambda(Y_a)Q\Lambda(Y_a) \right) Q. \tag{14.24}$$

We convert the first sum in (14.24) into matrix operators as follows:

$$\sum_{a=1}^{n} \Lambda(Y_a) = \sum_{a=1}^{n} \{\text{diag}[\tau(Y_a) - \tau(Y_a)\tau(Y_a)^{\mathsf{T}}]\} = \text{diag}(\tau_{1:n} 1_n) - \tau_{1:n} \tau_{1:n}^{\mathsf{T}},$$

where $\tau_{1:n}$ is the matrix $(\tau_1, \ldots, \tau_n)$, which is simply

$$L_Y [L_Y L_Y^{\mathsf{T}} + \varepsilon_X \lambda_{\max}(L_Y L_Y^{\mathsf{T}}) I_{n+1}]^{-1} L_Y^{\mathsf{T}}.$$

To convert the second sum in (14.24) into matrix operations, note that

$$\sum_{a=1}^{n} \Lambda(Y_a)Q\Lambda(Y_a)$$

$$= \sum_{a=1}^{n} \text{diag}(\tau(Y_a))Q\text{diag}[\tau(Y_a)] - \sum_{a=1}^{n} \text{diag}[\tau(Y_a)]Q\tau(Y_a)\tau(Y_a)^{\mathsf{T}} \tag{14.25}$$

$$- \sum_{a=1}^{n} \tau(Y_a)\tau(Y_a)^{\mathsf{T}} Q\text{diag}[\tau(Y_a)] + \sum_{a=1}^{n} \tau(Y_a)\tau(Y_a)^{\mathsf{T}} Q\tau(Y_a)\tau(Y_a)^{\mathsf{T}}.$$

Abbreviate the right-hand side by $A_1 - A_2 - A_2^\mathsf{T} + A_3$. By applying the definition $Q = I_n - 1_n 1_n^\mathsf{T}/n$, we can easily convert the sum in (14.25) into matrix operations as follows:

$$A_1 = \mathrm{diag}[(\tau_{1:n} \odot \tau_{1:n})1_n] - \tau_{1:n}\tau_{1:n}^\mathsf{T}/n,$$
$$A_2 = (\tau_{1:n} \odot \tau_{1:n})Q\tau_{1:n}^\mathsf{T},$$
$$A_3 = \tau_{1:n}[\mathrm{diag}^*(\tau_{1:n}^\mathsf{T}Q\tau_{1:n})]\tau_{1:n}^\mathsf{T},$$

where, in the third line, the notation $\mathrm{diag}^*(B)$ means setting all the off-diagonal entries of B to 0 — an operation that can be implemented by $\mathrm{diag}(\mathrm{diag}(B))$ in R. In summary, we have

$$\sum_{a=1}^{n}(Q/n - Q\Lambda(Y_a)Q)^2$$
$$= Q/n - (2/n)Q\left(\mathrm{diag}(\tau_{1:n}1_n) - \tau_{1:n}\tau_{1:n}^\mathsf{T}\right)Q$$
$$+ Q\left(\mathrm{diag}[(\tau_{1:n} \odot \tau_{1:n})1_n] - \tau_{1:n}\tau_{1:n}^\mathsf{T}/n - (\tau_{1:n} \odot \tau_{1:n})Q\tau_{1:n}^\mathsf{T}\right.$$
$$\left. - \tau_{1:n}Q(\tau_{1:n} \odot \tau_{1:n})^\mathsf{T} + \tau_{1:n}[\mathrm{diag}^*(\tau_{1:n}^\mathsf{T}Q\tau_{1:n})]\tau_{1:n}^\mathsf{T}\right)Q. \tag{14.26}$$

We now summarize the numerical procedure for GSAVE as the following algorithm.

Algorithm 14.1 GSAVE

1. Standardize $X_1,\ldots,X_n$ marginally and, if Y is a random vector, standardize $Y_1,\ldots,Y_n$ marginally.
2. Choose kernel functions κ_X and κ_Y. If the Gaussian radial basis function is chosen, then choose the tuning parameters γ_X and γ_Y using as described in Section 13.7. Also, choose ε_X and ε_Y as described in Section 13.7.
3. Compute the first d eigenvectors $v_1,\ldots,v_d$ of the matrix

$$(L_X Q L_X^\mathsf{T} + \eta_X I_{n+1})^{-1/2} L_X \sum_{a=1}^{n}(Q/n - Q\Lambda(Y_a)Q)^2 L_X^\mathsf{T}(L_X Q L_X^\mathsf{T} + \eta_X I_{n+1})^{-1/2},$$

where the sum $\sum_{a=1}^{n}(Q/n - Q\Lambda(Y_a)Q)^2$ is computed using (14.26).
4. The sufficient predictors are

$$v_i^\mathsf{T}(L_X Q L_X^\mathsf{T} + \eta_X I_{n+1})^{-1/2}[\ell_X(X_a) - E_n\ell_X(X)], \quad i = 1,\ldots,d,\ a = 1,\ldots,n.$$

In the following R-codes to implement GSAVE, the inputs are the same as those for the R-codes for GSIR, which were already explained. The main code, gsave, contains a function called ridgepower, which computes the matrix $(A + \varepsilon\lambda_{\max}(A)I_n)^c$ for some constant c. It has three inputs: a is the matrix A; e is the number ε; c is the power c to be raised. The main code also calls the function gram.gauss, which was given in the R-codes for GSIR. The code has options for categorical or continuous response, and the predictors can be evaluated at the training set or the test set. If it is evaluated at the training set, then x and x.new must be the same matrix. The kernel for x is the Gaussian radial basis function; the kernel for y is the Gaussian radial basis function if y is continuous, the discrete kernel if y is discrete.

R-codes for GSAVE

```
gsave=function(x,x.new,y,ytype,ex,ey,comx,comy,r){
n=dim(x)[1]
kx0=gram.gauss(x,x,comx);kx=rbind(1,kx0)
if(ytype=="scalar") {ky0=gram.gauss(y,y,comy);ky=rbind(1,ky0)}
if(ytype=="categorical") {ky0=gram.dis(y);ky=rbind(1,ky0)}
Q=diag(n)-rep(1,n)%*%t(rep(1,n))/n
kkx=kx%*%Q%*%t(kx)
kky=ky%*%t(ky)
if(ytype=="scalar") kkyinv=ridgepower(sym(kky),ey,-1)
if(ytype=="categorical") kkyinv=mppower(sym(kky),-1,1e-9)
piy=t(ky)%*%kkyinv%*%ky
sumlam=diag(apply(piy,1,sum))-piy%*%piy
a1=diag(apply(piy*piy,1,sum))-piy%*%piy/n
a2=(piy*piy)%*%piy-piy%*%diag(apply(piy,1,mean))%*%piy
a3=piy%*%diag(diag(piy%*%Q%*%piy))%*%piy
mid=Q/n-(2/n)*Q%*%sumlam%*%Q+Q%*%(a1-a2-t(a2)+a3)%*%Q
kx.new.0=gram.gauss(x,x.new,comx)
kx.new=rbind(1,kx.new.0)
n1=dim(kx.new)[2];Q1=diag(n1)-rep(1,n1)%*%t(rep(1,n1))/n1
kk=ridgepower(kx%*%Q%*%t(kx),epsx,-1/2)%*%kkx%*%Q
kk.new=ridgepower(kx%*%Q%*%t(kx),epsx,-1/2)%*%kkx.new%*%Q1
pred=t(kk.new)%*%eigen(sym(kk%*%mid%*%t(kk)))$vectors[,1:r]
return(pred)
}


ridgepower=function(a,e,c){
return(matpower(a+e*onorm(a)*diag(dim(a)[1]),c))}
```

14.4   Simulation Studies and an Application

As the theory for GSAVE suggested, its main use is for the situations where the central class is not complete, in which case GSIR is not guaranteed to recover it. As was shown in Section 14.2, in these cases GSAVE can recover a larger portion of the central class. For this reason, the models we used to investigate the performance of GSAVE all contain sufficient predictors that affect heteroscedasticity:

$$\begin{cases} \text{IV}: & Y = X_1\,\varepsilon \\ \text{V}: & Y = (1/50)\,(X_1^3 + X_2^3)\,\varepsilon \\ \text{VI}: & Y = (X_1/(1+e^{X_2}))\,\varepsilon \end{cases}$$

Again, we use these models along with the scenarios (A), (B), and (C) for the distribution of X used in Section 13.9. The specifications of n, m, N, p are the same as in Section 13.9.

In Table 14.1 we report the means and standard deviations of Spearman's correlations between the estimated and true sufficient predictors. The numbers in this table are taken from Lee et al. (2013), where the GSIR in that paper is the version described at the end of Section

Table 14.1 *Comparison of GSAVE and GSIR under Models IV, V, VI and Scenarios A, B, C*

model	method	scenario		
		A	B	C
IV	GSIR	0.41 ± 0.23	0.53 ± 0.25	0.26 ± 0.18
	GSAVE	0.89 ± 0.08	0.87 ± 0.08	0.76 ± 0.20
V	GSIR	0.20 ± 0.14	0.11 ± 0.08	0.12 ± 0.09
	GSAVE	0.88 ± 0.06	0.88 ± 0.06	0.82 ± 0.14
VI	GSIR	0.27 ± 0.17	0.64 ± 0.13	0.44 ± 0.17
	GSAVE	0.84 ± 0.09	0.76 ± 0.15	0.73 ± 0.15

13.6. We see that GSAVE performs substantially better than GSIR. The discrepancy can be explained by the fact that GSIR depends completely on $E[\mathrm{var}(f(X)|Y)]$, whereas GSAVE extracts further information from $\mathrm{var}[f(X)|Y]$.

As is the case for SAVE in the linear SDR setting, GSAVE works the best for extracting predictors affecting the conditional variance of the response, but often not so well for extracting predictors affecting the conditional mean. We now demonstrate this point by applying GSAVE to Models I, II, and III under Scenarios A, B, and C. For the tuning parameters, we applied the 10-fold cross validation procedure in Section 13.8 multiple samples in all models and scenarios, and in a majority of cases it selects the set

$$\varepsilon_x = \varepsilon_y = 0.001, \quad \rho_x = 1, \quad \rho_y = 0.2.$$

So, to simplify computation, we used this set of tuning parameters for this simulation. The results are reported in Table 14.2.

Table 14.2 *Performance of GSAVE under Models I, II, III and Scenarios A, B, C*

model	scenario		
	A	B	C
I	0.18 ± 0.11	0.25 ± 0.13	0.49 ± 0.11
II	0.83 ± 0.11	0.81 ± 0.07	0.69 ± 0.12
III	0.78 ± 0.19	0.88 ± 0.13	0.64 ± 0.26

We see that GSAVE performs rather poorly for Model I, but reasonably well for Models II and III. However, we expect that other second-order methods for linear SDR, such as Directional Regression in Li and Wang (2007) and the minimum discrepancy approach in Cook and Ni (2005), will be amenable to similar generalizations to nonlinear SDR, which, based on our experiences with linear SDR, might work better than GSAVE in recovering the sufficient predictors conditional mean. This issue is currently under investigation by the author and his students.

We applied GSAVE to the pen digit data with the same tuning parameters as those used for GSIR. While GSIR provides sufficiently clear separation of the digits, one of its disadvantages, as also shared by the classical SIR, is that when a categorical response has k classes, it can only provide $k-1$ meaningful sufficient directions. This is because the GSIR operator in this setting is of rank $k-1$. For example, if the response is binary, then it can only provide one sufficient predictor, whereas GSAVE can provide more than $k - 1$ predictors. To illustrate this point we apply GSAVE to the digits 6 and 9, using the same set of tuning parameters as in the pen digit application of GSIR: $\rho_x = 1$ and $\varepsilon_x = 10^{-6}$. Figure 14.1 (upper left panel) shows the first three GSAVE predictors, which give rather clear separation of the two groups. In the plots, "+" represents the digit 6; "·" represents the digit 6. As with the application of GSIR, the predictors are evaluated at the testing set. We also explored the degree of separation provided by the later predictors. For example, the upper-right panel shows the perspective plot for the first, second, and fourth GSAVE predictors, which show even stronger separation. The lower-left panel shows the first, third, and fourth GSAVE predictors; the lower-right panel shows the first, second, and fifth GSAVE predictors. In each perspective plot, the axis are chosen to show the best viewing angle. We see that these later predictors provide additional useful information for classification.

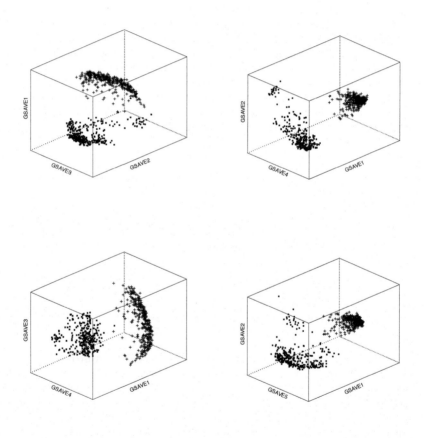

Figure 14.1 *First five sufficient predictors by GSAVE for the pen digit data.*

14.5   Relation between Linear and Nonlinear SDR

As we finish our exposition of nonlinear SDR, it is important to emphasize that, even though nonlinear SDR significantly broadened the scope and increased the flexibility and power of Sufficient Dimension Reduction, the place of linear SDR remains critically important, in the same way that linear regression and linear principal component analysis remain critically important, in spite of the advances and advantages of nonlinear regression and nonlinear principal component analysis.

Specifically, the importance of linear SDR can be seen from the following respects. First, the linear SDR has stronger interpretability: linear combinations such as $\beta^{\mathsf{T}}X$ tell us, for example, which components of X determine the first sufficient predictor. The same cannot be said of the inner product $\langle f, \kappa_X(\cdot, X) \rangle_{\mathscr{H}_X}$ for nonlinear SDR. Second, because linear SDR characterizes statistical relations through linear indices, with appropriate sparse regularization it can be used as a tool for variable selection, while taking into account the possible low-rank structure in the predictor. There is a growing literature about this, which will be further discussed in the next chapter. Third, as discussed in Li et al. (2011), a nonlinear SDR method can sometimes present the differences in variances of different classes as separation of locations in the nonlinear sufficient predictor, whereas second-order linear SDR methods, such as SAVE, Contour Regression and Directional Regression, often show difference in variances explicitly and clearly. This can be a significant advantage when the cause of variations among different classes are important. Fourth, the inverse regression methods discussed in Chapter 2 through Chapter 6 use the symmetry in the distribution of X to avoid high-dimensional smoothing, thus avoiding the sparsity of data in high dimension. Finally, the asymptotic theories and order determination methods are more mature and well developed for linear SDR than for nonlinear SDR, giving the former a richer tool box for applications.

In summary, the combined strength of nonlinear SDR in its ability to detect complex relations, and of linear SDR in its good interpretability and its transparent relation with components of the predictor, provides us with versatile and powerful means to gain insights into the hidden structures of high dimension data.

Chapter 15

The Broad Scope of Sufficient Dimension Reduction

15.1 Sufficient Dimension Reduction for Functional Data   253
15.2 Sufficient Dimension Folding for Tensorial Data   256
15.3 Sufficient Dimension Reduction for Grouped Data   259
15.4 Variable Selection via Sufficient Dimension Reduction   260
15.5 Efficient Dimension Reduction   262
15.6 Partial Dimension Reduction for Categorical Predictors   264
15.7 Measurement Error Problem   265
15.8 SDR via Support Vector Machine   267
15.9 SDR for Multivariate Responses   268

The growing scope of the momentous research on Sufficient Dimension Reduction — as well as the fast approaching due date of this book — mean that we have to be brutally selective about the topics to cover. This selection inevitably reflects the author's bias and his familiarity with the candidate subjects to consider. Many topics, albeit as important as those already presented, have to be left out of the discussions of the preceding chapters. The purpose of this final chapter is to survey some of these topics, laying out their basic ideas, important results, key applications, and references, so as to provide a road map for the interested readers should they wish to pursue these topics further.

In broad outlines these topics can be classified into the following categories: extending SDR to take into account the special structures in the data, such as functional, tensorial, or grouped structures, combining SDR with sparse penalty to assist variable selection, improving the efficiency of SDR using the techniques in optimal estimation and semiparametric inference, dealing with categorical predictors and predictors with measurement error, using the Support Vector Machine to perform dimension reduction, and the extension of the linear SDR methods to multivariate response.

15.1 Sufficient Dimension Reduction for Functional Data

In this section we survey the developments of extension of Sufficient Dimension Reduction to functional data, a term that refers to observations that are in the form of functions defined on an interval in $\mathbb{R}$, which typically represents time. Many modern applications produce data of this kind. For example, the Electroencephalography (EEG) data, the functional Magnetic Resonance Imaging (fMRI) data, the data collected by smart wearables that record vital signs or motion of the experimental subjects, the stock market data, and the economic data that trace the developments of countries over a period of time, are all special types of functional data. Functional data can occur as a single function or as a vector of functions; they can appear as predictors or responses.

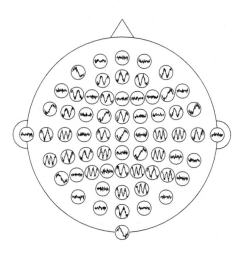

Figure 15.1 *A schematic representation of the functional data collected by EEG from a subject.*

To illustrate the idea, let us consider the structure of a typical EEG data, such as that used in Li et al. (2010a). As shown in Figure 15.1, the small circles indicate the locations of a set of 64 electrodes, and the curves in the small circles, which were artificially generated, illustrate the signals recorded by each electrode. The curves are recorded at equally-spaced 256 time points. A sample of 122 subjects participated in the test, with 77 alcoholic and 45 non-alcoholic. Thus, in this example, the predictors are 64-dimensional vectors whose components are functions recorded at 256 time points; the responses are the binary variable indicating whether the subject is alcoholic. We are interested in reducing the dimension of the 64 dimensional vector of functions to classify alcoholic and non-alcoholic subjects based on their EEG data.

The earliest papers on functional SDR, such as Ferré and Yao (2003), Ferré and Yao (2005) in SDR extends SIR to the functional setting. Suppose $X_1, \ldots, X_n$ are a sample of random functions, by which we mean X_i is a mapping from the sample space Ω to a Hilbert space $\mathscr{H}$ of functions defined on J, which is an interval in $\mathbb{R}$ representing time. That is, for each $\omega \in \Omega$, $X(\omega)$ is a function in $\mathscr{H}$. Usually, $\mathscr{H}$ can be a family of cubic splines defined on J.

The Sufficient Dimension Reduction problem is formulated as

$$Y \perp\!\!\!\perp X | \langle f_1, X \rangle_{\mathscr{H}}, \ldots, \langle f_d, X \rangle_{\mathscr{H}},$$

for a minimal set members $f_1, \ldots, f_d$ of $\mathscr{H}$. The goal is to find the subspace of $\mathscr{H}$ spanned by $f_1, \ldots, f_d$. We call this problem functional linear SDR. The variance $\mathrm{var}(X)$ is now defined as the linear operator on $\mathscr{H}$:

$$E[(X - EX) \otimes (X - EX)] = E(X \otimes X) - (EX) \otimes (EX),$$

where $\otimes$ is the same tensor product and the expectation of the random matrix $X \otimes X$ is the same as defined in Section 12.7. The expectation of X is defined via the Riesz representation of the linear functional

$$\mathscr{H} \to \mathbb{R}, \quad f \mapsto E(\langle f, X \rangle_{\mathscr{H}}).$$

As in the classical SDR setting, we split the support of Y into intervals, say $B_1, \ldots, B_m$, and calculate the conditional variances

$$\mathrm{var}(X | Y \in B_i) = E(X \otimes X | Y \in B_i) - E(X | Y \in B_i) \otimes E(X | Y \in B_i).$$

We then form the candidate operator

$$\Lambda_{\mathrm{FSIR}} = \sum_{i=1}^{m} P(Y \in B_i) \mathrm{var}(X | Y \in B_i)$$

where the subscript FSIR indicates "Functional Sliced Inverse Regression." It can be shown that, under a linear conditional mean assumption similar to that described in Chapter 3:

$$E(X | \langle f_1, X \rangle_{\mathscr{H}}, \ldots, \langle f_d, X \rangle_{\mathscr{H}}) \text{ is a linear function of } \langle f_1, X \rangle_{\mathscr{H}}, \ldots, \langle f_d, X \rangle_{\mathscr{H}},$$

we have that the range of the operator $\Sigma_{XX} \Lambda_{\mathrm{FSIR}}$ is a subset of the central subspace.

Under the same framework, several other estimators for the linear SDR were also extended to the functional setting. For example, Wang et al. (2013) extended Contour Regression, Lian and Li (2014) extended SAVE, to the functional setting, and Wang et al. (2015) proposed a hybrid estimator of functional SIR and functional SAVE. Also, concerned with the assumption that the sample path of X belongs to a Hilbert space might be too strong for some applications, Hsing and Ren (2009) proposed a more general framework of functional SDR that does not require the sample paths of X to reside in a Hilbert space, and proposed a method similar to SIR under this general framework.

Recently, Li and Song (2017) extended nonlinear SDR to the functional setting. They allow both X and Y to be functions, or vector of functions. To appropriately capture the nonlinear relations between X and Y, we need to create Hilbert spaces whose functions are defined on the domains $\mathscr{H}_X$ and $\mathscr{H}_Y$. To achieve this the authors introduced what they called the *nested Hilbert spaces*. Let $\mathscr{H}_X$ and $\mathscr{H}_Y$ be the Hilbert spaces in which X and Y reside. They construct second-level Hilbert spaces using the inner products from the first-level Hilbert spaces. Let us use the predictor X to illustrate the point. Let κ_X be a kernel function

$$\kappa_X : \mathscr{H}_X \times \mathscr{H}_Y \to \mathbb{R}$$

be a positive definite kernel generated by the inner product in $\mathscr{H}_X$. For example, the Gaussian radial basis function on $\mathscr{H}_X \times \mathscr{H}_X$ is defined as

$$\kappa_X(f_1, f_2) = e^{-\gamma \| f_1 - f_2 \|_{\mathscr{H}_X}^2},$$

where γ is a tuning constant; the polynomial kernel defined on $\mathscr{H}_X \times \mathscr{H}_X$ is

$$\kappa_X(f_1, f_2) = (\langle f_1, f_2 \rangle_{\mathscr{H}_X} + c)^k,$$

where $c \geq 0$ is a constant. The second-level space $\mathfrak{M}_X$ is defined as reproducing kernel Hilbert space generated by the kernel κ_X; that is, the inner product of $\mathfrak{M}_X$ is determined by

$$\langle \kappa_X(\cdot, f_1), \kappa_X(\cdot, f_2) \rangle_{\mathfrak{M}_X} = \kappa_X(f_1, f_2);$$

and $\mathfrak{M}_X$ is the completion of

$$\text{span}\{\kappa_X(\cdot, f) : f \in \mathscr{H}_X\}$$

in terms of the norm generated by $\langle \cdot, \cdot \rangle_{\mathscr{H}_X}$. The pair of Hilbert spaces, $\{\mathscr{H}_X, \mathfrak{M}_X\}$, are nested in the sense that the kernel of the second Hilbert space is uniquely determined by the inner product of the first Hilbert space. The first space need not be a reproducing kernel Hilbert space. It could, for example, be the space of splines in the time domain with an L_2 inner product. The authors developed functional GSIR and functional GSAVE under this framework, and derived the the asymptotic convergence of the functional GSIR.

15.2   Sufficient Dimension Folding for Tensorial Data

In many contemporary applications, the sampling units of the data are in the form of multi-dimensional arrays, or tensors. For example, an image is a two-dimensional array, a video clip is a three-dimensional array, a three-dimensional object in motion is a four-dimensional array. These raise the question of how to conduct sufficient dimension reduction on tensor-valued predictors for the purposes of classification and prediction. As an illustration, consider the situations where one might be interested in classifying a collection of youtube videos according to their contents, styles, genres, and so on, based on a training set, or classifying a collection of X-ray images as normal or abnormal based on a collection of images already classified by experts. In the first application, the predictor would be a 3-d tensor; in the second application, a 2-d tensor. The responses in both applications are categorical variables.

This problem was considered by Li et al. (2010a), who developed what they called the *dimension folding* method to reduce a high-dimensional tensor to a low-dimensional tensor. Let us begin with the special case where X is a random matrix in $\mathbb{R}^{p \times q}$. Of course, one can treat X as the random vector $\text{vec}(X)$ and employ the classical SDR framework

$$Y \perp\!\!\!\perp X | \beta^\mathsf{T} \text{vec}(X), \tag{15.1}$$

where β is a matrix in $\mathbb{R}^{(pq) \times d}$. However, in doing so, we lose the matrix structure of X. More-over, it would not be parsimonious modeling, because β involves many parameters, some of which may contain redundant information due to the matrix structure of X. A more reasonable way of posing this dimension reduction question seems to be to reduce a high-dimensional ma-trix into a low-dimensional one. Motivated by this, Li et al. (2010a) considered the following Sufficient Dimension Reduction problem: find $\alpha \in \mathbb{R}^{p \times d_L}$ and $\beta \in \mathbb{R}^{q \times d_R}$ such that

$$Y \perp\!\!\!\perp X | \alpha^\mathsf{T} X \beta \quad \text{or equivalently} \quad Y \perp\!\!\!\perp X | (\beta \otimes \alpha)^\mathsf{T} \text{vec}(X). \tag{15.2}$$

Here, d_L indicates the *left dimension*, and d_R, the *right dimension*. Note that the formulation (15.1) requires pq parameters; whereas the formulation (15.2) only requires $pd_L + qd_R$ pa-rameters, which is a substantial reduction of parameters because d_L and d_R are usually small numbers, such as 1, 2, or 3. The object of estimation in the new problem can be regarded as the subspace $\text{span}(\beta) \otimes \text{span}(\alpha)$ in $\mathbb{R}^{pq}$, which is defined as

$$\text{span}\{v \otimes u : v \in \text{span}(\beta), \, u \in \text{span}(\alpha)\}.$$

If this is the smallest subspace that satisfies (15.2), then it is called the central dimension folding subspace, or simply the *dimension folding subspace*, and is denoted by $\mathscr{S}_{Y|\circ X\circ}$. Note that since $\mathscr{S}_{Y|X}$ is the smallest subspace of $\mathbb{R}^{pq}$ that satisfies (15.2) and $\mathscr{S}_{Y|\circ X\circ}$ is a subspace of $\mathbb{R}^{pq}$ that satisfies (15.1), we have that $\mathscr{S}_{Y|\circ X\circ} \subseteq \mathscr{S}_{Y|X}$.

More generally, when X is a multi-dimensional array in $\mathbb{R}^{p_1 \times \cdots \times p_k}$, we consider the following dimension reduction problem

$$Y \perp\!\!\!\perp X \,|\, (\alpha_r \otimes \cdots \otimes \alpha_1)^{\mathsf{T}} \text{vec}(X),$$

where $\text{vec}(X)$ is a vector in $\mathbb{R}^{p_1 \cdots p_k}$ obtained by X by varying its last index first. The goal of this dimension reduction problem is to find

$$\text{span}(\alpha_k) \otimes \cdots \otimes \text{span}(\alpha_1),$$

which, if it is minimal, is called the *r-mode dimension folding subspace*. This subspace is denoted by $\mathscr{S}_{Y|X\circ r}$. The subspace $\text{span}(\alpha_i)$ — or α_i itself — is called the ith mode of the dimension folding subspace.

To estimate the dimension folding space, Li et al. (2010a) introduced the notion of a Kronecker envelope, a concept that was inspired by the envelope models developed by Cook et al. (2010). For any subspace $\mathscr{S} \subseteq \mathbb{R}^{p_1 \cdots p_r}$, its Kronecker envelope is the smallest subspace of the form $\text{span}(\mathscr{S}_r) \otimes \cdots \otimes \text{span}(\mathscr{S}_1)$ that contains $\mathscr{S}$, and is denoted by $\mathscr{E}^{\otimes}_{p_1, \ldots, p_r}$. That is,

$$\mathscr{E}^{\otimes}_{p_1, \ldots, p_r}(\mathscr{S}) = \cap\{\tilde{\mathscr{S}} : \tilde{\mathscr{S}} \in \mathfrak{A}\},$$

where

$$\mathfrak{A} = \{\mathscr{S}_r \otimes \cdots \otimes \mathscr{S}_1 : \mathscr{S}_1 \subseteq \mathbb{R}^{p_1}, \ldots, \mathscr{S}_r \subseteq \mathbb{R}^{p_r}\}.$$

Kronecker envelope connects the central subspace with the dimension folding subspace, and connects an estimator of the central subspace with an estimator of the dimension folding subspace. For example, it can be shown that

$$\mathscr{S}_{Y|X\circ u} \subseteq \mathscr{E}^{\otimes}_{p_1, \ldots, p_r}(\mathscr{S}_{Y|\text{vec}(X)}).$$

Moreover, if $T(F_n)$ is an unbiased estimate of the central subspace — that is, if $T(F_0) \subseteq \mathscr{S}_{Y|\text{vec}(X)}$, then it can be shown that

$$\mathscr{E}^{\otimes}_{p_1, \ldots, p_r}[T(F_0)] \subseteq \mathscr{S}_{Y|\circ X\circ}.$$

Via the mechanism of the Kronecker envelope, we can convert an estimator of the central subspace $\mathscr{S}_{Y|\text{vec}(X)}$, such as SIR, SAVE, and DR, into an estimator of the dimension folding subspace $\mathscr{S}_{Y|X\circ u}$. Using this idea, Li et al. (2010a) developed folded-SIR, folded-SAVE, and folded-DR to estimate the dimension folding subspace. The algorithms for these folded estimators consist essentially of iterative minimization of quadratic forms, where, at each iteration, the estimator of the ith mode of the dimension folding subspace, α_i, is updated. Moreover, each update can be expressed explicitly in the form of Ordinary Least Squares, which keeps the computation relatively simple. Under an additional Kronecker covariance structure on X,

Kim (2010), collaborating with her advisor Bing Li and Professor Naomi Altman at Pennsylvania State University, proposed an alternative method that computes the r-mode of the dimension folding subspace sequentially.

Sufficient dimension folding, as well as related regression methods with tensor-valued predictors or responses, have received increased attention and undergone momentous developments during the past few years, due to the high demands for statistical tools that process data in tensor form. In particular, Ding and Cook (2014) introduced two model-based approaches to dimension folding: the first is a generalization of the principal component analysis, where a tensor-valued X is modeled as

$$X = \mu + \Gamma_2 v \Gamma_1^\mathsf{T} + \sigma \varepsilon.$$

Here v is a latent random matrix of lower dimension, ε is an independent error, and Γ_1 and Γ_2 are nonrandom matrices. The second method is a generalization of the Principal Fitted Component introduced in Cook and Forzani (2008). One of the useful features of this model-based approach is that it inherits the asymptotic properties of Maximum Likelihood Estimation. Cook and Ding (2015) developed a different extension of SIR to r-mode tensor than those of Li et al. (2010a) and Kim (2010). Along a different direction, Xue and Yin (2014) extended the Outer Product of Gradient and the Minimal Average Variance Estimation (OPG and MAVE, see Chapter 7) to tensor-valued predictors, which are called the folded-OPG and folded-MAVE. They also made a distinction between the central dimension folding subspace and the central mean dimension folding subspace, and showed that these estimators target the central mean dimension folding subspace. Developing further along this direction, Xue et al. (2016) extended the ensemble estimator described in Chapter 11 to the tensor-predictor setting, so that, for example, the folded-OPG and folded-MAVE can be incorporated with an ensemble of functions of the response to recover the central dimension folding subspace. Zhong et al. (2015) considered the following SDR problem

$$Y \perp\!\!\!\perp X \,|\, \gamma_1^\mathsf{T} \mathrm{vec}(X), \ldots, \gamma_d^\mathsf{T} \mathrm{vec}(X),$$

where γ_i are vectors of the form $\beta_i^{(r)} \otimes \ldots \otimes \beta_i^{(1)}$.

There have also been substantial advances in the development of regression methods for tensor-valued predictors and responses. For example, Hung and Wang (2013) proposed a logistic regression model where the predictor is of the form

$$\mathrm{logit}[P(Y = 1 | X)] = \alpha + \alpha^\mathsf{T} X \beta,$$

where X is a matrix-valued predictor. Zhou et al. (2013) further developed generalized linear models of the form

$$g(\mu) = \alpha + \gamma^\mathsf{T} Z + (\alpha_r \otimes \cdots \otimes \alpha_1)^\mathsf{T} \mathrm{vec}(X),$$

where g is the link function, Z is a vector-valued predictor, and X is a tensor-valued predictor. The distribution of Y given X is assumed to belong to an exponential family. Zhou and Li (2014) introduced a regularized generalized linear model with a vector-valued and a matrix-valued predictor; that is,

$$g(\mu) = \gamma^\mathsf{T} Z + \langle B, X \rangle + J(B),$$

where g is the link function, $\langle B, X \rangle$ is the matrix-inner product $\mathrm{tr}(B^\mathsf{T} X)$, and $J(B)$ is a regularization term that encourages the small singular values of B to become 0.

Another recent development is extending Independent Component Analysis to tensor-valued data. This belongs to the realm of unsupervised dimension reduction. For example, Virta et al. (2017a) and Virta et al. (2017b) extended the two most commonly used methods in Independent Component Analysis — the Fourth Order Blind Identification (FOBI) and the Joint Approximate Diagonalization of Eigenmatrices (JADE) to tensor-valued data.

15.3 Sufficient Dimension Reduction for Grouped Data

In some applications, the predictors fall naturally into several groups. For example, in a global surface temperature reconstruction study (Mann et al. (2008), McShane and Wyner (2011)), the data contains 1209 climate proxies recorded yearly over the past 2000 years. One of the goals of the study is to assess whether the recent temperature rise is anomalous. Since instrumental climate records only became available in the mid-19th century, proxy series were used to reconstruct global surface temperatures for the past. These 1209 proxies fall into several groups: tree composites, tree rings, ice cores, cave deposits, lake sediments, and historical documentations. Clearly, it makes more sense to linearly combine predictors within each group (such as different ice-core predictors) than to linearly combine tree-ring predictors with ice-core predictors. Presence of such group information is becoming increasingly common in modern data collections. Incorporating this information into statistical analysis not only increases the interpretability, but also makes the model more parsimonious. Motivated by this, Li et al. (2010b) and Guo et al. (2015a) introduced the Groupwise Sufficient Dimension Reduction. See also Liu et al. (2017).

Groupwise Sufficient Dimension Reduction (groupwisd SDR) is formulated as follows. Suppose $\mathscr{S}_1, \ldots, \mathscr{S}_g$ are orthogonal subspaces of $\mathbb{R}^p$ that decompose $\mathbb{R}^p$. That is,

$$\mathscr{S}_1 \oplus \cdots \oplus \mathscr{S}_g = \mathbb{R}^p,$$

where $\mathscr{S}_1 \oplus \mathscr{S}_2$ simply means span$\{u_1 + u_2 : u_1 \in \mathscr{S}_1, u_2 \in \mathscr{S}_2\}$ and $\mathscr{S}_1 \perp \mathscr{S}_2$. These are known subspaces derived from the group information. For example, if X is 5 dimensional and $X_1, \ldots, X_5$ fall into two groups: $\{X_1, X_2\}$ and $\{X_3, X_4, X_5\}$, then $g = 2$, $\mathscr{S}_1 = \text{span}(e_1, e_2)$, and $\mathscr{S}_2 = \text{span}(e_3, e_4, e_5)$. In groupwise SDR we look for subspaces $\mathscr{T}_1 \subseteq \mathscr{S}_1, \ldots, \mathscr{T}_g \subseteq \mathscr{S}_g$ such that

$$Y \perp\!\!\!\perp X | P_{\mathscr{T}_1} X, \ldots, P_{\mathscr{T}_g} X, \quad \text{or equivalently,} \quad Y \perp\!\!\!\perp X | P_{\mathscr{T}_1 \oplus \cdots \oplus \mathscr{T}_g} X. \tag{15.3}$$

If the subspace $\mathscr{T}_1 \oplus \cdots \oplus \mathscr{T}_g$ is minimal, then it is called groupwise central subspace, and is denoted by $\mathscr{S}_{Y|X}(\mathscr{S}_1, \ldots, \mathscr{S}_g)$. The groupwise central subspace is the smallest sufficient dimension reduction subspace that respects the group structure in the predictor. Note that, by construction

$$\mathscr{S}_{Y|X}(\mathscr{S}_1, \ldots, \mathscr{S}_g) \supseteq \mathscr{S}_{Y|X}.$$

The subspace $\mathscr{S}_{Y|X}(\mathscr{S}_1, \ldots, \mathscr{S}_g)$ is the target of estimation.

Suppose $\mathscr{S}_i$ has dimension p_i and let $A_i \in \mathbb{R}^{p \times p_i}$ be a matrix such that $\text{span}(A_i) = \mathscr{S}_i$. Again, these matrices are known (they are not unique but this does not matter as long as their columns span $\mathscr{S}_i$). The above relation can be rewritten as

$$Y \perp\!\!\!\perp X | \beta_1^{\mathsf{T}} Z_1, \ldots, \beta_g^{\mathsf{T}} Z_g,$$

where $Z_i = A_i^{\mathsf{T}} X, i = 1, \ldots, g$.

Li et al. (2010b) and Guo et al. (2015a) proposed two approaches to tackle this problem.

The first (Li et al., 2010b) extends the MAVE and rMAVE in Chapter 11 by imposing a block diagonal structure on the matrix B in the objective function (11.11). The objective function can be minimized by iterating among different blocks of the matrix, where each step has an explicit least-squares-type solution. The groupwise rMAVE was developed similarly to rMAVE — that is, by iterating over steps of groupwise MAVE. This approach targets the groupwise central mean subspace, defined by

$$E(Y|X) = E(Y|P_{\mathscr{T}_1}X, \ldots, P_{\mathscr{T}_g}X),$$

instead of the conditional independence in (15.3). Of course, we can then use the idea of ensemble estimates in Chapter 11 to recover the groupwise central subspace. That is, replace Y by an ensemble of functions of Y, and apply the groupwise MAVE (or groupwise rMAVE) with the ensemble as the responses.

The second approach (Guo et al., 2015a) uses the idea of a direct-sum envelope in a way not unlike how the Kronecker envelope was used in dimension folding. Consider the class of all subspaces of $\mathbb{R}^p$ of the form

$$\mathfrak{B} = \{\mathscr{T}_1 \oplus \cdots \oplus \mathscr{T}_g : \mathscr{T}_1 \subseteq \mathscr{S}_1, \ldots, \mathscr{T}_g \subseteq \mathscr{S}_g\}.$$

The direct-sum envelope is the subspace

$$\cap\{\mathscr{S} \subseteq \mathfrak{B} : \mathscr{S}_{Y|X} \subseteq \mathscr{S}\}.$$

This subspace is written as $\mathscr{E}^{\oplus}(\mathscr{S}|\mathscr{S}_1, \ldots, \mathscr{S}_g)$. The direct-sum envelope serves a similar purpose as does the Kronecker envelope. That is, if $T(F_n)$ is an estimate of the central subspace $\mathscr{S}_{Y|X}$, then $\mathscr{E}^{\oplus}[T(F_n)|\mathscr{S}_1, \ldots, \mathscr{S}_g]$ is an estimate of $\mathscr{S}_{Y|X}(\mathscr{S}_1, \ldots, \mathscr{S}_g)$. Using this mechanism, Guo et al. (2015a) developed groupwise SIR (gSIR) and groupwise Directional Regression (gDR) to estimate the groupwise central subspace.

15.4   Variable Selection via Sufficient Dimension Reduction

Dimension reduction and variable selection are similar in both their goals and their means. They both aim at data reduction: the former seeks a low-rank core in the predictor, whereas the latter seeks a sparse core of the predictor. They are both achieved by conditional independence: Sufficient Dimension Reduction by $Y \perp\!\!\!\perp X|\beta^{\mathsf{T}}X$, and variable selection by $Y \in X|X_A$, where X_A stands for $\{X_i : i \in A\}$, A being a subset of $\{1, \ldots, p\}$ called the active set. Therefore, it is reasonable to combine SDR and variable selection to achieve low rank and sparsity at the same time.

A pioneering work in this direction was Cook (2004), which formulates a hypothesis to test that part of the predictor has no effect on regression. Usually, a subset of predictors X_A is selected if $Y \perp\!\!\!\perp X_{A^c}|X_A$. Sufficient dimension reduction offers us the extra information $Y \perp\!\!\!\perp X|\beta^{\mathsf{T}}X$, which implies the following equivalence

$$Y \perp\!\!\!\perp X_{A^c}|X_A \;\Leftrightarrow\; Y \perp\!\!\!\perp X|\beta_A^{\mathsf{T}}X_A,$$

as shown in Proposition 1 of Cook (2004). More generally, this problem can be formulated as testing

$$H_0 : \mathscr{S}_{Y|X} \perp \mathscr{H}$$

where $\mathscr{H}$ is a subspace of $\mathbb{R}^p$. Cook (2004) developed a test statistic based on the SIR estimate of $\mathscr{S}_{Y|X}$, as well as the asymptotic distribution of the test statistic. Note that the combination of SDR and the hypothesis test not only reduces X to X_A, but also projects X_A onto a lower dimensional subspace, thus achieving sparsity and low rank at the same time.

Li (2007) introduced the Sparse Sufficient Dimension Reduction. As we have seen in the previous chapters, many SDR methods can be formulated as the generalized eigenvalue problem $\mathrm{GEV}(\Lambda, \Sigma)$, where Λ is the candidate matrix, and Σ is the covariance matrix of X. Li (2007) observed that this generalized eigenvalue problem can always be reformulated as minimization of the objective function

$$\sum_{i=1}^{p} \|\Sigma^{-1} m_i - \alpha \beta^{\mathsf{T}} m_i\|_{\Sigma}^2 + \rho \operatorname{tr}(\beta^{\mathsf{T}} \Sigma \beta)$$

over the set $\{(\alpha, \beta) \in \mathbb{R}^{p \times d} \times \mathbb{R}^{p \times d}, \alpha^{\mathsf{T}} \Sigma \alpha = I_d\}$, where $m_1, \dots, m_p$ are the columns of $\Lambda^{1/2}$, $\tau > 0$ is a constant, and $\|u\|_{\Sigma}$ is the norm $(u^{\mathsf{T}} \Sigma u)^{1/2}$. With an objective at hand, Li (2007) added a regularization term, $\sum_{i=1}^{p} \sum_{j=1}^{d} \lambda_j |\beta_{ij}|$, to encourage sparsity of β.

Note that, although the above procedure ensures the sparsity of β, it does not actually select variables, because for a component of X to drop out we need an entire row of β to be 0, which cannot be achieved by penalizing individual entries of β. Bondell and Li (2009) proposed another regularized SDR method that does select variables. It uses a different objective function, which was developed by Cook and Ni (2005), as will be covered in the next section. The form of the objective function is unimportant for our discussion; the main point is the design of the penalty. Roughly speaking, Bondell and Li (2009) introduced the sparsity through an auxiliary parameter $\eta = (\eta_1, \dots, \eta_p)$ and use $\operatorname{diag}(\eta) \beta$ to replace β so that, when an η_i is encouraged to shrink to 0, it brings with it the entire ith row of β, and thereby eliminates the ith component of X.

Chen et al. (2010) introduced a different approach to variable selection through Sufficient Dimension Reduction. Observing that, for the purpose of variable selection, the penalty should be imposed at the row-level rather than at the element-level, they employed the following loss function

$$\rho(\beta) = \sum_{i=1}^{p} \lambda_i \|(\beta^{\mathsf{T}})_i\|_2.$$

where $\|\cdot\|$ is the Euclidean norm, and $(\beta^{\mathsf{T}})_i$ is the ith column of β^{T}, or the i row of β. Note that, within the elements of the same row, the penalty is the Euclidean norm, but between different rows, the penalty behaves like an L_1 norm. Thus the force of the penalty is exerted not on the individual elements of β, but on the rows of β.

Wang and Yin (2008) introduced the regularized MAVE to achieve sparse Sufficient Dimension Reduction. Instead of minimizing the MAVE objective function (11.11), they proposed to minimize

$$\sum_{i=1}^{n} \sum_{j=1}^{n} [Y_j - a_i - c_i^{\mathsf{T}} (\beta_1, \dots, \beta_m)^{\mathsf{T}} (X_j - X_i)]^2 W_h(X_i, X_j) + \lambda \|\beta_m\|_1,$$

where $\|\cdot\|_1$ is the L_1-vector norm. For each $m = 1, \dots, d$, where d is the assumed dimension of the central mean subspace, the above objective function is minimized iteratively between $\{(a_i, c_i) : i = 1, \dots, n\}$ and β_m until it converges. Then m is reset to $m + 1$ and the process is repeated until m reaches d. This process results in a matrix $B = (\beta_1, \dots, \beta_d)$ that is both low rank and sparse. For a related work on sparse single- and multiple-index model, see Alquier and Biau (2013).

15.5   Efficient Dimension Reduction

One way to improve efficiency of Sufficient Dimension Reduction is by combining a set of SDR estimators in some optimal way. For example, Cook and Ni (2005) developed a systematic way to optimally combine different SDR estimators, and Zhu et al. (2007) considered a class of a hybrid of SIR and SAVE.

Cook and Ni (2005)'s method, called minimum discrepancy approach (MDA), is akin to Generalized Method of Moments (GMM); see Hansen (1982). The idea underlying MDA can be briefly described as follows. Start with vectors $\hat{\zeta}_1, \ldots, \hat{\zeta}_h$ that estimate a set of vectors, say $\zeta_1, \ldots, \zeta_h$ in the central subspace. For example, these can be the slice means in SIR, or the estimates of $I_p - \text{var}(Z | Y \in J_i)$ in SAVE, where Z is the standardized predictor and J_i is the ith slice. Assume, for simplicity, $\zeta_1, \ldots, \zeta_h$ span the central subspace; that is, the collection of estimators are exhaustive. Let $\hat{\zeta}$ denote the matrix $(\hat{\zeta}_1, \ldots, \hat{\zeta}_h)$ and let ζ denote the matrix $(\zeta_1, \ldots, \zeta_h)$. Let β be a basis matrix of $\mathscr{S}_{Y|X}$. Then there is a matrix $\alpha \in \mathbb{R}^{d \times h}$ such that $\zeta = \beta \alpha$. So $\hat{\zeta} - \beta \alpha$, or equivalently $\text{vec}(\hat{\zeta} - \beta \alpha)$, can be viewed as a set of estimating equations that we can optimally combine using the idea of GMM. That is, we minimize the objective function

$$[\text{vec}(\hat{\zeta} - \beta \alpha)]^\top W_n [\text{vec}(\hat{\zeta} - \beta \alpha)] \tag{15.4}$$

where W_n is a positive definite weighting matrix. If

$$\sqrt{n} \text{vec}(\hat{\zeta} - \beta \alpha) \xrightarrow{\mathscr{D}} N(0, \Gamma),$$

then, the optimal weight matrix would be Γ^{-1}. Thus, given a sample estimate $\hat{\Gamma}$ of Γ, we can form the optimal objective function

$$[\text{vec}(\hat{\zeta} - \beta \alpha)]^\top \hat{\Gamma}^{-1} [\text{vec}(\hat{\zeta} - \beta \alpha)]. \tag{15.5}$$

Using the theory of optimal estimating equations it can be shown that the objective function (15.5) is optimal in the sense its minimizer has the smallest asymptotic variance among the minimizers of the objective functions of the form (15.4) with W_n being any positive definite matrices. This method can be applied systematically to combine several estimators optimally, or to combine a set of vectors derived from different slices optimally.

Another line of development toward improving the efficiency of SDR is to use a semiparametrically efficient estimator. This approach was used in Ma and Zhu (2012) and Ma and Zhu (2013a). The idea is to formulate SDR as a semiparametric problem. Suppose

$$\{f_{\theta,h} : \theta \in \mathbb{R}^p, h \in \mathscr{H}\}$$

is a family of probability density functions where θ is a Euclidean parameter and h is a function that varies in a class of functions $\mathscr{H}$, and consider the family of log likelihood

$$\{\log f_{\theta,h} : \theta \in \mathbb{R}^p, h \in \mathscr{H}\}.$$

According to the semiparametric theory, the projection of the score function with respect to the Euclidean parameter, $\partial \log f_{\theta,h} / \partial \theta$, on to the orthogonal complement of the tangent space of the family $\{f_{\theta_0,h} : h \in \mathscr{H}\}$, where θ_0 is the true value of θ, is the optimal estimating equation in the sense that its solution has the smallest asymptotic variance among the collection of all regular estimators. See Bickel et al. (1998).

In our context, the joint density of (X, Y) is $\psi(x, y)$ which, under the SDR assumption

$Y \perp\!\!\!\perp X | \beta^\mathsf{T} X$, is equal to $\lambda(x)\eta(y,\beta^\mathsf{T}x)$, where $\lambda(x)$ is the marginal density of X and $\eta(y,\beta^\mathsf{T}x)$ is the conditional density of Y given $\beta^\mathsf{T}X$. Thus our semiparametric family is

$$\left\{ \log[\lambda(x)] : \lambda(\cdot) \text{ is a density on } \Omega_x \right\}$$
$$+ \left\{ \log[\eta(y,\beta^\mathsf{T}x)] : \eta(\cdot,u) \text{ is conditional density } \Omega_Y \text{ for each } u \in \mathbb{R}^d, \beta \in \mathbb{G}^{p \times d} \right\}$$
$$\equiv \mathcal{M}_m + \mathcal{M}_c,$$

where the subscript m and c indicate "marginal" and "conditional" distributions, and Ω_X and Ω_Y are the supports of X and Y. The tangent space can be calculated by taking derivatives of the sub-models. For example, suppose λ_α is a parametric family where $\alpha = 0$ corresponds to the true marginal density λ_0. Then $\partial \log[\lambda_\alpha(x)/\partial \alpha]$ must satisfy the equation

$$\int_{\Omega_X} [\partial \log \lambda_\alpha(x)/\partial \alpha]_{\alpha=0} \lambda_0(x) dx = 0.$$

Thus, the tangent space of $\mathcal{M}_m$ is simply

$$\mathcal{T}_m = \{h(x) : E[h(X)] = 0\}.$$

By the same argument it can be deduced that the tangent space of $\mathcal{M}_c$ is

$$\mathcal{T}_c = \{h(y,\beta^\mathsf{T}x) : E[h(Y,\beta^\mathsf{T}X)|\beta^\mathsf{T}X] = 0\}.$$

The tangent space of $\mathcal{M}_m + \mathcal{M}_c$ is, then, $\mathcal{T}_m + \mathcal{T}_c$. Ma and Zhu (2012) went on to show that the orthogonal complement of $\mathcal{T}_m + \mathcal{T}_c$ is of the form

$$(\mathcal{T}_m + \mathcal{T}_c)^\perp = \left\{ h(x,y) - E[h(X,Y)|\beta^\mathsf{T}x,y] : E[h(Y,X)|x] = E[h(Y,X)|\beta^\mathsf{T}x] \right\}.$$

From here, Ma and Zhu's developments diverged into two directions. The first, as was done in Ma and Zhu (2012), was to develop reasonable estimating equations in the space $(\mathcal{T}_m + \mathcal{T}_c)^\perp$ by modifying existing inverse regression methods such as SIR, SAVE, and DR. The second, as was done in Ma and Zhu (2013a), projects the true score on to $(\mathcal{T}_m + \mathcal{T}_c)^\perp$ to produce the semiparametrically efficient score. Ma and Zhu (2012) also developed semiparametric approach to estimating the central mean subspace. See Ma and Zhu (2013b) for a general review that placed the semiparametric methods among the other SDR methods.

This semiparametric approach was further developed by Luo et al. (2014), which considered the estimation of a general statistical function including the conditional mean. Suppose we are interested a conditional quantity $T(Y|X)$, such as the conditional mean, the conditional quantile, the conditional variance. Here, the notation $T(Y|X)$ is meant to be a function of X alone, in the same way that $E(Y|X)$ and $\text{var}(Y|X)$ are functions of X, not Y. In this general framework the SDR problem was stated as

$$T(Y|X) \text{ is measurable with respect to } \sigma(\beta^\mathsf{T}X). \tag{15.6}$$

Luo et al. (2014) defined the space spanned by the lowest dimensional β the T-central subspace, and denote it by $\mathcal{S}_{T(Y|X)}$. The question raised by Luo et al. (2014) is, suppose, instead of the semiparametric family specified by

$$\psi(x,y) = \lambda(x)\eta(y,\beta^\mathsf{T}x),$$

as used in Ma and Zhu (2012), we consider the family of densities $\psi(x,y)$ that satisfies (15.6), then what is the semiparametrically efficient estimator? Obviously, (15.6) is a weaker condition than the conditional independence condition $Y \perp\!\!\!\perp X | \beta^\mathsf{T} X$, so that $\mathscr{S}_{T(Y|X)} \subseteq \mathscr{S}_{Y|X}$. This means our estimation target is focused on the specific aspects of the conditional distribution we are interested in, such as mean regression, quantile regression, and estimation of heteroscedasticity. Luo et al. (2014) developed the semiparametric efficient score for this problem, as well as the methods to implement it.

15.6   Partial Dimension Reduction for Categorical Predictors

Many data analysis problems contain categorical predictors, which need to be appropriately accounted for by Sufficient Dimension Reduction. Specifically, the subjects in the sample may be divided into groups — such as male and female — and the Sufficient Dimension Reduction subspaces may well be different from group to group. The situation is analogous to Analysis of Covariance, where both the slopes and the means of different groups in the population may be different. This problem is different from groupwise Sufficient Dimension Reduction as described in Section 15.3, as the latter is concerned with the group structure among the predicting variables, rather than subjects.

Suppose, in addition to random variables (X, Y), we have a set of categorical predictors. Since a set of categorical variables can always be collapsed into a single categorical variable, we assume, without loss of generality, that we have a single categorical variable W. Let then $\{1, \ldots, c\}$ be all the possible values of W, which can be interpreted as c categories. Our goal is to reduce the dimension of the continuous random vector X for predicting Y, in the presence of W. Chiaromonte et al. (2002) formulated the Sufficient Dimension Reduction problem as follows: find a matrix $\beta \in \mathbb{R}^{p \times d}$ such that

$$Y \perp\!\!\!\perp X | \beta^\mathsf{T} X, W. \tag{15.7}$$

We call the subspace of $\mathbb{R}^p$ spanned by the columns of β a partial Sufficient Dimension Reduction subspace. Furthermore, if $\mathscr{S}$ is the intersection of all span(β) where β satisfies (15.7), then we call $\mathscr{S}$ the partial central subspace, and denote it by $\mathscr{S}_{Y|X}^{(W)}$. This subspace is our target of estimation.

The partial central subspace can be estimated by the following mechanism. For each $w \in \{1, \ldots, r\}$, let (X_w, Y_w) be the random element whose distribution is the conditional distribution of $(X, Y)|W = w$. Chiaromonte et al. (2002) showed that the following equality holds:

$$\mathscr{S}_{Y|X}^{(W)} = \mathscr{S}_{Y_1|X_1} + \cdots + \mathscr{S}_{Y_c|X_c}. \tag{15.8}$$

Based on this relation, we can first estimate $\mathscr{S}_{Y_w|X_w}$, $w = 1, \ldots, c$, using the data in the wth category by any SDR method, and then assemble them together by solving a combined eigenvalue problem. Specifically, let $\hat{M}_1, \ldots, \hat{M}_c$ be candidate matrices for estimating $\mathscr{S}_{Y_w|X_w}$, $w = 1, \ldots, c$, and, without loss of generality, assume $\hat{M}_w$ to be symmetric and positive semidefinite. Let n_w be the sample size of the wth category, and let n be the total sample size. We merge the candidate matrices by

$$\hat{M} = \sum_{w=1}^{c} \frac{n_w}{n} \hat{M}_w.$$

The partial central subspace can then be estimated by the first d eigenvectors of the matrix $\hat{M}$.

Let Z_w represent the standardized version of X_w — standardized the conditional mean and conditional variance matrix for category w. When the true covariance matrices Σ_w of the categories are the same, it is more efficient to estimate the standardized central subspaces $\mathscr{S}_{Y_w|Z_w}$

by, say, candidate matrices $\hat{N}_1, \ldots, \hat{N}_c$, and then use $\hat{\Sigma}_{\text{pool}}^{-1/2} \hat{v}_1, \ldots, \hat{\Sigma}_{\text{pool}}^{-1/2} \hat{v}_d$ to estimate $\mathscr{S}_{Y|X}^{(W)}$, where $\hat{\Sigma}_{\text{pool}}$ is the pooled variance matrix from the c categories, and $\hat{v}_i$'s are the leading eigenvectors of

$$\hat{N} = \sum_{w=1}^{c} \frac{n_w}{n} \hat{N}_w.$$

This was the approach proposed by Chiaromonte et al. (2002). In that paper, $\hat{N}_w$ are the SIR candidate matrix derived from the data in the wth category.

Li et al. (2003a) developed partial Sufficient Dimension Reduction for conditional mean, which is defined by the relation

$$E(Y|X,W) = E(Y|\beta^{\mathsf{T}} X, W).$$

The smallest subspace span(β) where β satisfies the above relation is called the partial central mean subspace, and is denoted by $\mathscr{S}_{E(Y|X)}^{(W)}$. Similar to (15.8), Li et al. (2003a) showed that

$$\mathscr{S}_{E(Y|X)}^{(W)} = \mathscr{S}_{E(Y_1|X_1)} + \cdots + \mathscr{S}_{E(Y_c|X_c)}.$$

We can then apply any SDR methods that target for the central mean subspace to the wth category to estimate $\mathscr{S}_{E(Y_w|Z_w)}$, and then combine them in the similar way as we did for the central partial subspace.

Partial dimension reduction was further developed by a number of authors to enhance its efficiency and to broaden its applications. For example, Wen and Cook (2007) proposed to use the Minimum Discrepancy Approach described in Section 15.5 to combine estimators obtained from different categories to achieve optimal estimation. This approach can be applied to estimating either the partial central subspace $\mathscr{S}_{Y|X}^{(W)}$ or the partial central mean subspace $\mathscr{S}_{E(Y|X)}^{(W)}$. Shao et al. (2009) developed the partial SAVE method to estimate $\mathscr{S}_{Y|X}^{(W)}$. In some semiparametric regression models we would like to keep a set of predictors intact while performing Sufficient Dimension Reduction, even if those variables are not categorical. For example, in a partially linear model, or a partially linear single index model, a subset of the predictors are modeled linearly, and should be left intact when we model the rest of the parameters nonlinearly. See, for example, Heckman (1986) and Carroll et al. (1997). Feng et al. (2013) proposed to a discretization-expectation method, which first discretizes the variables to be shielded from dimension reduction, and then perform partial dimension reduction on the rest of variables, treating the discretized variables as the categorical predictor W. In some genetic applications, as described in Liu et al. (2017), both the subjects and the predictor variables fall naturally into groups. To handle these situations, Liu et al. (2017) proposed a structured ordinary least squares method — a combination of groupwise SDR and partial SDR — to take into account both group structures.

Another approach to handling categorical variable in Sufficient Dimension Reduction was taken by a recent work by Bura et al. (2016), which considered an inverse regression model where the predictor given the response has a multivariate exponential family distribution. Under this model, the authors derived the minimal sufficient predictor $R(X)$ that satisfies $Y \perp\!\!\!\perp X | R(X)$, and used it as sufficient predictor. In this setting, $R(X)$ can be either continuous or categorical, or partly continuous and partly categorical.

15.7 Measurement Error Problem

In regression analysis, the predictor X is sometimes observed with error. Or, to put it another way, the predictor X cannot be directly observed; it is only observed through surrogate random vector W. Auxiliary data or information allows us to estimate certain aspects of the relation between X and W. To assist understanding, we can view W as measurements obtained by a cheap instrument, and X as measurements by an expensive instrument. We observe a large sample of (Y,W), but also have a sample of (X,W) that can be used to calibrate W. How to estimate the regression coefficients of Y versus X — even though X is not observed — is the measurement error regression problem. See Fuller (1987), Carroll and Stefanski (1990), and Carroll et al. (1995).

Specifically, suppose that the regression model of Y on X and the relation between X and W are specified by

$$Y = \beta^{\mathsf{T}}X + \varepsilon, \quad W = \gamma + \Gamma^{\mathsf{T}}X + \delta,$$

where $X \perp\!\!\!\perp \varepsilon$ and $\delta \perp\!\!\!\perp (X,\varepsilon)$. Let Σ_w denote the covariance matrix $\mathrm{var}(W)$. Let $U = \Sigma_{XW}\Sigma_w^{-1}W$, and call this random vector the projected surrogate predictor. Here, we assume that Σ_{XW} can be estimated by the calibration sample. It can be shown that

$$\Sigma_U^{-1}\Sigma_{UY} = \Sigma_X^{-1}\Sigma_{XY} = \beta.$$

That is, with the knowledge of Σ_{XW}, and the observations on (Y,W), we can estimate β even without observing X.

It is then natural to ask, in the Sufficient Dimension Reduction setting, is it possible to use the surrogate predictor to derive an unbiased estimator of the central subspace? This statement, if true, is both more general and stronger than the above statement about linear models, because the relation between Y and X is unspecified except we know that it is a function of the linear index $\beta^{\mathsf{T}}X$. Nevertheless, this is possible. Carroll and Li (1992) showed, under the linear conditional mean assumption, that if one performs SIR on Y versus U, then the process leads to an unbiased estimator of central subspace $\mathscr{S}_{Y|X}$. Furthermore, under both the linear conditional mean and the constant conditional variance assumptions, Lue (2004) showed that the same phenomenon holds for pHd: if one applies the principal Hessian directions to Y versus U, then the process again leads to an unbiased estimator of $\mathscr{S}_{Y|X}$.

Li and Yin (2007) pursued this question further: are the above two cases coincidences or is there a deeper reason behind them? What about other estimators, such as SAVE, DR (DR was developed around the same time Li and Yin (2007) was written), Iterative Hessian Transformations, OPG, and MAVE? Under the assumption that $X \sim N(\mu_X, \Sigma_X)$, $\delta \sim N(0, \Sigma_\delta)$ and $\delta \perp\!\!\!\perp (X,Y)$, Li and Yin (2007) proved the following relations

$$\mathscr{S}_{Y|X} = \mathscr{S}_{Y|U}, \quad \mathscr{S}_{E(Y|X)} = \mathscr{S}_{E(Y|U)}. \tag{15.9}$$

In other words, we can apply *any* unbiased estimator on (U,Y) to produce unbiased estimator of $\mathscr{S}_{Y|X}$. It is as if the measurement error were transparent, which cannot impede the true relation between Y and X, as long as X and δ are Gaussian random vectors. The proof of (15.9) in Li and Yin (2007) was long and tortuous, but Yin and Li (2010) came up with an alternative, much simpler proof. Li and Yin (2007) refer to this property as the invariance law for the measurement error problem, because it means replacing the true predictor by the projected surrogate predictor does not change the central subspace.

Note that Li and Yin (2007)'s general relation (15.9) relies on the multivariate Gaussian assumption on the predictor and the measurement error. This Gaussian assumption can be weakened if we only wish to show some specific estimators based on the surrogate predictors are unbiased. The above-mentioned results of Carroll and Li (1992) and Lue (2004) are two examples. Along these lines, Zhang et al. (2014) showed that, under the linear conditional mean assumption, the cumulative mean estimator developed by Zhu et al. (2010), when applied to (Y,U), produces an unbiased estimator of the central subspace.

15.8 SDR via Support Vector Machine

Support Vector Machine (SVM, Vapnik, 1998) is a classification method that draws an optimal hyperplane to separate two labeled clusters. Let X be a predictor in $\mathbb{R}^p$ and let Y be a binary response, labeled as $Y = -1, 1$. Let $(X_1, Y_1), \ldots, (X_n, Y_n)$ be a sample from (X, Y). The goal of SVM is to find a hyperplane $\{\psi^\mathsf{T}(x - \bar{X}) = t : x \in \mathbb{R}^p\}$ whose normal vector is as much as possible aligned with Y. The ideal situation is that $[\psi^\mathsf{T}(X_i - \bar{X}) - t]Y_i$ being always greater 1, which means the two clusters are positioned completely outside the two margin hyperplanes $[\psi^\mathsf{T}(X_i - \bar{X}) - t]Y_i = \pm 1$, and at the same time, the distance between the two margins, $2/\|\psi\|$, is maximized. When the two clusters cannot be perfectly separated by a hyperplane, we maximize the distance between the two margins while penalizing the cases of X_i falling within the two margins. This leads to the following minimization criterion at the population level:

$$\psi^\mathsf{T}\psi + \lambda E\{1 - Y[\psi^\mathsf{T}(X - EX) - t]\}^+, \tag{15.10}$$

where the function x^+ stands for $\max(0, x)$, and is called the hinge function, and $\lambda > 0$ is a tuning parameter. The hinge function exerts no force when $1 - Y[\psi^\mathsf{T}(X - EX) - t] \leq 0$, which corresponds to the case where X lies outside the two margins, but contributes to the objective function being minimized when X falls within the two margins. The function $\psi^\mathsf{T}\psi$ is a decreasing function of the distance between the two margins, and thus minimizing it amounts to maximizing the distance between the two margin hyperplanes.

This idea can be extended to the nonlinear case using the reproducing kernel Hilbert space. Let $\kappa : \Omega_X \times \Omega_X \to \mathbb{R}$ be a positive definite kernel, and let $\mathscr{H}$ be the reproducing kernel Hilbert space generated by κ. Let μ be the mean element $E[\kappa(\cdot, X)]$. We replace the objective function (15.10) by

$$\langle \phi, \phi \rangle_{\mathscr{H}} + \lambda E\{1 - Y[\langle \psi, \kappa(\cdot, X) - \mu \rangle_{\mathscr{H}} - t]\}^+. \tag{15.11}$$

Li et al. (2011) observed that SVM can be used as a natural way to construct contours as employed by the Contour Regression. It is more versatile than constructing contours based on $X_i - X_j$ with small $|Y_i - Y_j|$ in that it allows for the possibility of nonlinear contours, thus opening up a way for nonlinear Sufficient Dimension Reduction. For linear SDR, Li et al. (2011) proposed to divide the space of X into slices, as we did in the Sliced Inverse Regression, and apply a modified SVM to different pairs of slices to obtain the contour hyperplanes to extract their normal vectors. We then combine these normal vectors by principal component analysis to form vectors in the central subspace. The modified version of the SVM objective function is

$$\psi^\mathsf{T}\Sigma\psi + \lambda E\{1 - Y[\psi^\mathsf{T}(X - EX) - t]\}^+, \tag{15.12}$$

where Σ is the covariance matrix $\mathrm{var}(X)$, and Y represents the labels of a pair of slices; that is, $Y = 1$ if X belongs to one slice, and $Y = -1$ if X belongs to the other slice. Li et al. (2011) showed that this method is unbiased under the linear conditional mean condition. They also developed the asymptotic distribution of this estimator.

For nonlinear Sufficient Dimension Reduction, Li et al. (2011) modified the objective function to

$$\langle \phi, \Sigma_{XX}\phi \rangle_{\mathscr{H}} + \lambda E\{1 - Y[\langle \psi, \kappa(\cdot, X) - \mu \rangle_{\mathscr{H}} - t]\}^+, \tag{15.13}$$

where Σ_{XX} is the covariance operator

$$E\{[\kappa(\cdot, X) - \mu] \otimes [\kappa(\cdot, X) - \mu]\}.$$

Li et al. (2011) showed that, if $\mathscr{H}$ is dense in $L_2(P_X)$, then the minimizer of the modified objective function (15.13) is a member of the central class. We perform the modified nonlinear SVM on each pair of slices, and assemble them together by principal component analysis. This method of performing linear or nonlinear Sufficient Dimension Reduction by linear or kernel SVM is called principal SVM or kernel principal SVM.

Principal SVM and kernel principal SVM have been further developed by several authors. For example, Artemiou and Dong (2016) extended them using an L_q-SVM; that is, by replacing the function u^+ in by $(u^+)^q$ in (15.12) and (15.13) for some constant $q > 0$. The advantage of doing so is that the resulting objective function is strictly convex; whereas the L_1-SVM objective functions (15.12) and (15.13) are not strictly convex. Shin et al. (2017) extended principal SVM using weighted SVM (see also Shin et al., 2014). For example, in the linear case, replace (15.12) by

$$\psi^\mathsf{T}\Sigma\psi + \lambda E\{W_\pi(Y)[1 - \psi^\mathsf{T}(X - EX) - t]^+\}, \tag{15.14}$$

where $W_\pi(-1) = 1 - \pi$ and $W_\pi(1) = \pi$ for some $\pi \in (0,1)$. When the response variable is binary, methods such as SIR and principal SVM can only recover one direction in the central subspace, and the same can be said of GSIR and kernel principal SVM. An advantage of Shin et al. (2017)'s approach is that, by changing the values of π, we can potentially recover more vectors in the central subspace. For other extensions and related work, see Artemiou and Shu (2014), Artemiou and Tian (2015), Zhou and Zhu (2016), and Shin and Artemiou (2017).

15.9   SDR for Multivariate Responses

The methods described in Chapters 3 through 8 and Chapter 11 are restricted to a univariate response Y. In this section we outline the developments to extend the methods to multivariate responses.

When the goal is to estimate the central mean subspace, the problem is relatively easy to solve. As shown in Cook and Setodji (2003), if Y is a q-dimensional response $(Y_1,\ldots,Y_q)^\mathsf{T}$, then the central mean subspace $\mathscr{S}_{E(Y|X)}$ satisfies the following relation

$$\mathscr{S}_{E(Y|X)} = \operatorname{span}\{\mathscr{S}_{E(Y_1|X)},\ldots,\mathscr{S}_{E(Y_q|X)}\}.$$

This means we can simply perform Sufficient Dimension Reduction for conditional means $E(Y_i|X)$ and then assemble the subspaces together by the Principal Component Analysis. All the methods for estimating the central mean subspace described in Chapter 8 and Chapter 11 can serve this purpose.

For estimating the central subspace, however, the matter is more complicated and the development diverges into several different directions. One extension is slicing the space of Y into hypercubes, which underlies the approach proposed by Li et al. (2003b). Realizing that multivariate slicing in a high-dimensional space may result in the curse of dimensionality, the authors introduced the *most predictive variates* (MP variates), which is defined as the linear combinations $\theta^\mathsf{T}Y$ that maximize the ratio

$$\operatorname{var}[E(\theta^\mathsf{T}Y|X)]/\operatorname{var}(\theta^\mathsf{T}Y). \tag{15.15}$$

This optimization can be reformulated as a generalized eigenvalue problem. After extracting the first few MP variates, say $\tilde{Y}_1 = \theta_1^\mathsf{T}Y, \ldots, \tilde{Y}_k = \theta_k^\mathsf{T}Y$, we perform Sliced Inverse Regression on the low dimensional response $(\tilde{Y}_1,\ldots,\tilde{Y}_k)^\mathsf{T}$. Once we obtain the dimension-reduced X, say $\beta^\mathsf{T}X$, we substitute it back into (15.15) to update $\tilde{Y}_1,\ldots,\tilde{Y}_k$, perhaps with further dimension reduction. The conditional expectation at each step is done by multivariate slicing.

Another approach is called the pooled marginal slicing (Saracco, 2005). Again, we illustrate this approach at the population level using SIR. Let $\mathscr{S}_{\text{SIR}}$ be the subspace spanned by $\text{var}[E(X|Y)]$ where Y is a random vector. Let $\mathscr{S}_{\text{SIR}}^{(i)}$ be the subspace spanned by $\text{var}[E(X|Y_i)]$, which is the SIR matrix for the ith component of Y. Then, it can be shown that

$$\text{span}\{\mathscr{S}_{\text{SIR}}^{(1)}, \ldots, \mathscr{S}_{\text{SIR}}^{(q)}\} \subseteq \mathscr{S}_{\text{SIR}}.$$

Thus, we can first perform SIR on each individual response Y_i, and then assemble them together using Principal Component Analysis. Note that this approach is not guaranteed to recover the space $\mathscr{S}_{\text{SIR}}$ fully.

Yin and Bura (2005) proposed a method based on the higher conditional moments of Y. Yin and Cook (2002) generalized the central mean subspace of Cook and Li (2002) to central kth moment subspace via the relations

$$E(Y^i|X) = E(Y^i|\beta^\mathsf{T}X), \quad i = 1, \ldots, k.$$

If β is a matrix that satisfies the above relations with lowest column rank, then $\text{span}(\beta)$ is called the central kth moment subspace, written as $\mathscr{S}_{E(Y_k|X)}$. Yin and Cook (2002) showed

$$\mathscr{S}_{E(Y|X)} \subseteq \mathscr{S}_{E(Y^2|X)} \subseteq \cdots \subseteq \mathscr{S}_{E(Y^k|X)} \subseteq \mathscr{S}_{Y|X}.$$

Yin and Bura (2005) further generalized the central kth moment subspace to the multivariate case by the relations

$$E(Y^{\otimes i}|X) = E(Y^{\otimes i}|\beta^\mathsf{T}X), \quad i = 1, \ldots, k,$$

where $Y^{\otimes k}$ is the k-fold Kronecker product $Y \otimes \cdots \otimes Y$. In the more general setting we still have

$$\mathscr{S}_{E(Y|X)} \subseteq \mathscr{S}_{E(Y^{\otimes 2}|X)} \subseteq \cdots \subseteq \mathscr{S}_{E(Y^{\otimes k}|X)} \subseteq \mathscr{S}_{Y|X}.$$

Yin and Bura (2005) extended the ideas of OLS and PHD to this setting: in particular, they showed that, under the linear conditional mean and the constant conditional variance assumptions, the spaces spanned by

$$E[Z \otimes (Y^{\otimes i})^\mathsf{T}], \quad E[ZZ^\mathsf{T} \otimes (Y^{\otimes i})^\mathsf{T}], \quad i = 1, \ldots, k,$$

are all contained in $\mathscr{S}_{E(Y^{\otimes k}|Z)}$, where Z is the standardized X. Based on this result they developed sample estimates for the central kth moment space. Note that, again, this method is not guaranteed to fully recover the central subspace.

Li et al. (2008) introduced a highly flexible SDR estimator for multivariate responses called the Projective Resampling estimator. It does not suffer from the curse of dimensionality (maintaining $\sqrt{n}$-convergence rate regardless of the dimension q), is relatively easy to implement, and can be applied in conjunction with any SDR method for univariate responses. Furthermore, if the underlying SDR method for univariate response is exhaustive, then so is the corresponding Projective Resampling estimator.

The idea is the following. Suppose T is a continuous random vector defined on $\mathbb{R}^q$, say $T \sim N(0, I_q)$, that is independent of (X, Y). For each t, let $M(t)$ be a positive semi-definite matrix spanning the central subspace of the scalar response $t^\mathsf{T}Y$ versus X; that is, $\text{span}(M(t)) = \mathscr{S}_{t^\mathsf{T}Y|X}$. Then, under mild conditions,

$$\text{span}\{EM(T)\} = \mathscr{S}_{Y|X}.$$

At the sample level, suppose $\hat{M}(t) = M((X_1, t^\mathsf{T}Y_1), \ldots, (X_n, t^\mathsf{T}Y_n))$ is a candidate matrix derived from any SDR method the univariate response $t^\mathsf{T} Y$. We draw an i.i.d. sample $T_1, \ldots, T_{m_n}$ of T independently from $(X_1, Y_1), \ldots, (X_n, Y_n)$ and form the matrix $\sum_{i=1}^{m_n} M(T_i)$. We then estimate the central subspace $\mathscr{S}_{Y|X}$ by performing a generalized eigenvalue problem on this matrix. Since the sample $(X_1, Y_1), \ldots, (X_n, Y_n)$ is used repeatedly via the random projections, not unlike the way we perform bootstrap estimation, we call this method the Projective Resampling method. Li et al. (2008) showed that, if m_n is of the order n, then the estimator maintains the parametric convergence rate of $n^{-1/2}$.

For the ensemble estimate in Chapter 11 with ensemble $\mathfrak{F}_C$, the idea of Projective Resampling is in some sense embedded in the ensemble procedure: we simply replace e^{itY} by $e^{it^\mathsf{T} Y}$ to handle the multivariate cases. Also, nonlinear SDR can be applied to multivariate responses without modification because the kernel for Y applies to random vectors.

Bibliography

S. Akaho. Proceedings of international meeting on psychometric society. *A kernel method for canonical correlation analysis*, pages 1–48, 2001.

P. Alquier and G. Biau. Sparse single-index model. *Journal of Machine Learning Research*, 14:243–280, 2013.

A Artemiou and Y. Dong. Sufficient dimension reduction via principal lq support vector machine. *Electronic Journal of Statistics*, 10:783–805, 2016.

A. Artemiou and M. Shu. A cost based reweighted scheme of principal support vector machine. *Topics in Nonparametric Statistics*, pages 1–12, 2014.

A. Artemiou and L. Tian. Using sliced inverse mean difference for sufficient dimension reduction. *Statistics & Probability Letters*, 106:184–190, 2015.

F. R. Bach and M. I. Jordan. Kernel independent component analysis. *Journal of Machine Learning Research*, 3:1–48, 2002.

R. R. Bahadur. Sufficiency and statistical decision functions. *Annals of Mathematical Statistics*, 25:423–462, 1954.

P. M. Bentler and J. Xie. Corrections to test statistics in principal hessian direction. *Statistics and Probability Letters*, 47:381–389, 2000.

A. Berlinet and C. Thomas-Agnon. *Reproducing kernel Hilbert spaces in probability and statistics*. Kluwer Academic, 2004.

D. P. Bertsekas. *Nonlinear Programming, Second Edition*. Athena Scientific, 1999.

P. Bickel and D. Freedman. Some asymptotic theory for the bootstrap. *The Annals of Statistics*, 9:1196–1217, 1981.

P. J. Bickel, C. A. J. Klaassen, Y. Ritov, and J. A. Wellner. *Efficient and Adaptive Estimation for Semiparametric Models*. Springer, 1998.

P. Billingsley. *Probability and Measure, Third Edition*. John Wiley & Sons, 1995.

H. D. Bondell and L. Li. Shrinkage inverse regression estimation for model-free variable selection. *Journal of the Royal Statistical Society, Series B*, 71:287–299, 2009.

D. D. Boos and R. J. Serfling. A note on differentials and the clt and lil for statistical functions with applications to m-estimates. *Annals of Statistics*, 8:618–624, 1990.

G. E. Box and D. R. Cox. An analysis of transformations. *Journal of the Royal Statistical Society, Series B*, 2:211252, 1964.

E. Bura and R. D Cook. Estimating the structural dimension of regressions via parametric inverse. *Journal of the Royal Statistical Society, Series B*, 63:393–410, 2001.

E. Bura and B. J. Yang. Dimension estimation in sufficient dimension reduction: A unifying approach. *Journal of Multivariate Analysis*, 102:130–142, 2011.

E. Bura, S. Duarte, and L. Forzani. Sufficient reductions in regressions with exponential family inverse predictors. *Journal of the American Statistical Association*, 111:1313–1329, 2016.

C. Carmeli, E. De Vito, A. Toigo, and V. Umanita. Vector valued reproducing kernel hilbert spaces and universality. *Analysis and Applications*, 08:19–61, 2010.

R. J. Carroll and K.-C. Li. Measurement error regression with unknown link: dimension reduction and data visualization. *Journal of the American Statistical Association*, 87:1040–1050, 1992.

R. J. Carroll and L. A. Stefanski. Approximate quasi-likelihood estimation in models with surrogate predictors. *Journal of the American Statistical Association*, 85:652–663, 1990.

R. J. Carroll, D. Ruppert, and L. A. Stefanski. *Measurement Error in Nonlinear Models*. Chapman & Hall, London, 1995.

R. J. Carroll, J. Fan, I. Gijbels, and M. P. Wand. Generalized partially linear single-index models. *Journal of the American Statistical Association*, 92:477–489, 1997.

C.-H. Chen and K.-C. Li. Can sir be as popular as multiple linear regression? *Statistica Sinica*, 8:289–316, 1998.

X. Chen, C. Zou, and R. D. Cook. Coordinate-independent sparse sufficient dimension reduction and variable selection. *The Annals of Statistics*, 38:3696–3723, 2010.

F. Chiaromonte, R. D. Cook, and B. Li. Sufficient dimension reduction in regressions with categorical predictors. *The Annals of Statistics*, 30:475–497, 2002.

J. B. Conway. *A Course in Functional Analysis, Second Edition*. Springer, 1990.

R. D. Cook. Using dimension-reduction subspaces to identify important inputs in models of physical systems. *In 1994 Proceedings of the Section on Physical and Engineering Sciences. American Statistical Association, Alexandria, VA.*, pages 18–25, 1994.

R. D. Cook. *Regression Graphics: Ideas for Studying Regressions through Graphics*. John Wiley & Sons, Inc., 1998.

R. D. Cook. Testing predictor contributions in sufficient dimension reduction. *The Annals of Statistics*, 32:1062–1092, 2004.

R. D. Cook. Fisher lecture: dimension reduction in regression. *Statistical Science*, 22:1–40, 2007.

R. D. Cook and F. Critchley. Identifying regression outliers and mixtures graphically. *Journal of American Statististical Association*, 95:781794, 2000.

R. D. Cook and S. Ding. Tensor sliced inverse regression. *Journal of Multivariate Analysis*, 133:216–231, 2015.

R. D. Cook and L. Forzani. Principal fitted components for dimension reduction in regression. *Statistical Science*, 23:485–501, 2008.

R. D. Cook and L. Forzani. Likelihood-based sufficient dimension reduction. *Journal of American Statistical Association*, 104:197–208, 2009.

R. D. Cook and B. Li. Dimension Reduction for Conditional Mean in Regression. *The Annals of Statistics*, 30(2):455–474, 2002.

R. D. Cook and B. Li. Determining the dimension of iterative hessian transformation. *The Annals of Statistics*, 32:2501–2531, 2004.

R. D. Cook and L. Ni. Sufficient dimension reduction via inverse regression a minimum discrepancy approach. *Journal of the American Statistical Association*, 108:410–428, 2005.

R. D. Cook and C. M. Setodji. A model-free test for reduced rank in multivariate regression. *Journal of the American Statistical Association*, 98:328–332, 2003.

R. D. Cook and S. Weisberg. Sliced Inverse Regression for Dimension Reduction: Comment. *Journal of the American Statistical Association*, 86(414):328–332, 1991.

R. D. Cook, B. Li, and F. Chiaromonte. Dimension reduction in regression without matrix inversion. *Biometrika*, 94:569–584, 2007.

R. D. Cook, B. Li, and F. Chiaromonte. Envelope models for parsimonious and efficient multivariate linear regression. *Statistica Sinica*, 20:927–960, 2010.

A. P. Dawid. Conditional independence in statistical theory. *Journal of the Royal Statistical Society. Series B (Methodological)*, pages 1–31, 1979.

S. Ding and R. D. Cook. Dimension folding pca and pfc for matrix-valued predictors. *Statistica Sinica*, 24:463–492, 2014.

Y. Dong and B. Li. Dimension reduction for non-elliptically distributed predictors: second-order methods. *Biometrika*, 97:279–294, 2010.

D. L. Donoho and I. M. Johnstone. Ideal spatial adaptation by wavelet shrinkage. *Biometrika*, 81:425–455, 1994.

M. L. Eaton. A characterization of spherical distributions. *Journal of Multivariate Analysis*, 20:272–276, 1986.

J. Fan. Local linear regression smoothers and their minimax efficiencies. *The Annals of Statistics*, 21:196–216, 1993.

J. Fan and I. Gijbels. Variable bandwidth and local linear regression smoothers. *The Annals of Statistics*, 20:2008–2036, 1993.

J. Fan, Q. Yao, and H. Tong. Estimation of conditional densities and sensitivity measures in nonlinear dynamical systems. *Biometrika*, 83:189–206, 1996.

K.-T. Fang and L.-X. Zhu. Asymptotics for kernel estimate of sliced inverse regression. *The Annals of Statistics*, 24:1053–1068, 1996.

Z. Feng, X. Wen, Z. Yu, and L. Zhu. On partial sufficient dimension reduction with applications to partially linear multi-index models. *Journal of the American Statistical Association*, 108:237–246, 2013.

L. T. Fernholz. *von Mises Calculus for Statistical Functionals*. Springer, 1983.

L. Ferré and A. F. Yao. Functional sliced inverse regression analysis. *Statistics*, 37(6):475–488, November 2003.

L. Ferré and A. F. Yao. Smoothed functional inverse regression. *Statistica Sinica*, 15:665–683, 2005.

A. A. Filippova. Mises theorem on the asymptotic behavior of functionals of empirical distribution function and its statistical applications. *Theory Prob. Appl.*, 7:24–57, 1962.

M. Forina, R. Leardi, C. Armanino, and S. Lanteri. *PARVUS - An Extendable Package of Programs for Data Exploration, Classification and Correlation*. Elsevier, Amsterdam, 1988.

K. Fukumizu, F. R. Bach, and M. I. Jordan. Dimensionality reduction for supervised learning with reproducing kernel Hilbert spaces. *Journal of Machine Learning Research*, 5:73–99, 2004.

K. Fukumizu, F. R. Bach, and A. Gretton. Statistical consistency of kernel canonical correlation analysis. *The Journal of Machine Learning Research*, 8:361–383, 2007.

K. Fukumizu, F. R. Bach, and Michael I. Jordan. Kernel dimension reduction in regression. *The Annals of Statistics*, 37, 2009.

W. A. Fuller. *Measurement Error Models*. Wiley, New York, 1987.

W. K. Fung, X. He, L. Liu, and P. Shi. Dimension reduction based on canonical correlation. *Statistica Sinica*, 12:1093–1113, 2002.

R. D. Gill. Non- and semi-parametric maximum likelihood estimators and the von Mises method, part 1. *Scandinavian Journal of Statistics*, 16:97–128, 1989.

G. H. Golub, M. Heath, and G. Wahba. Generalized cross-validation as a method for choosing a good ridge parameter. *Technometrics*, 21(2):215–223, 1979.

Z. Guo, L. Li, W. Lu, and B. Li. Groupwise dimension reduction via envelope method. *Journal of the American Statistical Association*, 110:1515–1527, 2015a.

Zifang Guo, Lexin Li, Wenbin Lu, and Bing Li. Groupwise sufficient dimension reduction via envelope method. *Journal of American Statistical Association*, 110:1515–1529, 2015b.

P. Hansen. Large sample properties of generalized method of moments estimators. *Econometrica*, 50:1029–1054, 1982.

W. Hardle and T. M. Stoker. Investigating smooth multiple regression by method of average derivatives. *Journal of the American Statistical Association*, 84:986–995, 1989.

W. Hardle, P. Hall, and H. Ichimura. Optimal smoothing in single-index models. *Annals of Statistics*, 21:157–178, 1993.

N. Heckman. Spline smoothing in a partly linear model. *Journal of the Royal Statistical Society, Series A*, 48:244–248, 1986.

J. Hoffmann-Jorgensen. *Probability with a View Towards Statistics*. Chapman & Hall, 1994.

R. A. Horn and C. R. Johnson. *Matrix Analsysis*. Cambridge University Press, 1985.

M. Hristache, A. Juditsky, J. Polzehl, and V. Spokoiny. Structure adaptive approach for dimension reduction. *The Annals of Statistics*, 29:1632–1640, 2001.

T. Hsing and R. Eubank. *Theoretical Foundations of Functional Data Analysis, with an Introduction to Linear Operators*. Wiley, 2015.

T. Hsing and H. Ren. An rkhs formulation of the inverse regression dimension-reduction problem. *The Annals of Statistics*, 37(2):726–755, 2009.

H. Hung and C. C. Wang. Matrix variate logistic regression model with application to eeg data. *Biometrics*, 14:189–202, 2013.

H. Ichimura. Semiparametric least squares (sls) and weighted sls estimation of single-index models. *Journal of Econometrics*, 58:71–120, 1993.

T. Kato. *Perturbation Theory for Linear Operators*. Springer, 1980.

J. Kelley. *General Topology*. D. Van Nostrand Company, Inc., 1955.

M. K. Kim. On dimension folding of matrix or array valued statistical objects. *Ph.D. thesis, Pennsylvania State University*, 2010.

K.-Y. Lee, B. Li, and F. Chiaromonte. A general theory for nonlinear sufficient dimension reduction: formulation and estimation. *The Annals of Statistics*, 41, 2013.

K.-Y. Lee, B. Li, and H. Zhao. Variable selection via additive conditional independence. *Journal of the Royal Statistical Society: Series B*, 78:1037–1055, 2016.

E. L. Lehmann. An interpretation of completeness and basus theorem. *Journal of the American Statistical Association*, 76:335–340, 1981.

E. L. Lehmann and G. Casella. *Theory of Point Estimation, Second Edition*. Springer, 1998a.

E. L. Lehmann and G. R. Casella. *Theory of Point Estimation, Second Edition*. Springer, 1998b.

B. Li. Linear operator-based statistical analysis: A useful paradigm for big data. 2017. *Canadian Journal of Statistics*, to appear.

B. Li and Y. Dong. Dimension reduction for nonelliptically distributed predictors. *The Annals of Statistics*, 37:1272–1298, 2009.

B. Li and E. Solea. A nonparametric graphical model for functional data with application to brain networks based on fMRI. To appear in *Journal of the American Statistical Association*, 2017.

B. Li and J. Song. Nonlinear sufficient dimension reduction for functional data. *The Annals of Statistics*, 45:1059–1095, 2017.

B. Li and S. Wang. On directional regression for dimension reduction. *Journal of the American Statistical Association*, 102:997–1008, 2007.

B. Li and X. Yin. On surrogate dimension reduction for measurement error regression: an invariance law. *The Annals of Statistics*, 35:2143–2172, 2007.

B. Li, R. D. Cook, and F. Chiaromonte. Dimension reduction for the conditional mean in regressions with categorical predictors. *The Annals of Statistics*, 31:1636–1668, 2003a.

B. Li, H. Zha, and F. Chiaromonte. Contour regression: A general approach to dimension reduction. *The Annals of Statistics*, 33(4):1580–1616, 2005.

B. Li, S. Wen, and L. Zhu. On a projective resampling method for dimension reduction with multivariate responses. *Journal of the American Statistical Association*, 103:1177–1186, 2008.

B. Li, M. K. Kim, and N. Altman. On dimension folding of matrix-or array-valued statistical objects. *The Annals of Statistics*, pages 1094–1121, 2010a.

B. Li, A. Artemiou, and L. Li. Principal support vector machines for linear and nonlinear sufficient dimension reduction. *The Annals of Statistics*, 39:3182–3210, 2011.

K.-C. Li. On principal hessian directions for data visualization and dimension reduction: Another application of stein's lemma. *Journal of the American Statistical Association*, 87:1025–1039, 1992.

K.-C. Li and N. Duan. Regression analysis under link violation. *The Annals of Statistics*, 17:1009–1052, 1989.

K.-C. Li, Y. Aragon, K. Shedden, and C. T. Agnan. Dimension reduction for multivariate response data. *Journal of the American Statistical Association*, 98:99–109, 2003b.

Ker-Chau Li. Sliced inverse regression for dimension reduction. *Journal of the American Statistical Association*, 86(414):316–327, 1991.

L. Li. Sparse sufficient dimension reduction. *Biometrika*, 94:603–613, 2007.

L. Li, B. Li, and L.-X. Zhu. Groupwise dimension reduction. *Journal of American Statistical Association*, 105:1188–1201, 2010b.

H. Lian and G. Li. Series expansion for functional sufficient dimension reduction. *Journal of Multivariate Analysis*, 124:150–165, 2014.

R. Y. Liu, K. Singh, and S. H. Lo. On a representation related to the bootstrap. *Sankhyā: The Indian Journal of Statistics*, 51:168–177, 1989.

Y. Liu, F. Chiaromonte, and B. Li. Structured ordinary least squares: A sufficient dimension reduction approach for regressions with partitioned predictors and heterogeneous units. *Biometrics*, 73:529539, 2017.

H.-H. Lue. Principal hessian directions for regression with measurement error. *Biometrika*, 91:409–423, 2004.

W. Luo and B. Li. Combining eigenvalues and variation of eigenvectors for order determination. *Biometrika*, 103:875–887, 2016.

Wei Luo, Bing Li, and Xiangrong Yin. On efficient dimension reduction with respect to a statistical functional of interest. *Annals of Statistics*, 42:382–412, 2014.

Y. Ma and L. Zhu. A semiparametric approach to dimension reduction. *Journal of the American Statistical Association*, 107:168–179, 2012.

Y. Ma and L. Zhu. Efficient estimation in sufficient dimension reduction. *The Annals of Statistics*, 41:250–268, 2013a.

Y. Ma and L. Zhu. A review on dimension reduction. *International Statistical Review*, 81: 134–150, 2013b.

M. E. Mann, Z. Zhang, M. K. Hughes, R. S. Bradley, S. K. Miller, Rutherford S., and F. Ni. Proxy-based reconstructions of hemispheric and global surface temperature variations over the past two millennia. *Proceedings of the National Academy of Sciences*, 105:13252–13257, 2008.

P. McCullagh. *Tensor Methods in Statistics*. Chapman & Hall/CRC, 1987.

P. McCullagh and J. A. Nelder. *Generalized Linear Models*. Chapman & Hall, 1989.

B. B. McShane and A. J. Wyner. A statistical analysis of multiple temperature proxies: Are reconstructions of surface temperatures over the last 1000 years reliable? *The Annals of Applied Statistics*, 5:5–44, 2011.

W. Parr. The bootstrap: some large sample theory and connections with robustness. *Statistics & Probability Letters*, 3:97–100, 1985.

J. Pearl and T. Verma. *The logic of representing dependencies by directed graphs*. University of California (Los Angeles). Computer Science Department, 1987.

J. Pearl, D. Geiger, and T. Verma. Conditional independence and its representations. *Kybernetika*, 25(7):33–44, 1989.

C. R. Rao and S. K. Mitra. Generalized inverse of a matrix and its applications. *Sixth Berkeley Symposium*, pages 601–620, 1972.

J. A. Reeds. *On the definition of von Mises functionals*. Ph.D. dissertation, Harvard University, Cambridge, MA., 1976.

D. Ruppert and M. P. Wand. Multivariate locally weighted least squares regression. *The Annals of Statistics*, 22:1346–1370, 1994.

J. Saracco. Asymptotics for pooled marginal slicing estimator based on sir_α approach. *Journal of Multivariate Analysis*, 96:117–135, 2005.

B. Scholkopt, A. Smolar, and K.-R. Muller. Nonlinear component analysis as a kernel eigenvalue problem. *Neural Computation*, 10(5):1299–1319, 1998.

J. R. Schott. *Matrix Analysis for Statistics*. Wiley, 1997.

Y Shao, R. D. Cook, and S. Weisberg. Marginal tests with sliced average variance estimation. *Biometrika*, 94:285–296, 2007.

Y. Shao, R. D. Cook, and S. Weisberg. Partial central subspace and sliced average variance estimation. *Journal of Statistical Planning and Inference*, 139:952–961, 2009.

S. J. Shin and A. Artemiou. Penalized principal logistic regression for sparse sufficient dimension reduction. *Computational Statistics & Data Analysis*, 111:48–58, 2017.

S. J. Shin, Y. Wu, H. H. Zhang, and Y. Liu. Probability enhanced sufficient dimension reduction in binary classification. *Biometrics*, 70:546–555, 2014.

S. J. Shin, Y. Wu, H. H. Zhang, and Y. Liu. Probability enhanced sufficient dimension reduction in binary classification. *Biometrika*, 104:67–81, 2017.

E. Solea and B. Li. Copula gaussian graphical models for functional data. Submitted to *The Annals of Statistics*, 2016.

B. K. Sriperumbudur, A. Gretton, K. Fukumizu, B. Scholkopt, and G. R. G. Lanckriet. Hilbert space embeddings and metrics on probability measures. *Journal of Machine Learning Research*, 11:1517–1561, 2010.

M. Stone. Cross-validatory choice and assessment of statistical prediction. *Journal of the Royal Statistical Society, Series B*, 36:111–147, 1977.

J. B. Tenenbaum, V. de Silva, and J. C. Langford. A global geometric framework for nonlinear dimensionality reduction. *Science*, 290:2319–2323, 2000.

V. N. Vapnik. *Statistical Learning Theory*. Wiley, New York, 1998.

J. Virta, B. Li, K. Nordhausen, and H. Oja. Independent component analysis for tensor-valued data. *Journal of Multivariate Analysis*. To appear., 2017a.

J. Virta, B. Li, K. Nordhausen, and H. Oja. Jade for tensor-valued observations. *Journal of Computational and Graphical Statistics*. To appear., 2017b.

R von Mises. On the asymptotic distribution of differentiable statistical functions. *Annals of Mathematical Statistics*, 18:309–348, 1947.

G. Wang, N. Lin, and B. Zhang. Functional contour regression. *Journal of Multivariate Analysis*, 116:1–13, 2013.

G. Wang, Y. Zhou, X.-N. Feng, and B. Zhang. The hybrid method of fsir and fsave for functional effective dimension reduction. *Computational Statistics and Data Analysis*, 91:64–77, 2015.

H. Wang and Y. Xia. Sliced regression for dimension reduction. *Journal of the American Statistical Association*, 103:811–821, 2008.

Q. Wang and X. Yin. A nonlinear multi-dimensional variable selection method for high dimensional data: Sparse mave. *Computational Statistics and Data Analysis*, 52:4512–4520, 2008.

Y. Wang. Nonlinear dimension reduction in feature space. *PhD Thesis, The Pennsylvania State University*, 2008.

X. Wen and R. D. Cook. *Journal of Statistical Planning and Inference*, 137:1961–1978, 2007.

S. Wold, M. Sjostrom, and L. Eriksson. Pls-regression: a basic tool of chemometrics. *Chemometrics and Intelligent Laboratory Systems*, 58:109–130, 2001.

H. M. Wu. Kernel sliced inverse regression with applications to classification. *Journal of Computational and Graphical Statistics*, 17(3):590–610, 2008.

Y. Xia. A constructive approach to the estimation of dimension reduction directions. *The Annals of Statistics*, 35:2654–2690, 2007.

Y. Xia, H. Tong, W. K. Li, and L.-X. Zhu. An adaptive estimation of dimension reduction space. *Journal of Royal Statistical Society, Series B*, 64:363–410, 2002.

Y. Xue and X. Yin. Sufficient dimension folding for regression mean function. *Journal of Computational and Graphical Statistics*, 23:1028–1043, 2014.

Y. Xue, X. Yin, and X. Jiang. Ensemble sufficient dimension folding methods for analyzing matrix-valued data. *Computational Statistics & Data Analysis*, 103:193–205, 2016.

Z. Ye and R. E. Weiss. Using the bootstrap to select one of a new class of dimension reduction methods. *Journal of the American Statistical Association*, 98:968–979, 2003.

Y.-R. Yeh, S.-Y. Huang, and Y.-Y. Lee. Nonlinear dimension reduction with kernel sliced inverse regression. *IEEE Transactions on Knowledge and Data Engineering*, 21:1590–1603, 2009.

X. Yin and E. Bura. Moment based dimension reduction for multivariate response regression. *Journal of Statistical Planning and Inference*, 136:3675–3688, 2005.

X. Yin and R. D. Cook. Dimension reduction for the conditional kth moment in regression. *Journal of the Royal Statistical Society, Series B*, 64:159–175, 2002.

X. Yin and B. Li. A note on the invariance law for surrogate dimension reduction. *Communications in Statistics*, 39:2721–2724, 2010.

X. Yin and B. Li. Sufficient dimension reduction based on an ensemble of minimum average variance estimators. *The Annals of Statistics*, 39:3392–3416, 2011.

X. Yin, B. Li, and R.D. Cook. Successive direction extraction for estimating the central subspace in a multiple-index regression. *Journal of Multivariate Analysis*, 99:1733–1757, 2008.

P. Zeng. Determining the dimension of the central subspace and central mean subspace. *Biometrika*, 95:469–479, 2008.

J. Zhang, L. Zhu, and L.-P. Zhu. Surrogate dimension reduction in measurement error regressions. *Statistica Sinica*, 24:1341–1363, 2014.

W. Zhong, X. Xing, and K. Suslick. Tensor sufficient dimension reduction. *Wiley Interdisciplinary Reviews: Computational Statistics*, 7:178–184, 2015.

H. Zhou and L. Li. Regularized matrix regression. *Journal of the Royal Statistical Society: Series B*, 76:463–483, 2014.

H. Zhou, L. Li, and H. Zhu. Tensor regression with applications in neuroimaging data analysis. *Journal of the American Statistical Association*, 108:540–552, 2013.

J. Zhou and L. Zhu. An integral transform method for estimating the central mean and central subspaces. *Journal of Multivariate Analysis*, 101:271–290, 2010.

J. Zhou and L. Zhu. Principal minimax support vector machine for sufficient dimension reduction with contaminated data. *Computational Statistics & Data Analysis*, 94:33–48, 2016.

L. Zhu, B. Miao, and H. Peng. On sliced inverse regression with high-dimensional covariates. *Journal of the American Statistical Association*, 101:630–643, 2006.

L. Zhu, M. Ohtaki, and Y. Li. On hybrid methods of inverse regression-based algorithms. *Computational Statistics and Data Analysis*, 51:2621–2635, 2007.

L. P. Zhu, L. X. Zhu, and Z. H. Feng. Dimension reduction in regressions through cumulative slicing estimation. *Journal of the American Statistical Association*, 105:1455–1466, 2010.

Y. Zhu and P. Zeng. Fourier methods for estimating the central subspace and the central mean subspace in regression. *Journal of the American Statistical Association*, 101:1613–1651.

Index

T-central subspace, 263
π-λ Theorem, 18
k-fold Cross Validation, 223
r-mode sufficient dimension folding subspace, 257

adjoint operator, 194
affine equivariance, 99
analytic continuation theorem, 208
Arc Software, 33
Average Derivative Estimation, 160

Banach space, 10
basis matrix, 49
BIC-type criteria, 141
bilinear form, 194
bootstrapped eigenvector variation, 147
bounded linear functional, 193
Box-Cox transformation, 89

candidate matrix, 26
candidate operator, 209
canonical correlation; CANCOR, 37
cardinality, 67
central mean subspace, 97
central solution subspace, 94
central subspace, 23
complete dimension reduction class, 208
complete sub σ-field, 206
conditional independence, 17
constant conditional variance, 48
Contour Regression, 65
coordinate descent, 90
coordinate mapping, 199
coordinate of a function, 12
coordinate of covariance operator, 200
coordinate representation, 12
coordinate representation of a linear operator, 12
copula, 89
cumulant generating function, 5

curse of dimensionality, 16

dense modulo constants, 196
density MAVE; dMAVE, 178
density OPG; dOPG, 173
direct-sum envelope, 260
Directional Regression, 70

e-affine invariant, 113
efficient dimension reduction, 262
Electroencephalography; EEG, 254
elliptical distribution, 27, 83
empirical directions, 64
empirical distribution, 4
Ensemble Estimator, 180
envelope model, 104
estimating equation, 94
exhaustiveness, 25, 56, 69, 77
exponential family, 5

Fisher consistency, 25
forward regression, 159
Fourth Order Blind Identification; FOBI, 259
Frobenius norm, 222
functional data, 253
functional Magnetic Resonance Imaging; fMRI, 254
Functional Sliced Inverse Regression: FSIR, 255

Gateaux derivative, 108
Gauss-Seidel algorithm, 90
generalized cross validation; GCV, 220
generalized eigenvalue problem, 3
Generalized Linear Model, 5
Generalized Sliced Average Variance Estimation; GSAVE, 233
Generalized Sliced Inverse Regression; GSIR, 211
Gram matrix, 12
Grassman manifold, 26

groupwise central subspace, 259
groupwise MAVE, 260
groupwise rMAVE, 260
Groupwise Sufficient Dimension Reduction, 259

Hadamard product, 243
heteroscedastic conditional covariance operator, 233
Hilbert space, 10
Hilbert-Schmidt operator, 199
homoscedastic conditional variance operator, 235

idempotent, 10
Independent Component Analysis, 259
influence function, 109
inner product, 9
inner product space, 9
invariant subspace, 104
inverse regression equation, 94
Iterative Hessian Transformation, 104

Kernel Canonical Correlation Analysis, 191
Kernel Inverse Regression, 42
kernel of linear operator, 10
kernel Principal Component Analysis: kernel PCA, 202
kernel principal SVM, 268
kernel Sliced Inverse Regression, 220
kernel SVM, 268
Kronecker product, 110

ladel estimator, 152
linear conditional mean, 14, 28
linear manifold, 8
linear operator, 10
linear subspace; subspace, 8
local linear regression, 160
Louwner's ordering, 175

mean element, 195
measurement error problem, 265
Minimum Average Variance Estimate; MAVE, 167
minimum discrepancy approach; MDA, 262
Monte Carlo approximation, 130
Moore-Penrose inverse, 40

nested Hilbert spaces, 256

nonlinear Sufficient Dimension Reduction; nonlinear SDR, 191

order determination, 141
Ordinary Least Squares, 100
Outer Product of Gradients; OPG, 160

Parametric Inverse Regression; PIR, 37
partial dimension reduction, 264
partial least squares, 104
perturbation theory, 115
Principal Component Analysis, 2
Principal Hessian Direction, 101
principal SVM, 268
projection, 10
projection in Euclidean space, 11
Projective Resampling, 269

range of a linear operator, 10
refined dMAVE; rdMAVE, 178
refined dOPG; reOPG, 173
refined MAVE; rMAVE, 170
refined OPG; rOPG, 170
regression operator, 212
relative universality, 216
Reproducing Kernel Hilbert Space; RKHS, 192
Riesz representation theorem, 193

sample moment, 2
second-moment operator, 195
self adjoint, 10
semigraphoid axioms, 19
semiparametrically efficient estimator, 262
sequential test, 107
sequential test for SIR, 118
sequential test for DR, 132
sequential test for PHD, 124
sequential test for SAVE, 126
Single Index Model, 22
singular value, 130
singular value decomposition, 130
Sliced Average Variance Estimate; SAVE, 47
Sliced Inverse Regression; SIR, 27
Sliced Regression, 180
sparse Sufficient Dimension Reduction, 261
Spearman's correlation, 35
spherical distribution, 83
statistical functional, 107
stochastic ordering, 107

structural dimension, 23
Sufficient Dimension Folding, 256
sufficient dimension reduction σ-field, 204
sufficient dimension reduction class, 208
sufficient dimension reduction subspace, 21
sufficient predictor, 31
Support Vector Machine, 267

tangent space, 262
tensor product, 194
tensorial data, 256
Tychonoff regularization, 217

unbiasedness, 25
universality, 196

variance operator, 195
von Mises expansion, 107

Ye-Weiss method, 147